Preface

Antibiotics, sulfonamides, vaccines, serums, and other measures that can stop infections in their tracks or prevent them altogether—all these are taken for granted by Americans today. Yet at the beginning of this century practically none of the infectious diseases could be prevented or cut short once the illness had begun. The story of how this change came about involves not only the discoveries made by medical scientists and their application by practicing physicians and public health officials, but also the socioeconomic forces that made all this possible. The abundance of effective measures now available for the cure and prevention of infections would have been impossible without the improvements in education, the advances in industrial technology, and the increasing concern for human welfare which developed in the United States and elsewhere.

The story of the control of infectious diseases in the last seventy-five years has been told in bits and pieces—in articles or books on the development of individual drugs or vaccines, in histories of individual diseases, and in biographies of some of the participants. Yet most of it is still immured within the covers of technical books and journals intended for members of the health professions. The story needs to be told as a whole because the parts are interdependent. Knowledge about one disease was used to understand and control other diseases, the results of basic research in one field were transferred to other fields, the discovery of one remedy led to the finding of others, and at every turn enough popular support had to be generated to mobilize resources behind the enterprise. The history of the control of disease, here as elsewhere, is always intimately connected with the history of the society in which it occurs. Having lived through most of this period and been a practicing physician for more than half of it, having carried out research on several of the infectious diseases and studied the effects of a number of serums, vaccines, sulfonamides, and antibiotics, having known many of the principal investigators in the field and been a consultant to several government agencies, and having collaborated with several pharmaceutical companies, I have witnessed the advances in the control of infectious diseases from many angles and have ventured to put this fascinating story together.

Because progress has been so extensive and involved so many areas during these seventy-five years, it has been necessary to leave out certain diseases. Infections other than those caused by bacteria and viruses are not discussed, even though this decision has meant the omission of malaria, a disease that was intimately bound up with the earlier history of North America. Nor is vaccination for smallpox discussed in detail since the principles of vaccination against this disease had been elucidated and the important battles for immunization of the population had been won by 1900. Further progress in the twentieth century has consisted mainly of refinement of techniques and the extension of control procedures throughout all the states. Infections that are found mainly in tropical or semitropical countries and are only occasionally imported into this country are also excluded. They deserve their separate history. I have told the story mainly from the viewpoint of people in the United States because this is how I saw it and because the experience of this country has been typical of that of the economically advanced nations in general, but I have not hesitated to relate incidents that occurred in other countries when they were important to the development of the story. Omitted also are many significant discoveries that have not as yet produced practical measures for the prevention or therapy of a disease. These are the province of future historians.

It has been my desire to make this book understandable to the inquisitive layman without misplacing emphasis, sacrificing accuracy, or omitting details that are important to medical historians and interested physicians. For this reason technical terms are defined, where necessary, and at the same time documentation is supplied.

I owe a great debt to many colleagues, alongside of whom I have studied infectious diseases and with whom I have shared opinions—especially my preceptor, Dr. Maxwell Finland; my coinvestigators for many years, Drs. Mark H. Lepper and George G. Jackson; and the many fellows and trainees with whom I have been associated at George Washington University and the University of Illinois, especially my first fellow, Dr. Harry A. Feldman. Dr. Martin Cummings and his staff at the National Library of Medicine have been consistently helpful in making the facilities of that outstanding collection available to me. Dr. John B. Blake and his associates in the Division of the History of Medicine, especially Dorothy Hanks,

Fighting Infection

This volume is published as part of a long-standing cooperative program between Harvard University Press and the Commonwealth Fund, a philanthropic foundation, to encourage the publication of significant and scholarly books in medicine and health.

Fighting Infection
Conquests of the Twentieth Century

Harry F. Dowling

A Commonwealth Fund Book
Harvard University Press
Cambridge, Massachusetts, and London, England
1977

To my colleagues from whom I learned much while we
were studying infectious diseases together,

Maxwell Finland
Harry A. Feldman
Mark H. Lepper
George G. Jackson

Printed in the United States of America
Library of Congress Cataloging in Publication Data
Dowling, Harry Filmore.
  Fighting infection: conquests of the twentieth century.
  "A Commonwealth Fund book."
  Includes index.
    1. Communicable diseases—History. 2. Communicable diseases—
United States—History. I. Title.
[DNLM: 1. Communicable diseases—Prevention and control.
2. Communicable diseases—Therapy. 3. History of Medicine, 20th
century. 4. History of medicine, 20th century.— U. S.
WC11AA1 D747s]
RA643.D64   616.9   77–8307
ISBN 0–674–30075–0

have smoothed my path in many ways. I wish to thank also Catherine M. Brosky, librarian of the Graduate School of Public Health, University of Pittsburgh, for making available to me the Thomas Parran Papers.

I am indebted to Professor James Harvey Young of Emory University, Dr. James H. Cassedy of the National Library of Medicine, and Dr. Lisle A. Rose of the Historical Office, Department of State, who read the entire manuscript and together with Dr. Blake gave advice and help unstintingly. I wish to thank many others who read one or more chapters, including Edgar A. Cooper, H. Filmore Dowling, Jr., Manfred J. Waserman, and Drs. Clifford Bachrach, Fred M. Davenport, John N. Dowling, René Dubos, Harry A. Feldman, H. Corwin Hinshaw, Dorothy M. Horstmann, Alexander D. Langmuir, John T. Litchfield, Peter D. Olch, Sidney Olansky, Karl H. Pfeutze, Marjorie Pyle, Charles H. Rammelkamp, Thomas B. Turner, Theodore E. Woodward, and Hyman J. Zimmerman. My thanks to Kathleen M. Danz and Maria Regina for ferreting out innumerable references and to Mary Elizabeth Rose, who edited the manuscript and made many valuable suggestions.

This work was supported by a grant to the University of Illinois from the Commonwealth Fund.

Contents

Figures

Table

Fighting Infection

1.   *The Field of Battle*

On December 13, 1799, after a day of hard work and exposure on his farm at Mount Vernon, ex-President George Washington was seized with pain in the throat and could hardly speak. That night he began to gasp for breath. Three doctors were called. They blistered his throat, gave him inhalations of vinegar and water, gargles of vinegar and sage tea, and an enema, and bled him three times, the last time taking a quart of blood. In spite of these measures—some would say on account of them—he died less than forty-eight hours after his initial symptoms. This was a common fate of the great, as well as the not-so-great, who were unlucky enough to have a serious infection at the beginning of the nineteenth century. Doctors had no medicines to prevent Washington's death from an infection of the throat or larynx, nor even the slightest clue as to how to find any.[1]

During the next hundred years the science of medicine was placed on a firmer foundation by intensive study of the pathology of diseased organs, tissues, and finally of cells. The newer knowledge of physiology and pathology led to a better understanding of how the body worked normally and how illness interfered with normal function. Yet this knowledge, which eventually provided the basis for determining how remedies acted, did not at first lead to the production of new drugs. A few years before Washington's illness, doctors had been trying to relieve suffocation from swelling of the larynx by opening the trachea just below it. Washington's physicians decided against this procedure, however, because the operation had failed more often than it had succeeded. A century later tracheotomy, as the operation came to be called, had been perfected and would have been used to relieve the ex-President's distressed breathing. Also by that time bleeding had been almost completely discarded as treatment for an infection. But these were the only significant changes between 1799 and 1899 in the therapy of an infection such as the one that killed Washington.

1

As an example of the futility of most therapy at the end of the nineteenth century, James B. Herrick described his emotions as an intern in Chicago's Cook County Hospital when he found that he could do nothing for a poor patient who was in the terminal stages of a severe infection. "The high temperature, the rapid pulse, the chattering delirium, and the picking at the bedclothes were like a nightmare that drove away my sleep." Many of the patients whom Herrick was later called to see as a consultant were suffering from complications of pneumonia, typhoid fever, gonorrhea, and syphilis; for them, too, he could do little but make the diagnosis, shake his head, and depart. At that time specific remedies for infectious diseases could be counted on the fingers of one hand. It had been known for centuries that the bark of the cinchona tree would stop certain chills and fevers; and quinine, the concentrated drug extracted from this bark, was in the medicine bag of every physician by the late nineteenth century. Yet not until 1880, when Alphonse Laveran of France first identified the parasites that cause malaria, did doctors learn precisely when to use quinine—in malaria, where it was effective, rather than in the other fevers, where it was useless. Doctors had also known for centuries that mercury sometimes helped patients who had syphilis, although it was not effective in every case, and when it did work, improvement was slow. Ipecac was the time-honored remedy for the type of dysentery caused by amoebas, but differentiation of the amoebic form from the more common diarrheas caused by bacteria could not be made until after the pathogenic, or disease-producing, amoebas were described in 1875 and methods for growing and identifying bacteria were developed.[2] The other two specific remedies available in 1900, diphtheria and tetanus antitoxins, were the first to be made by man instead of nature. Although they had been introduced during the previous decade, they were not yet universally accepted by doctors or patients, nor were they always given properly. As early fruits of the new science of bacteriology, however, they gave promise of more to come.

It is impossible to say when people first realized that some diseases were infectious. The ancient Egyptians and Jews had an inkling when they perceived that leprosy and other diseases could spread from person to person and established rules to prevent their spread. One milestone in the knowledge of infections was a work by the Italian physician, Girolamo Fracastoro, in 1546, which explained that the "seeds" or germs of contagious diseases were carried from person to person and described the characteristics of many infectious diseases. But bacteriology was not really born until the seventeenth century when Anthony von Leeuwenhoek of Holland built over two hundred microscopes and, peering through one of them, saw protozoa in water and "more animals [presumably bacteria] living . . . on the teeth in a man's mouth, than there are men in a whole kingdom."[3]

The use of improved microscopes led in 1839 to the formation by a German anatomist, Theodor Schwann, of the theory that the cell was the primary, fundamental unit of living matter. This theory lent credence to the concept that single-celled organisms, such as bacteria, could live independently, reproduce, and cause disease. Support came from another source when in 1835 an Italian, Agostino Bassi, first demonstrated that a specific living agent was responsible for a particular disease by showing that muscardine, a disease of silkworms, was caused by a fungus. This observation stimulated investigators to search for the agents that caused other diseases. Some reputable scientists protested that there were no such specific agents, contending that microorganisms were generated spontaneously in putrefying matter and were incidental to disease rather than the cause of it. Louis Pasteur's studies on fermentation, conducted between 1857 and 1860, convinced him that this process could not occur without the presence of living yeasts. In 1861 he reported that he had been unable to demonstrate the spontaneous generation of microorganisms; rather, microorganisms grew in his sterile flasks only after he had introduced contaminated objects or unsterile air. Over the next few years scientists turned away from the idea of spontaneous generation and began instead to look for the microorganism that might be the cause of each infection.[4]

In 1840 the German physiologist Jacob Henle had laid down the principle that before a living agent could be considered the cause of a disease, it would have to be identified, grown apart from other agents, and then used to produce the disease it was supposed to cause. The procedures necessary to accomplish this were later developed by the German bacteriologist Robert Koch and are followed today under the name of Koch's postulates in order to prove the relationship of a living agent to a disease. In 1868 Casimir Joseph Davaine, a French physician, first transferred a bacterial disease by inoculating healthy animals with blood from diseased ones. In this blood Davaine could see long, slender, rodlike bacteria which had first been described by a German physician, Franz Pollender, in 1855. These were anthrax bacilli, so named after the disease that they produced in animals and sometimes in humans. In the next decade a number of bacteria were identified, transmitted to animals, and cultured in various artificial media, but it was not until 1881 that Koch reported his simplified techniques of growing bacteria on plates of sterile, solidified gelatin, from which a single colony of a particular species of microorganism could be removed and grown in pure culture. Also, by perfecting the staining of bacteria with colored dyes, he made identification much easier. Thus, the scientists of the mid-nineteenth century went far beyond their predecessors. They demonstrated conclusively that bacteria comprised a distinct group of organisms and that some could cause disease. They found out how to grow bacteria on artificial media and how to isolate a strain in pure culture. And they learned to identify microorganisms by their appearance under the microscope and by the way that they grew on artificial media and in animals. The next step was to use this knowledge in the prevention of disease, in other words, to put the germ theory to work.[5]

In 1880 Pasteur announced his success in preventing infections by the use of a new bacteriological method. He had found that cultures of the bacillus of chicken cholera lost much of their virulence when kept for some time and that injection of these weakened or attenuated bacilli protected fowls against subsequent infections of virulent bacilli. Turning to anthrax bacilli, he succeeded in weakening them by cultivating them for a few days at high temperature. By giving animals progressively increasing doses of these attenuated bacilli, he produced an immunity that protected them against ordinarily lethal doses of virulent bacilli. Later he found that the microorganism that causes swine erysipelas could be made less virulent by still another method, passing it from rabbit to rabbit. Edward Jenner, when he introduced vaccination against smallpox in 1796, had used a virus already modified by nature so that it produced a much milder disease in humans than the smallpox virus and yet was able to immunize against it. In contrast, Pasteur had used laboratory methods to attenuate microorganisms so that they produced no disease while retaining their immunizing properties. Since then many other methods have been used for weakening microorganisms to make vaccines from them, but the principle established by Pasteur is the basis of them all.[6]

It remained to apply this principle to the prevention of human disease. Pasteur began to work on rabies, a disease of the nervous system transmitted by the bites of infected animals. He removed the spinal cords of rabbits that had died of rabies, dried them for a number of days to weaken the virus, and made a vaccine from them.

Dogs were successfully protected against rabies with vaccines made in this way, and on July 6, 1886, Pasteur gave the first injection of a similar vaccine to a boy who had been severely bitten by a rabid dog. Three months later he announced that the child was well and flourishing.[7]

But Pasteur was not the first to use the new science of bacteriology in the prevention of disease. Joseph Lister, a surgeon of Glasgow, had been trying to figure out why wounds so seldom healed without the formation of pus. Acting on Pasteur's investigations, which showed that the air contained "minute particles . . . [,] the germs of various low forms of life," which were not "merely accidental concomitants of putrescence but . . . its essential cause," Lister began in 1865 to apply carbolic acid, now called phenol, to wounds and to surgical incisions during and after operations. The effect of this procedure was striking; in case after case he found that wounds remained free of infection, or that infection often disappeared if it had already started. In addition, patients convalescing in his ward did not develop serious infections of the skin and the tissues beneath it, as in the past. Later Lister also had phenol sprayed into the air of the operating room. Despite the damp and foggy atmosphere thus created in the room and the frequent changes of complicated dressings that were required, antisepsis gradually caught on. In the 1880s another technique was introduced called asepsis, which consisted of sterilizing all objects coming in contact with a wound, rather than killing microorganisms with a chemical compound after they had reached the wound. Asepsis eventually replaced antisepsis because it was just as effective and much kinder to the tissues of patients and to the skins of doctors and nurses.[8]

Meanwhile the causative agents of infectious diseases were being discovered at a rapid rate. Between 1880 and 1900 investigators identified twenty-one microorganisms as the specific causes of human diseases. Among these were the bacilli that cause typhoid fever, tuberculosis, and diphtheria; the pneumococcus and the meningococcus, responsible for the commonest forms of pneumonia and meningitis; and the streptococcus, which produces septic sore throat, scarlet fever, and puerperal sepsis, as well as infections in other organs and the bloodstream. It was also found that certain bacteria, particularly diphtheria and tetanus bacilli, remain localized in one area while producing toxins which spread widely to cause serious illness and death. The identification of the different bacteria was greatly aided by a discovery of Christian Gram, a Dane, who reported in 1884 that when bacteria were treated with gentian violet, followed by an iodine solution and then by alcohol, some would retain the color of the violet stain while others would lose it. This procedure is now universally known as Gram's stain. Bacteria which retain the gentian violet are designated as gram-positive, while those that are decolorized are called gram-negative.[9]

Another far-reaching discovery, the result of the work of many individuals rather than a particular one, was that animals infected with bacteria manufacture substances called antibodies, which enable them to combat bacteria of the same species and thus aid in recovery. When blood from infected animals is removed and freed of its cells, the remaining portion, the serum, contains the protective antibodies, and when given to other lower animals or to humans, this serum helps the new host fight the disease. The phenomenon of antibody production was also helpful in diagnosis, for antibodies cause bacteria to agglutinate, or aggregate in clumps. Agglutination and other tests for demonstrating the presence of antibodies could be employed to determine whether a person was ill from, or had previously had, a disease caused by a particular infectious agent.[10]

During these years, when investigators were boldly cutting swaths into previously untouched fields of knowledge, the harvest consisted mostly of a better understanding of how bacteria produced infections in humans and lower animals. Little of the knowledge was of benefit to the patient suffering from an infectious disease. Some progress was being made, however, in the prevention of infections. At the end of the nineteenth century smallpox was in the process of being eliminated by vaccination. In Massachusetts, where mortality statistics are available for these decades, smallpox had accounted for 27 deaths per 100,000 population in 1860 and 67 in 1872, whereas in 1900 it caused only one death per 1,000,000 population. By 1900 cholera had been practically excluded from this country as a result of sanitary and quarantine measures. Declines in death rates from other diseases were likewise substantial, although less spectacular. The mortality rate from typhoid fever in Massachusetts ranged between 72 and 134 per 100,000 during the 1860s; in 1900 it was down to 22 per 100,000 and still falling, mainly because of improved sanitation. The death rate from tuberculosis of the respiratory system fell from 365 per 100,000 in 1861 to 190 in 1900, a decline that can be attributed mostly to improvements in the standard of living and in personal hygiene. Yet infectious diseases still exacted a high toll from the American people. Among the ten leading causes of death in 1900 were pneumonia, tuberculosis, infections of the gastrointestinal tract, and diphtheria. Together these diseases were responsible for 590 deaths per 100,000 population, or more than one-third of all deaths. Other infections causing an impressive number of deaths included whooping cough, measles, scarlet fever, meningitis, and typhoid fever.[11] In fact, despite the gains, the level of health of many Americans at the end of the nineteenth century was generally low.

Efforts to prevent infections by public health measures were hampered in part by inadequate knowledge as to how they were spread. From the start of the nineteenth century control measures had gyrated back and forth in response to a bitter debate between the contagionists and the anticontagionists. The former contended that infectious diseases were transmitted by living agents and pointed to the obvious contagiousness of smallpox, measles, and syphilis, but they had difficulty explaining the spread of certain other diseases now recognized as infectious. For instance, because the role of mosquitoes as transmitters was unknown, they were unable to understand how yellow fever traveled from person to person. Also the origin of many cases of scarlet fever, diphtheria, and pneumonia was obscure because it was not realized that these diseases could be conveyed to susceptible persons by healthy carriers.

Since the simplistic theory of direct contagion did not fit all the known facts, most doctors by the middle of the nineteenth century had embraced the miasmatic theory, which blamed epidemics on some peculiar change in the state of the atmosphere, produced in turn by poor sanitary conditions. Public health measures based on the contagionist theory, which included isolation and quarantine, often failed to stop the spread of a disease, while measures based on the miasmatic theory, even though wrong in its basic principle, were successful. The reason was that these measures involved the removal of refuse, garbage, and offal from streets and the draining of marshes and stagnant pools, which thus got rid of the places where bacteria grew or flies and mosquitoes bred. To carry out these procedures, American cities one by one established health departments, set up sanitary requirements, and built community water and sewage systems.[12]

The proper direction and control of public health measures are impossible unless the leaders know what they are fighting against; this requires collecting accurate information on the numbers of persons born and dying in a given period of time and the causes of illness and death—in other words, vital statistics. America lagged behind England in the collection of such data, partly because of its widely scattered population and partly because, to a practical people, collecting statistics did not seem as important as tilling the soil, mining coal, and building houses and factories. In 1880 the Death Registration Area, those states and cities considered by the United States Bureau of the Census to be 90 percent complete and accurate in the reporting of deaths, included only two states, the District of Columbia, and nineteen cities from other states, representing only 17 percent of the population of the nation. Not until 1933 were all states represented. Another designation, the Birth Registration Area, was established in 1915, comprising ten states and the District of Columbia. This area did not include all states until 1933.[13]

Deficiencies in the collection of vital statistics were one reflection of the poor coordination of public health efforts. Massachusetts established the first effective board of health in 1869, and by 1901 sanitary organizations of some kind had been established for all but five states. Yet they were mainly advisory boards and were expected to interfere as little as possible in local affairs. There was even less direction of public health on the federal level. At the turn of the century the functions of the only federal agency, the Marine Hospital Service, were limited to the medical care of seamen and to preventing contagious diseases from getting into the United States or spreading from state to state. Health affairs within states were left almost completely to local and state authorities. The first formal conference of federal, state, and local public health officials was not held until 1902. Illustrative of the changing functions of the federal unit were the alterations in its name. In 1902 it became the Public Health and Marine Hospital Service, but not until 1912 did Congress stress its major role by calling it the Public Health Service.[14]

Although many cities and states were slow in providing effective public health services, a few were highly active. Among the latter were Providence, Rhode Island, and the state of Michigan, which by 1888 had established public health laboratories, mainly for the analysis of water and food. In 1887 Joseph J. Kinyoun of the Marine Hospital Service set up a one-room laboratory for research, which became the forerunner of the National Institutes of Health. The most aggressive campaign for the people's health was mounted in New York City where a small bacteriologic laboratory for the diagnosis of infectious diseases was installed in 1893. Under the leadership of Hermann M. Biggs and William H. Park it became a pioneer in developing methods for the control of infectious diseases and an example to laboratories in this country and throughout the world. Where public health measures were actively utilized, most of the motivation and hard work came from a few doctors and other public-spirited citizens and a handful of dedicated public officials. The general attitude at the turn of the century was illustrated by the remark of a Pennsylvania state senator who said he took the chairmanship of the Public Health Committee "because there is nothing to do in public health."[15]

At the turn of the century social and economic factors also contributed to the poor control of infectious diseases in the United States. The last hundred years had seen the population grow from 5 million to 76 million and its territory stretch from the Atlantic to the Pacific. Nearly half the people now lived in cities, three of which had more than a million inhabitants. To the old impersonal forces of nature, against which man had always struggled, were added the impersonal forces of an industrial economy. It was hard to salvage good health from a life of soot and smoke outdoors, of underventilation and overwork in the factory or sweatshop, and of drabness, dirt, and hunger in the tenement. In spite of William Cowper's claim that "God made the country and man made the town," rural Americans were not much better off. Poverty prepared the soil for infections in the country as well as in the city. The flood of new immigrants compounded the health problems of the cities; the country was beset by the ebb and flow of migrant workers. Organized efforts to fight disease were uneven, inefficient, and hamstrung in the cities; they were practically nonexistent in the country.[16]

A considerable obstacle to improved health care was the low level of medical education. Although the first medical school had been established in the United States in 1765 and others soon followed, much of medical care was given by doctors who had learned under the apprentice system. Education in the medical schools lagged behind that available in Europe. Although a few of the better schools were attempting to add laboratories and lengthen the curriculum, the inferior ones either gave diplomas after a short course of lectures or required of the applicant only that he mail them the stated fee. In such a climate quackery flourished both within and without the profession.[17]

A final obstacle to good medical care was the public's skepticism of what even the best doctors were able to accomplish. The results obtained with the few worthwhile drugs available to physicians in the nineteenth century were overshadowed in some people's minds by the discomforts, the serious complications, and the deaths produced by the many other drugs that were ladled out in profusion and in heroic doses. Small wonder that medical sects, such as homeopathy and osteopathy, were often more attractive, and that in 1900

Americans spent $60 million for proprietary medicines, that is, drugs with a trade marked name, advertised to the public and sold without a prescription.[18]

But improvements were in the making in both medical science and medical education. The American Medical Association, formed in 1847 with the avowed purpose of combating quackery and improving medical education, was beginning to influence the quality of the doctors who were permitted to practice medicine. After 1875 states began to set up licensing requirements, and by 1895 21 states required examination and 14 others permitted only the graduates of certain medical schools to practice. The opening of the Johns Hopkins School of Medicine in 1893 established a model of rigorous training in the basic medical sciences plus broad, practical experience with patients which is still being followed today. The deficiencies of the inferior schools were glaringly exposed in 1910 in a report by Abraham Flexner, which finally forced them either to close down or to make the necessary improvements.[19]

Meanwhile, in the first part of the nineteenth century chemists had begun to extract the active ingredients from the crude drugs then in use, such as morphine from opium, quinine from cinchona bark, and atropine from belladonna. Later they learned how to synthesize aspirin and other drugs from coal tar, a by-product of the chemical industry. Finally, at the end of the century the new science of bacteriology contributed the first serums and vaccines. The rise of medical research was manifested by the appearance in this country of journals devoted exclusively to that subject—the *Journal of Medical Research* and the *Journal of Experimental Medicine* in 1896 and the *Journal of Infectious Diseases* in 1904. There was no doubt that medical science, as well as medical education, was ready for the great step forward in the twentieth century.[20]

William Osler, the leading teacher of medicine at the turn of the century, wrote in 1905 that more had been accomplished for the health of the human race in the nineteenth century than at any period in the past.[21] Yet the new sciences of bacteriology and pharmacology, the beginnings of synthetic chemistry, the increasing acceptance by the public of the techniques of public health, and the rising stature of the medical profession promised even greater progress in the next century. The American people, looking back on their phenomenal record of growth in so many areas in the nineteenth century, had roseate dreams for the century to come, but even they could hardly have envisioned the progress toward the conquest of disease that would be made before three-quarters of the twentieth century had passed.

2. *Control by Public Health Measures*

On the evening of February 15, 1898, the battleship *Maine* blew up in Havana harbor, and six days later the United States plunged into war with Spain, that "splendid little war," according to Secretary of State John Hay. But it could as well have been called "a miserable little war," for ignorance and mismanagement led to a supply shortage, spoiled food, filth, and disease. At one point the invading forces were so decimated by illness that it seemed they would be forced to abandon Cuba, and during the year of the war 5,438 men died of disease, compared to only 968 dying from wounds received in battle, a ratio of 5.6 to 1. Among the troops encamped in the United States the major killer was typhoid fever, which developed in 90 percent of the volunteer regiments within eight weeks of entering camp and in one-fifth of all soldiers in camp during the course of the year. The frightful morbidity and mortality from typhoid in this hasty war merely reflected the wide prevalence of the disease throughout the country. Yet the picture was changing, mostly through the application of public health measures.[1]

American medicine was in transition at the turn of the century. Doctors had a twofold task: to import basic knowledge and techniques from Europe and to learn to apply them to the control of infectious diseases. It would be some time, because of the paucity of both trained investigators and adequate facilities for research, before they would contribute much toward a third objective, the addition of new knowledge. The cause of typhoid fever had been discovered and its method of spread elucidated in Europe. Thus, the story of the control of typhoid fever bridges the gap between European and American medicine and between basic science and its successful application.

Typhoid fever is caused by a short, plump, rod-shaped bacterium, which produces infection only when it enters the alimentary tract of man. To start up a new infection, the bacillus must travel from the urine or feces of an infected person into the mouth of another by means of contaminated food, fingers, or intermediary objects. From the gastrointestinal tract it enters the bloodstream and localizes, particularly in the lymph nodes, the spleen, the bone marrow, and the lymphatic tissue of the small intestine and the gall bladder. The disease usually lasts for three or more weeks and is characterized by chills, fever, headaches, general aching, lassitude, stupor, and sometimes coma or delirium. Gastrointestinal symptoms, such as abdominal pain, bloating, and diarrhea are common. An evanescent pink or red rash is often seen. Death occurs in 10 to 15 percent of patients, the usual causes being general toxicity, bleeding from ulcers in the wall of the small intestine, or perforation of one of the ulcers into the abdominal cavity.

Because the signs and symptoms may be present in various combinations and any of them may be absent in a particular patient, doctors had a hard time distinguishing typhoid fever from other diseases before the causative agent was discovered. Fever had been recognized as a major feature of certain diseases from ancient times. The Greek physician Hippocrates in the fourth century B.C. tried to differentiate fevers by the signs and symptoms that accompanied them. He delineated a group of patients with continuous fever, vomiting, diarrhea, abdominal distention and pain, a red rash, and sometimes coma or delirium—a description that could well fit typhoid fever, especially since these cases were seen mostly in the autumn, when that disease is especially prevalent. Many other writers recorded similar long-continued fevers through the centuries, including those attacking the early settlers in Jamestown in 1607 and in New Amsterdam in 1658.[2]

Recognition of typhoid fever as a distinct disease began in the early nineteenth century when French clinicians correlated the presence of certain ulcers in the small intestines with one type of continued fever. In England, America, and some other countries another fever was described in which a rash, delirium, and stupor or coma were also prominent features. Its name, typhus, denoted a confused mental state with a tendency to stupor. The first clear differentiation between typhus and typhoid fevers was made in 1836 and 1837 by H. C. S. Lombard of Geneva and William Gerhard of Philadelphia. They separated the two diseases on clinical grounds and also recognized that the intestinal ulcers characteristic of typhoid were not found in patients with typhus. The pathological changes characteristic of typhus were not identified until the second decade of the twentieth century. They consisted of microscopic changes in the walls of the smaller arteries, particularly in the brain and skin, which were strikingly different from the changes in the lympathic tissue found in typhoid fever, particularly those in the intestinal walls that gave rise to ulcers. These pathological differences explained why mental symptoms were more pronounced and eruption more consistent in typhus, while abdominal symptoms were more characteristic of typhoid.[3]

The clearcut separation of typhoid and typhus fevers not only helped doctors to diagnose them more accurately but also made it easier to find out how they spread. Whereas typhus fever was eventually shown to be transmitted by the body louse or the rat flea, which accounted for its frequency wherever people were crowded together under unsanitary conditions, as in ships and jails and among troops in wartime, the involvement of the small intestine in typhoid fever explained why that disease was conveyed by discharges from the alimentary tract. Jail and ship fevers were highly contagious, but typhoid did not always spread to the patient's family or attendants. In fact, some observers contended that it was not infectious at all. In the middle of the nineteenth century, when the miasmatic theory of the spread of disease prevailed, some doctors held that typhoid fever could originate in putrefying material, while a greater number believed that something in the feces putrefied and then carried the disease.

In 1856 William Budd, a physician in Bristol, England, published the first of a series of articles in which he demonstrated that in the stools of patients with the disease later known as typhoid was an agent that carried the infection to other persons. By painstaking studies he showed that epidemics of the disease occurred when intestinal discharges of typhoid patients were transmitted to well persons who soon thereafter became ill with the disease. He evolved a system for preventing the spread of typhoid which included boiling contaminated linen, chemical disinfection of the discharges from typhoid patients, and the washing of hands by attendants. He extended disinfection to the cistern in which sewage from the households or institution was collected because "the sewer may be looked upon . . . *as a direct continuation of the diseased intestine.*" He also recommended that water and milk be boiled during an epidemic and urged others to use these measures because "with this last step . . . science passes into DUTY." But Budd was only a small-town practitioner, and although his closeness to his patients was ideal for detecting the spread of an infectious disease, he did not have the prestige of a big-city physician. Thus, his observations were for the most part ignored while the medical profession and public tended to follow the views of Charles Murchison, the leading authority on typhoid, who dogmatically held that putrefaction was the essential factor.[4]

In the early 1880s doctors were using the crude methods of that period to stain patients' tissues and secretions and to search them for the microorganisms that caused infections. One of the first to be found was the causative agent of typhoid fever. The typhoid bacillus, or *Salmonella typhi* as it is now called, was described in 1880 by Carl Joseph Eberth of Zurich, confirmed the following year by Koch, and grown in pure culture, free of other microorganisms, by George Gaffky in 1884. Within four years the bacillus had been cultivated from the blood, stools, and gall bladders of patients with the disease and from the urine and stools of convalescing patients. Finally, in 1904 Karl Wilhelm Drigalski found typhoid bacilli in the feces of individuals who had apparently never suffered from the disease and recognized that these healthy carriers could cause epidemics.[5]

Around this same time another discovery was made which was destined to be valuable for the diagnosis of typhoid fever and many other infections. In 1896 Max Gruber of Vienna and Herbert Edward Durham, an Englishman working in Gruber's laboratory, reported that the serum of animals which had recovered from a typhoid infection, when mixed with typhoid bacilli, caused them to agglutinate. They suggested that the serums of patients could be tested by this method to determine whether they were infected with the typhoid bacillus. Later in the same year Georges Widal in Paris showed that this procedure could be used for the diagnosis of typhoid fever, and the test bears his name.[6] Thus, by the beginning of the twentieth century the clinical picture and pathology of typhoid fever were understood, the causative agent and the method of its spread were known, and the diagnosis could be made with certainty. Now this knowledge had to be applied to the control of the disease in America.

The death rates from typhoid were extremely high in many American cities in 1900, ranging up to 144 per 100,000 population in Pittsburgh and 155 in Troy, New York. This kind of sacrifice did not go unheeded. Bacteriological studies had incriminated water as a source both of typhoid fever and of other diseases, such as cholera and dysentery, and sanitary surveys had shown how rivers became contaminated. In 1896, for example, Josiah Hartzell announced that the cities of Pittsburgh and Allegheny "drink their own sewage" and that "in the Mississippi River, eight cities alone deposited, during the past year, 152,675 tons of garbage, manure, and offal, 108,250 tons of night soil, and 3,765 dead animals." Local wells were no more dependable as a water supply than rivers and lakes because they were too easily contaminated by waste from privies. The first step in purification was to filter the water. In 1880 about thirty thousand people in American cities were supplied with filtered water; by 1900 this figure had increased to nearly two million. In 1908 chemical purification was introduced as a further safeguard with the addition of chlorine to a public water supply—a mixed blessing in the opinion of many who drink the heavily chlorinated water in some large cities today.

Another step in the elimination of disease-producing microorganisms was the proper disposal of human and animal waste. Even though some cities provided a public system for the disposal of sewage by the mid-nineteenth century and others followed, these facilities were seldom adequate. As late as 1907 only three-fourths of the houses of Richmond, Virginia, were served by sewers. A doctor wrote of Baltimore in the 1890s: "one's bath flowed shamelessly and soapily into the gutters of the cobbled streets, over which were stepping-stones such as are preserved to be wondered at in Pompeii." Satisfactory sewage facilities were not achieved for most of the United States until the early decades of the twentieth century, and even today the final treatment of sewage is not adequate in many cities and towns.[7]

As communities provided safe water and proper sewage disposal, waterborne epidemics decreased. But they did not stop altogether. Public water systems still became contaminated through mechanical failure or human error, and private systems, often less rigorously maintained, accounted for a disproportionately high number of outbreaks.[8]

Milk is also an effective spreader of a number of infectious diseases, among others tuberculosis, streptococcal sore throat, diphtheria, typhoid, and the summer diarrheas of infants. Epidemics of typhoid conveyed by contaminated milk were detected as early as 1857 in England, and a number were traced to milk in America toward the end of the nineteenth century. Two different methods were advocated for keeping milk free of disease-producing bacteria, certification and pasteurization. In the United States in 1893 a campaign began to certify milk on the basis of a low bacterial count, achieved by handling the milk under sanitary conditions from cow to consumer. Although vigorously promoted in some localities, this method was too difficult to monitor to be feasible on a large scale. The other method could be traced back to Pasteur's announcement in 1864 that keeping wine for a few minutes at a high temperature would kill the bacteria that caused the wine to turn sour. In 1886 a German chemist, Franz von Soxhlet, recommended that this procedure be used for milk, and the method came to be known as pasteurization. Its value as a public health measure was shown in 1893 when a New York philanthropist, Nathan Straus, established a system of milk stations where pasteurized milk was dispensed in nursing bottles with instructions to mothers for feeding their babies.[9]

The movement for pasteurized milk gained momentum slowly. At the turn of the century most dealers were not pasteurizing their milk. As late as 1906 bacteriologists investigating typhoid in the District of Columbia concluded that the milk supplied was "for the most part, too old, too dirty and too warm." Even the most progressive health departments were depending upon inspection of dairies and education of dairymen to lower the bacterial count. But this was not enough. Finally, in 1908 Chicago became the first city in the world to require pasteurization, and others followed. A proper milk control program came to include inspection of dairy farms to see that milk was gathered under clean conditions, followed by pasteurization and refrigeration until it reached the consumer. As a result of such programs, milkborne disease became limited mostly to those rural areas where control was inadequate. As typhoid was brought under control by a variety of measures, other bacteria became relatively more important as a source of milkborne epidemics. These epidemics, though less and less frequent, served as a continued warning to health officials and public that milk was an excellent purveyor of disease if not closely controlled.[10]

As the cities began to take public health measures seriously, typhoid fever became primarily a rural disease. In New York the mortality rate for the city first fell below that for the rest of the state in 1910; in many states this occurred later. In rural America, the battle against typhoid and other diseases transmitted by the fecal-oral route was carried on vigorously by means of education and the enforcement of sanitary regulations. Yet from 1959 to 1961 typhoid was still more frequent in rural than in metropolitan areas in all sections of the United States except the Middle Atlantic states. The discrepancy was greatest in the Pacific states, where the average annual incidence was 11.9 per 100,000 population in rural as opposed to 2.1 in metropolitan areas.[11]

In the period when various bacteria were being cultured from patients with infectious diseases and becoming identified as the causative agents, they were sometimes found in healthy persons also. At first little attention was paid to these findings, but during the last decade of the nineteenth century the idea gradually gained ground that well persons who were carriers of pathogenic micro-organisms were responsible for a significant proportion of new cases of the disease. As so often happens in the history of science, the idea arose independently in two different places. Koch had found that patients who recovered from cholera carried the micro-organism in their feces for some time afterward, and in 1893 he emphasized their role in spreading disease. In the same year Park and Alfred L. Beebe of the New York City Health Department showed that many persons who were in contact with diphtheria patients were carriers of the diphtheria bacilli. Although typhoid bacilli had been cultured from the urines and stools of convalescing patients, the concept that they were important sources of infection was not widely adopted until 1902 when Koch stressed the necessity of finding them and disinfecting their excreta. A model program was established in Germany, where eleven scientific institutes, under Koch's aegis, both assisted the local health officers in tracing each epidemic to its source and made bacteriologic examinations on specimens of blood and excreta obtained from patients or suspected carriers.[12]

A similar aggressive campaign was mounted by the Health Department of New York City where bacteriologists found that 5 percent of patients convalescent from typhoid were carriers. Their work also highlighted the importance of the chronic carrier, one who excretes pathogenic bacteria over a period of months or years. The most famous of these was an Irish cook, Mary Mallon, who was dubbed Typhoid Mary because of the trail of cases she left behind her. George Soper of the Health Department picked up her track in 1906 when he traced six cases of typhoid which had occurred in one household to a cook who had left abruptly after a stay of three weeks. When he finally located her, she had been the cause of seven epidemics in six years. When she would not allow her stools to be examined, the Health Department placed her in a hospital where numerous examinations showed that her feces were teeming with typhoid bacilli. Astonished, Soper wrote, "The cook was virtually a living culture tube, in which the germs of typhoid multiplied." After three years of incarceration she was released on the promise that she would not work where she handled food and would report at regular intervals. When she disappeared from sight and suspicious epidemics began popping up, it appeared obvious that she had broken both promises. She was not located again for five years, and this time she was kept in a hospital until she died 23 years later. Altogether 59 cases and three deaths were traced to her, though doubtless there were many more. When Mary sued for release from the hospital, the courts and much of the press sustained the Health Department's actions. The name Typhoid Mary became known around the world as a symbol of the chronic carrier of disease.[13]

There was one way that Typhoid Mary might have gained her freedom. Doctors had found that typhoid bacilli lodged particularly in the gall bladder in chronic carriers and that when this organ was removed, the stools often remained permanently free of the bacilli.[14] Although Mary refused the operation, other chronic carriers welcomed it. As waterborne and milkborne typhoid became less common in the United States, the cases of the disease that did occur were mostly traceable to food prepared by carriers. Health officials became adept at hunting carriers down and persuading them either to refrain from handling food or to have their gall bladders removed. As a result, epidemics of typhoid originating from a carrier became less and less frequent, although occasionally one still occurs, particularly when someone, unaware of being a carrier, prepares food on an out-of-the-ordinary occasion, such as a picnic or church supper, where facilities for proper handling and storage are inadequate.

The experience with typhoid demonstrated that control measures can be readily applied to fecal-oral carriers because their excreta are easily cultured and most carriers are willing to take one of the two protective steps. The problem is more complex in diseases transmitted by the respiratory route. Cultures taken from the nose and throat are not as consistently positive, the conditions of transmission are not as easily identified, and the carrier cannot be expected to shun all human company for more than a few days. The challenges and problems of controlling respiratory carriers are illustrated by the other disease that was intensively studied during the early part of the twentieth century, diphtheria. It is caused by a small, club-shaped bacillus named *Corynebacterium diphtheriae*, infection with which produces a false membrane in the throat, less often in the nose or larynx. Specifically, when these bacteria lodge on a membrane, they kill the cells, whereupon the blood vessels pour out fluid, which clots and covers the surface, trapping the bacilli along with white blood cells and the remains of the surface cells. This thin, tough layer, which adheres to the surface of the infected organ, is called a false membrane or pseudomembrane. When the membrane extends into the larynx, it may close this narrow channel, cutting off the supply of oxygen to the lungs and killing the patient by suffocation. Hence the most gruesome name for the disease: *morbus strangulatorius.* Diphtheria bacilli usually remain and multiply around the site of the pseudomembrane, while manufacturing a toxin which spreads throughout the body, causing fever, extreme prostration, injury to the nerves governing various muscles, and especially damage to the muscle of the heart which is often severe enough to cause sudden death.

The ancient writers did not distinguish the different kinds of sore throat, but Aretaeus, the Cappadocian, in the second century A.D. was probably describing diphtheria when he wrote about ulcers on the tonsils that were "foul and covered with a white, livid, or black concretion." He reported that some patients died of the infection in the throat, adding, "if it spread to the windpipe, it occasions death by suffocation within the space of a day." Although later writers described illnesses that could have been diphtheria, it was not until 1826 that the French physician Pierre Bretonneau clearly differentiated this disease from other inflammations of the throat by emphasizing the primary significance of the pseudomembrane and named the disease *diphtherite,* which means skin or membrane. Bretonneau was also the first successfully to bypass the obstruction by tracheotomy, which enabled the patient to breathe until the inflammation had subsided.[15]

From the study of epidemics of diphtheria Bretonneau became convinced that the disease was contagious, and he believed that a particular agent acted on the tissues to create the characteristic false membrane. This was proved in 1869 by a German surgeon, Friedrich Trendelenburg, who produced similar membranes in animals by opening the trachea and putting in small pieces of diphtheritic tissue from patients. Amid the general search for bacteria that might be causing diseases, the German pathologist Edwin Klebs reported in 1883 that he had seen a club-shaped bacillus in the diphtheritic membranes of some patients. It is uncertain whether this microorganism was actually *Corynebacterium diphtheriae,* but the following year another German, Friedrich Loeffler, settled the matter by identifying similar bacteria in the membranes of diphtheria patients and also culturing them in artificial media. One name for the bacillus, the Klebs-Loeffler bacillus, honors both discoverers. When Loeffler inoculated animals with the diphtheria bacilli, they proliferated only in the local area, even though tissues in distant organs were damaged and the animals died. He guessed correctly that a toxin produced by the bacteria was responsible for the remote areas of injury. Although Loeffler cautiously refrained from claiming that the bacillus he described was the cause of diphtheria, other investigators rapidly piled up evidence that diphtheria was a single disease, no matter where the false membrane might be situated, and that it could be diagnosed by finding the Klebs-Loeffler bacillus in the membrane.[16]

While investigators all over Europe eagerly tackled the problems of identifying and growing the diphtheria bacillus, not many Americans were in a position to do so. Yet a few doctors, who were either self-trained or had spent time in European laboratories, were able to set up small bacteriological laboratories of their own. One of the first of these was T. Mitchell Prudden of New York. In 1889 he reported that he had cultivated streptococci, rather than the bacilli described by Loeffler, from the throats of 22 children whom clinicians had diagnosed as having diphtheria as a complication of measles or scarlet fever. In retrospect, it is obvious that the diagnosis of diphtheria was erroneous and that these cases were examples of the difficulty of making the diagnosis of diphtheria in the days before the science of bacteriology had enabled doctors to understand the disease. At any rate, two years later William H. Welch, professor of pathology at the Johns Hopkins Medical School, reported that he had been able to culture the typical bacilli from the pseudomembranes of every one of eight patients with diphtheria, and soon other Americans, including Prudden himself, were finding the bacillus consistently in cases of diphtheria.[17]

Then in 1892 an important event in the history of infectious diseases occurred when the New York City Health Department established a small diagnostic laboratory and Biggs, the director, placed Park in charge of the bacteriology of diphtheria. The two men immediately moved to make the laboratory a practical help to clinicians. Not content with offering to examine for diagnosis material brought to the laboratory from suspected cases of diphtheria, they placed culture tubes in pharmacies throughout the city and arranged for collection after doctors had inoculated them. The medical profession responded so well that in 1894 the Health Department was in a position to require that three negative cultures be obtained before a patient was allowed out of isolation. In this way it hoped to break the chain by which the disease was transmitted. When Biggs explained to Koch the diagnostic procedures offered by the New York City Health Department, the latter confessed, "You put us to shame in this work."[18]

Loeffler had been perplexed by one finding in his original study, that he had grown the diphtheria bacillus from the throat of a healthy child, and others made the same observation. Park's interest was aroused when the only child in a family developed diphtheria soon after a nurse came to work in that household immediately after leaving a position in a diphtheria ward. Park cultured the same type of bacillus from the throats of the child and the nurse. Following up this observation with the help of Beebe, he found that while diphtheria bacilli usually disappeared from the throats of patients within ten days after the beginning of the disease, in a few patients they remained for periods up to two months or more. Furthermore, other members of a diphtheria patient's family often harbored the bacilli in their throats. Finally, virulent diphtheria bacilli were cultured from the throats of eight of 330 healthy persons who had not been in contact with diphtheria patients. These and other studies alerted the medical and public health professions to the significance of the healthy carrier of diphtheria bacilli. It became clear that diphtheria bacilli could remain in the throats of patients for weeks or months after they had recovered from the disease and could also be found in the throats of many healthy persons who had been in contact with diphtheria patients, as well as in some who had not.[19]

Knowledge of the ubiquity of the diphtheria bacillus led to vigorous campaigns to isolate patients, quarantine contacts, and disinfect anything that had been in the vicinity of a patient with the disease. Some of these measures produced results; others did not. In Providence, Rhode Island, Charles V. Chapin, the public health director, devised a method for determining the transmission of acute respiratory diseases, based on the secondary attack rate, that is, the number of cases occurring within a specified time in an attacked family, exclusive of the primary case. Using this procedure, he found that in Providence in 1903, among 1,420 persons in families of diphtheria patients who had been removed to the hospital, 55 subsequently developed diphtheria. From attack rates in families where the patient had not been taken to the hospital, he calculated that 120 of the 1,420 would have come down with the disease if they had remained at home. Thus, strict isolation of patients with diphtheria paid off. One important by-product of observations like these was the enlargement and upgrading of special hospitals for patients with contagious diseases or special buildings associated with municipal hospitals. Another measure, however, called terminal disinfection, consisting of fumigation of the patient's quarters after recovery or death, along with the burning of some objects and the chemical disinfection of others, turned out to be of negligible value.[20]

The problem of what to do with healthy carriers was more perplexing. Their number was far greater than were typhoid carriers, especially during the winter months. In Baltimore in November and December 1921, for example, 3 percent of 1,587 schoolchildren examined were harboring virulent diphtheria bacilli, and at times 25 to 50 percent or more of healthy staff and inmates of public institutions in Providence were carriers. Health departments did their best to stress to doctors and the public that healthy carriers could transmit the disease; yet preventing the spread of diphtheria bacilli by culturing the throats of the entire population and quarantining all the carriers was an impossibility.[21]

Thus, while sanitary measures and the control of carriers materially reduced the frequency of typhoid fever, these procedures had little effect in preventing new cases of diphtheria, or indeed of most infections spread by the respiratory route. The reason is obvious; we need not drink or bathe in our neighbor's excreta, but we cannot avoid washing our respiratory passages with the air he breathes out. And since there is no easy method of purifying air before we inhale it, respiratory infections must be prevented in some other way.

3. *Prevention by Immunization*

One way to protect oneself from an infectious disease is to run away from it, as Boccaccio's group of storytellers fled to the country to wait out a visitation of the plague. This method is seldom feasible. Another way is to have an attack of the disease, since it has been recognized for centuries that some infections seldom repeat themselves in the same individual. Yet this method is often precarious, because the first attack may be severe enough to kill or maim the victim. Man has found a way to improve upon nature by producing a mild or modified infection which protects the individual against another attack of the disease. The person who is infected naturally or artificially develops immunity. His body recognizes the difference between "self" and "not-self" and produces antibodies which neutralize or destroy the foreign substances. The method of bringing about protection in this way is called immunization and can be used when there is enough time before the disease strikes. But if a patient already has symptoms of the disease, the method will not work. Instead, the patient can be injected with serum taken from an animal previously rendered immune. The antibodies manufactured by the animal will combat the invaders in the same manner as antibodies that the patient would have produced himself. During the first third of the twentieth century rapid progress was made in the techniques of immunization and serum therapy and the knowledge underlying both procedures.

Early in the eighteenth century inoculation of healthy persons with material from smallpox pustules had been introduced into Europe and America. The person inoculated ordinarily suffered only a mild inflammation, which prevented him from contracting the disease on later contact. But sometimes this procedure precipitated a severe case of smallpox, occasionally leading to death. Consequently it was abandoned after an English physician, Edward Jenner, reported in 1798 that cowpox would produce a similar immunity against smallpox with much less risk. Jenner's method was called vaccination after "vaccinia," the name for cowpox; later the term came to mean the administration of any microorganism or its products for the purpose of producing immunity to that microorganism.

In the eighteenth and nineteenth centuries doctors in various countries tried inoculating material from patients with other infections, such as measles, gonorrhea, and syphilis. But these attempts were tentative and sporadic, producing no reliable method of immunization against infections. Because the nature of the diseases was unknown, such experiments were only intermittent stabs in the dark. Light began to dawn on the subject of immunity when Pasteur found that he could protect animals by first injecting attenuated microorganisms. Using this method, he succeeded in protecting humans against rabies. Yet because techniques were still crude and based on scanty knowledge of basic processes, safe and effective vaccines composed of live microorganisms were not forthcoming for the prevention of other infections.[1]

A way out of the dilemma was found in 1886 when Daniel E. Salmon and Theobald Smith of the federal Bureau of Animal Industry reported that pigeons had been protected against hog cholera by a heat-killed vaccine. A few years later Almroth Wright of the British Army Medical School received a visit from a Russian bacteriologist, Waldemar Haffkine, who claimed that he had immunized guinea pigs with the heat-killed microorganisms of cholera. Soon after this, Wright began to work on vaccination against typhoid. In 1896 he injected two persons with heat-killed typhoid bacilli, and a few months later two Germans, Richard Pfeiffer and Wilhelm Kolle, upon repeating this experiment, found agglutinating antibodies against typhoid bacilli in the serums of the persons vaccinated. Wright then made a bold assumption: because both guinea pigs and humans had developed agglutinins following typhoid infections, and since agglutinins had appeared in the blood of the few persons who had been given killed typhoid bacilli, killed bacilli would immunize against the disease. The following year Wright vaccinated 84 persons during an epidemic of typhoid in a mental hospital. No one who had been vaccinated developed typhoid; yet the results were inconclusive, because only four cases had occurred among the 118 unvaccinated persons. The epidemic was apparently on the wane.[2]

These results—the first of many equivocal ones that would be obtained with typhoid vaccine—nevertheless convinced Wright that the absence of typhoid in the immunized persons was proof of the vaccine's effectiveness. A short time later, when sent to India to investigate plague, he seized the opportunity to immunize with typhoid vaccine any soldier who would give his consent. By 1898, when word of his unauthorized activities had reached England and he was ordered to stop, he had injected over 4,000 soldiers. This was the beginning of a prolonged, acrimonious, and at times virulent controversy between Wright, who believed that he had produced evidence that every soldier should be inoculated forthwith, and the Medical Advisory Board of the army and others, who were not convinced that Wright's vaccine would prevent typhoid.[3]

According to Wright's figures, of 4,502 men vaccinated in India, 44 had developed typhoid, compared to 657 among the 25,851 uninoculated men, or 0.98 percent versus 2.54 percent. These results looked impressive, but his critics pointed out several flaws. First, the unvaccinated soldiers were not true controls but merely persons who had declined to volunteer for vaccination after his explanatory lecture. They may or may not have been at greater risk of developing typhoid than those who volunteered; there was no way of knowing. It was possible, since they were not interested in vaccination, that they may have been less likely to follow rules of personal sanitation than those in the vaccinated group. At any rate, they did not constitute a valid control group. Furthermore, the frequency of typhoid in vaccinated soldiers varied considerably in the different locations where the vaccine was given. This probably resulted from differences in the potency of the batches of vaccine, which were made up at different times. If so, it called for further improvements in the method of making the vaccine rather than for wholesale, immediate vaccination. Finally, Wright's vaccine often caused severe reactions, such as pain at the site of injection, fever, and malaise, sometimes lasting several days, another indication of the need for a better vaccine. About this time Great Britain became involved in a war in South Africa, and over 14,000 soldiers were vaccinated. Again only those who volunteered were injected, and the unvaccinated troops did not represent valid controls. The Army Council decided, upon recommendation of the Medical Advisory Board—and from a statistical viewpoint they were correct—that the protective value of vaccination was still unproved. Accordingly, they ordered that a more thorough study be carried out.[4]

The echoes of this controversy reached the public, some of whom opposed Wright, especially the antivaccinationists, a well-organized group of zealots who had been fighting vaccination against smallpox and who were happy to jump into the battle against typhoid vaccination also. Consequently, even though subsequent studies in large numbers of soldiers showed that typhoid occurred three to ten times more frequently in the unvaccinated than in the vaccinated and that the death rate in the uninoculated was seven to twenty times higher, when World War I began, vaccination against typhoid was not compulsory in the British army. When it became apparent that a law requiring all troops to be vaccinated could not be gotten through Parliament, Lord Kitchener cut the Gordian knot by ordering that no soldier be sent abroad unless he had been inoculated for typhoid. In Germany, where much of the basic work leading to the development of typhoid vaccine had been done, no controversy over typhoid vaccine developed. Perhaps for this reason, at the beginning of World War I German troops had not been immunized against typhoid, and vaccination was not made compulsory until many of their soldiers became ill.[5]

As it turned out, the Americans produced the best evidence for the value of vaccine in preventing typhoid. Mindful of the toll taken by the disease among American soldiers in the Spanish-American War, the army introduced typhoid vaccination in 1909 on a voluntary basis and made it compulsory in 1911. The number of cases of typhoid dropped dramatically, from 303 per 100,000 troops in 1909 to 26 per 100,000 troops in 1912. Except for minor rises in 1916 during the mobilization on the Mexican border and in 1931 because of a small milkborne epidemic in one unit, the incidence of typhoid in the army has since remained low. The large mobilizations for World War I and II caused practically no increase in the disease, a phenomenon unheard of in earlier wars. Although improvements in sanitary measures played a large part in lowering the incidence of typhoid in the army, as in civilian life, the sudden decrease in cases of the disease could not have been the result of such improvements alone, for they had been in progress for many years. Besides, the rates for admission to army hospitals for other diarrheal diseases, which should have been equally affected by sanitary measures, did not decrease correspondingly with the sharp drop in typhoid rates.[6]

Additional evidence of the value of the vaccine was obtained during localized epidemics in which only a portion of the exposed persons had been vaccinated. In 1917 an epidemic of typhoid occurred in Hawaii when part of the water supply in Schofield barracks became contaminated with typhoid bacilli. Among 4,000 immunized persons using this water, mostly soldiers, there were 55 cases of typhoid, as compared with 800 cases among 8,000 unimmunized persons who also drank the water. An even better illustration of the effectiveness of the vaccine occurred in an epidemic transmitted by orange juice in a residential hotel occupied by 181 women enlisted in the United States Coast Guard and 194 civilian women. Both groups were engaged in clerical work and were of comparable ages. Typhoid developed in 17, or 12 percent, of the 140 civilians who had neither been vaccinated nor had suffered a previous attack of the disease, but in only one, or 0.06 percent, of the vaccinated enlisted women.[7]

The experience with typhoid vaccination, besides showing the protective value of the vaccine, demonstrated that the immunity provided was not always complete, and that some individuals either did not build up an immunity when vaccinated or succumbed to the disease because they had ingested overwhelming doses of typhoid bacilli. This principle holds for all vaccines, although greater protection can be obtained against some diseases than against others. In fact, the immunity conferred by an attack of typhoid is not absolute either, and second attacks of typhoid have long been known to occur, although they are infrequent. In a particularly severe epidemic of typhoid in 1950, 34 percent of the exposed persons developed the disease. Of the 54 persons who had contracted typhoid fever in an epidemic three months earlier, 20 percent developed typhoid again during the second epidemic.[8]

In 1966 Richard Hornick and Theodore Woodward at the University of Maryland reported that, when 1,000 typhoid bacilli were instilled in the intestines of volunteers, none developed typhoid fever. The proportion of subjects coming down with the disease rose progressively as larger numbers of bacilli were instilled, reaching 95 percent when 1 billion were given. Approximately half of the volunteers came down with the disease when challenged with 10 million bacilli, and previous immunization gave no protection. In contrast, when 100,000 bacilli were given, only 28 percent of the unvaccinated volunteers and 8 percent of the vaccinated became ill. These experiments verified the observations made during epidemics that the vaccine did not protect against an overwhelming dose of typhoid bacilli. Fortunately such large doses are seldom encountered outside the experimental laboratory.[9]

Many improvements have been made in typhoid vaccines. Perhaps the most extensive experiments were carried out by Joseph F. Siler and his associates at the Army Medical School in Washington, now the Walter Reed Army Institute of Research. In comparing the immunizing capacity of strains of typhoid bacilli gathered from all over the world, they found to their surprise that the strain contained in the vaccine used in this country for the previous thirty years was a relatively poor producer of immunity. The most effective strain they tested had been isolated from a carrier in Panama. A vaccine made from it was used in the Civilian Conservation Corps beginning in January 1937, following which the incidence of typhoid fell from 7.8 per 100,000 persons in 1936, the last year in which the old, less effective strain was used, to 3.4 in 1939. The new strain has since become the standard one for vaccines made in this country. In strictly controlled experiments carried out under the auspices of the World Health Organization in countries where typhoid was still prevalent, a vaccine like the one developed at the Walter Reed Institute prevented 79–90 percent of cases of typhoid, compared to the controls. Such experiments helped prove the proposition that Wright thought he had answered: killed typhoid bacilli will prevent typhoid in many of the persons who are given a potent enough vaccine. In practice, typhoid vaccine is widely used to immunize persons who are likely to be exposed to the disease. These include not only military personnel but also persons who are traveling or living where food or water supplies are not carefully controlled, who are involved in catastrophes such as floods, where water is likely to become contaminated, and who are attending patients with typhoid fever.[10]

In comparison with vaccination against typhoid, immunization against diphtheria turned out to be more effective and became the principal method of bringing this disease under control. Methods for immunizing against diphtheria were developed in close connection with those for immunizing against tetanus. In a typical case of tetanus, several days after receiving a wound contaminated with dirt, the patient is aware of stiffness in various muscles, which becomes worse within a few hours or days. Often rigidity of the muscles of the jaw makes opening the mouth painful or impossible; hence, a common name for the disease is lockjaw. In more severe cases, painful convulsions occur following a slight stimulus, such as a noise, a light, or a simple jarring of the patient's bed. Such a frightening illness was bound to be recognized by the earliest doctors. One of Hippocrates's aphorisms was, "Spasm, supervening on a wound is fatal."[11]

In 1884 Arthur Nicolaier of Göttingen reported that he had produced spasms typical of tetanus in animals by inoculating them with specimens of soil. At the site of inoculation he found "long narrow bacilli." Animals injected with previously heated soil remained healthy. Since he did not find the bacilli throughout the body, Nicolaier concluded that the disease was caused by a toxic substance produced by the bacteria and carried elsewhere. Shortly afterward Emile Roux and Alexander Yersin of the Pasteur Institute, in a search for the diphtheria toxin, produced paralyses in animals by injecting bacteria-free filtrates of cultures of diphtheria bacilli, which proved that they had isolated the toxin. In 1889 Shibasaburo Kitasato, working in Koch's laboratory, managed to culture the tetanus bacillus, now called *Clostridium tetani*, in an atmosphere deficient in oxygen. He produced tetanus by inoculating animals with these bacilli, and when he found the bacilli only at the site of inoculation, he too concluded that the disease was caused by a toxin. One year later he and Emil Behring demonstrated that the toxin manufactured by tetanus bacilli produced immunity when animals which they had previously injected with sublethal amounts of the toxin were able to survive the injection of a lethal amount. More important still, when the serum of the immunized animals was injected into other animals, it in turn protected them against tetanus. During the same period Behring showed that the toxin of diphtheria bacilli also produced immunity and that the serum from immunized animals could protect other animals against diphtheria.[12]

Thus tetanus and diphtheria, diseases that differ greatly in their symptoms, were found to be caused by a similar mechanism. In each case, bacteria multiply in a local area, manufacturing a toxin which is called an exotoxin because it is secreted by living bacteria into the surrounding medium. In diphtheria the toxin is carried by the blood to various muscles where it causes paralysis or, in the case of the heart, irregularity or cessation of the heartbeat. In tetanus the toxin travels along the nerves to the central nervous system. Later, other bacilli, classified along with *Clostridium tetani* as anaerobes, or microorganisms that grow in an oxygen-deficient atmosphere, were found to produce similar lethal toxins. Among these were bacteria that produce gas gangrene when they grow in deep wounds and *Clostridium botulinum*, which grows in imperfectly sterilized foods, secreting a toxin that causes paralysis of nerves.

Behring and his coworkers perfected the tetanus and diphtheria antitoxins so that they could be used for the treatment of patients. In controlling the spread of diphtheria from person to person, however, the antitoxin was of little value. True, a doctor could sometimes prevent the disease from occurring by giving antitoxin to a person who had been in contact with a diphtheria patient, but the antibodies formed in animals and conveyed in their serum lasted only a few weeks after injection and the individual might come in contact with diphtheria bacilli at a later date. What was needed was a method that enabled people to produce their own antibodies. For a long time investigators were unable to immunize humans with pure diphtheria toxin because the toxin could be lethal. But in 1907 Theobald Smith reported that immunity could be produced in guinea pigs by mixing toxin with enough antitoxin to make it harmless, and he suggested that this method be used in the immunization of humans. Behring followed Smith's suggestion and in 1913 reported that he had given such a mixture to humans and that the antitoxin portion had blocked the adverse reactions while the toxin had produced a good antibody response. A practical method of immunizing against diphtheria had been found.[13]

Park and his associates in New York City promptly took up the new procedure and used it effectively in conjunction with a test that had been devised a few years before by Bela Schick of Vienna. The Schick test consisted of the injection of a small quantity of diphtheria toxin under the skin. In patients with no circulating antitoxin, and hence no evidence of immunity, an inflammatory reaction appeared at the site of the injection, indicating a positive test. In 1919 Park initiated a program of testing every child in the schools of New York City whose parents would give permission and immunizing those with positive tests, again making that city a leader in the practical application of bacteriologic principles.[14]

Yet toxin-antitoxin mixtures were not without hazards. If the toxin were not properly neutralized, serious symptoms and even death could occur. Catastrophes in various parts of the world from the injection of improperly made toxin-antitoxin mixtures spurred the search for a better immunizing agent. Gaston L. Ramon in France and Alexander T. Glenny in England found that toxin could be altered by the addition of formalin, a 40 percent solution of formaldehyde, so that it no longer caused adverse reactions while it retained its immunizing properties. During the 1920s they led a movement to displace toxin-antitoxin with this formalin-altered toxin, called toxoid. But here, as so often, the good already experienced was the enemy of the untried best. Since diphtheria was being successfully prevented with toxin-antitoxin immunization, and mortality rates were falling noticeably wherever it was used intensively, some were reluctant to give it up. Not until the early 1930s did the first state health department change over entirely to the use of toxoid, and it was several more years before toxoid displaced toxin-antitoxin everywhere.[15]

A further extension of immunization practices was made by the Chicago Health Department. After adopting the system of immunizing schoolchildren that had proven so successful in other cities, it discovered that 30 percent of the cases of diphtheria in Chicago now occurred in children under five years of age. By examining the records of births in the city over the past several years, the department was able to locate most of the children in this age group and thus to send nurses to their homes to test and immunize them. As a result, deaths from diphtheria fell from 12 per 100,000 population in 1930 to less than two in 1932. By sharing experiences of this kind, state and local health officials, with the active help of practicing physicians, were able progressively to lower the mortality from diphtheria. The experience in Omaha, Nebraska, was typical. The death rate ranged between 34 and 118 per 100,000 population from 1880 to 1896, when antitoxin came into use, and through 1920 it still reached as high as 21 in epidemic years. Toxin-antitoxin immunization was introduced in Omaha in 1923, and a vigorous campaign against diphtheria was launched in 1927. The mortality rate fell to 3.4 per 100,000 population in 1930 and 1.2 in 1940 and remained below 1.0 after 1945.[16]

Despite such examples, the slogan with which one city launched its campaign in 1928, "Banish diphtheria from Cincinnati," has not been fulfilled, except temporarily, in any large city, and certainly not in the United States as a whole. In only two years thus far has the number of reported cases in this country fallen below 200, and in 1974 there were 272 cases. Small outbreaks and epidemics continue to occur, especially in communities containing a number of unimmunized or incompletely immunized persons. These are usually the poorer communities and those with a high percentage of foreign-born inhabitants. Poverty, ignorance, and unconcern remain obstructions to the application of knowledge here as elsewhere.[17]

For many years the prevention of tetanus took a different direction from the prevention of diphtheria. The course of the disease was such that it could be prevented by the injection of antitoxin. If a small dose was given soon after an injury occurred in which there was contamination by soil, tetanus usually did not develop. Even in deep or crushing wounds where much tissue was destroyed, prompt surgical removal of dirt, debris, and dead tissue, along with the injection of antitoxin, almost always eliminated the danger of tetanus. The antitoxin neutralized the toxin as it was formed and before it could unite with nerve tissue.

In civilian life, where tetanus-producing injuries were infrequent, the adoption of antitoxin as a preventive was slow. Deaths from tetanus decreased in the Death Registration Area of the United

States from approximately 35 per million population during 1901–1905 to 7.2 during 1935–1939. Yet much of this improvement could be attributed not to the use of antitoxin but to the diminishing proportion of the population working in rural areas and to the increased precautions against accidents, such as a ban on the sale of fireworks.[18] Clearer proof of the value of antitoxin came from the military forces.

Explosive projectiles leave a trail of torn, ragged tissue and impacted dirt. Together they form an ideal nidus for the growth of tetanus bacilli. The painful spasms of tetanus made a nightmare of the last days of many a wounded soldier, especially those who had fought in cultivated fields. For example, nearly one of every 100 British wounded developed tetanus in the early months of World War I when the armies were fighting in Flanders. Thus there was a compelling reason to make sure that every soldier wounded in battle received a prophylactic dose of tetanus antitoxin. By the time the United States entered World War I these injections were routine, so that only 13 cases of tetanus developed for every 100,000 wounded men, as compared with 380 per 100,000 in the German army, where routine injections were not the practice.[19]

Despite these excellent results, tetanus antitoxin had its drawbacks as a prophylactic. For one reason or another the patient did not always receive the necessary injection, or it was given too late. Also, tetanus antitoxin, like other serums, caused allergic reactions, which were sometimes fatal, in an appreciable proportion of recipients. Active immunization, inducing the production of antibodies by the patient himself, was more desirable. But tetanus toxin is so powerful—less than one-billionth of a gram of dried toxin will kill a mouse, and pound for pound man is even more susceptible—that it cannot be neutralized by antitoxin with any assurance of safety, as can diphtheria toxin. Accordingly, when Ramon found out how to make toxoid, he predicted that this process would be even more valuable for tetanus than for diphtheria, and it was. Immunity from a course of toxoid injections was found to protect adequately against tetanus for as long as twenty-one years if another "booster" dose was given after the injury was received.[20]

Immunization with toxoid was made obligatory in the French army in 1936, as a result of which no case of tetanus occurred in the campaigns of 1939 and 1940 among immunized soldiers. The real test, however, was made by the United States military forces. Beginning in 1941 they relied on toxoid alone as a prophylactic without injecting antitoxin in addition after an injury. They gave a basic course of toxoid injections to all troops and a booster injection when an injury occurred. As a consequence, during American involvement in World War II from 1942 to 1945, when 3.6 million soldiers and sailors were admitted to hospitals as a result of battle wounds or accidents, the complete standard regimen failed to prevent tetanus in only seven persons.[21]

Tetanus and diphtheria toxoids and other vaccines were combined in a single injection in the French army as early as 1936. They were advocated for civilian use after studies had showed that the levels of tetanus and diphtheria antitoxin in the blood of children who received the combined toxoids were at least as high as in those who had been injected separately. The principal method for obtaining widespread protection has been the immunization of children, especially through the schools and clinics. As a result, by 1969 over half of the inhabitants of the United States under twenty years of age had received complete immunization against diphtheria and tetanus and only 6 percent had received no immunization at all against these infections. From 1950, when tetanus became a reportable disease, the number of cases dropped from about 500 to a low of 101 cases in 1973.[22]

The early promise of success from vaccinating against typhoid stimulated investigators to try other vaccines made with whole bacteria. Pneumonia, as a leading cause of death, was a prime target. The pneumococcus, cause of the most common variety of pneumonia, was identified with that disease in 1881, and beginning in 1891 a number of investigators found that immunity could be produced in animals by injecting pneumococci. In 1911 Wright went to South Africa, where pneumonia was common and highly lethal among the natives working in the gold and diamond mines, to start the mass immunization of these workers against pneumonia. On the basis of a few thousand immunizations, he and his associates claimed that they had decreased the frequency of pneumonia by 50–60 percent as compared with unvaccinated controls. Critics questioned this conclusion, however, because of a faulty system of allotting controls and other statistical fallacies. Apparently here, as in vaccination against typhoid, Wright's enthusiasm had run ahead of his facts.[23]

Other investigators who tried vaccinating Africans against pneumonia also believed that they had prevented many cases from occurring. As a consequence, starting in 1917, all Africans were immunized as soon as they came to work in the mines. Following this, the death rate for pneumonia fell from 12.3 per 1,000 workers in 1911 to 2.65 in 1917. At the same time, however, the percentage of workers who came from the tropical regions and hence were presumably more susceptible to pneumonia decreased correspondingly. And when inoculations were stopped in three mines, the death rate from pneumonia did not rise.[24]

In the United States attention was focused on pneumococcal vaccines during the severe epidemic of influenza that began in 1918. Russell Cecil and his associates vaccinated troops in several army camps with pneumococci and reported that they had reduced the frequency and fatality rate of pneumonia by about 50 percent as compared with controls. Roughly similar results were obtained by others over the next decade, although a few studies showed little or no protection from the vaccine. At the end of the 1920s the inoculation of thousands of persons in South Africa, the United States, and other countries had not settled the question of whether vaccination prevented pneumonia. Because the vaccines available at that time, which were prepared from whole pneumococci, produced uncomfortable and sometimes severe reactions and because attention became focused on treatment of pneumonias with serums and then with other agents, interest in vaccination declined and did not revive until recent years.[25]

Vaccination was also tried in an attempt to protect against the meningococcus, a frequent cause of meningitis, or inflammation of the meninges, the membranes that cover the brain. In the period before World War I Abraham Sophian of the New York City Health Department popularized a vaccine made of whole meningococci, but his results were not sufficiently convincing to excite much interest.[26] Sporadic trials of vaccines in the next three decades were no more successful, and again the stimulus to develop an effective vaccine disappeared when effective therapy became available, only to rise again in recent years.

One outgrowth of the success of vaccines as preventives of infections was an attempt to use them also for treatment. A bacteriologist in 1911 listed sixty-five diseases for which vaccines had been used therapeutically. Some of the vaccines used for therapy contained bacteria known to cause the disease under treatment, such as pneumococci for the treatment of pneumonia. Sometimes the vaccine was prepared from microorganisms cultured from the patient's own infection. Other vaccines were mixtures of common bacteria. For instance, vaccines made of microorganisms ordinarily found in the nose and throat of healthy persons were used in the treatment of chronic infections of the sinuses and bronchial tubes. Frequently mixtures of bacteria were recommended for diseases where the cause was unknown, such as arthritis, or where the results of other therapy were poor, as in recurrent abscesses of the skin. Since many of the trials were hit-or-miss and practically all were conducted without proper controls, they generated much hope and confusion while producing no proof of the value of vaccine therapy. The field was thus opened wide for those who could profit by deceiving, and the charlatans took full advantage of the opportunity. Gradually, as diagnosis became more accurate and more diseases were arrested by specific therapies, the ineffectiveness of vaccines became evident and they lost their appeal. Today it is generally accepted that vaccines will neither arrest nor cut short an infection once the symptoms of the disease have appeared.[27]

In summary, the principle of preparing vaccines in the laboratory that could be used to immunize persons against infections, which had been introduced by Pasteur, was effectively used in the first part of the twentieth century in preventing a few diseases. Most of the pioneer work on immunization was done in England, France, and Germany, while Americans were more active in the development of organized programs for immunizing large numbers of people, both military and civilian. Before effective vaccines could be developed for the prevention of many other infectious diseases, especially those caused by viruses, greater knowledge and more advanced techniques were needed. In the meantime, progress speeded up in another field, the therapy of infections with specific serums.

4. Treatment with Serum

People are told repeatedly that the prevention of disease benefits everyone. But to the average person in good health, disease and its prevention remain a remote abstraction. Only when a person actually gets sick is he likely to become concerned. He immediately begins to ask his doctor, with his eyes if not his lips, "What are you going to do to get me well?" Up through the nineteenth century, for most infections, the doctor responded by treating symptoms, such as lessening pain and bringing down fever, and by surgical procedures to remove diseased tissue or drain off pus if it appeared. The handful of specific remedies that had been gleaned from nature's abundant stock merely pointed up the absence of such specifics for the host of other infections.

The first specific remedies devised by man were reported by Behring and Kitasato in 1890. When they found that animals injected with diphtheria and tetanus toxins produced antitoxins which protected them from a later injection, they collected blood from the animals, separated off the serum which contained the antitoxin, and used it to protect other animals. The next step was to give antitoxins to patients with these diseases. The main emphasis at first was on diphtheria because of its high frequency. The first patient was treated late in 1891, and early the following year the firm of Meister, Lucius and Bruning began immunizing animals so as to produce the antitoxin commercially.[1]

Yet there was still a long way to go to produce effective antitoxins, for the ones first produced by Behring had only weak protective properties. A colleague at the Robert Koch Institute in Berlin, Paul Ehrlich, found the way to produce more potent antitoxins by injecting the toxin in progressively increasing doses. In the process he worked out a precise method for measuring the potency of diphtheria serums, which became the basis for evaluating all later serums. More important than Ehrlich's direct contributions to serum therapy was his theory about the interaction between antibodies and anti-

36

gens, the substances in microorganisms that stimulate the host to produce antibodies. He speculated that side chains on the antibodies attach to other side chains on the antigens much as a key fits into a lock. Through the chemical bond thus formed, the antibody exerts its neutralizing effect on the microorganism. This theory, which has continued to stand up as knowledge of the chemical and molecular structure of cells and antibodies has unfolded, has formed a solid basis for the work of thousands of investigators, not only in infectious diseases but also in fields as diverse as pharmacology, allergy, cancer immunity, and organ transplantation.[2]

Unfortunately, the original collaboration between Behring and Ehrlich that led to the production of an effective diphtheria serum did not have a happy ending. Both men were to receive a royalty on the profits from diphtheria antitoxin, but Behring talked Ehrlich out of his share in return for a promise that Ehrlich would be appointed director of the State Serum Institute for Research, a promise which Behring could not or would not fulfill. Ehrlich never forgave him, especially when he found out that Behring had built a castle on the Rhine from his share of the profits. Research and commercial exploitation, like oil and water, did not mix any better then than they do today.[3]

While the producers of antitoxin were having their problems, those interested in its use in patients were having other problems. Physicians were skeptical of the value of the serum at first, partly because they were not sure of Behring's abilities. They did trust both the skill and honesty of Roux, however, and when he reported in 1894 that he too had produced antitoxin and obtained excellent results with it in the treatment of diphtheria, an American observer later wrote that "hats were thrown to the ceiling, grave scientific men arose to their feet and shouted their applause in all the languages of the civilized world. I had never seen and have never since seen such an ovation displayed by an audience of scientific men."[4]

The clinical success of diphtheria antitoxin acted as a stimulus for the development of pharmaceutical firms as well as of government laboratories in the United States. In 1894 the New York City Health Department started inoculations in horses, the animals that yielded the largest quantities of antitoxin, and early in 1895 they treated their first patients with locally produced serum. The problem of establishing and enforcing standards, which was destined to plague makers and prescribers of serums, indeed of all drugs, as the number of remedies increased, was especially difficult for the early producers of antitoxin. They had to find out by trial and error which animals produced the best antitoxin; they had to master the techniques required to maintain strict asepsis; they had to devise proper receptacles for the serum so that it could be kept sterile and yet be easily available; they had to learn the injurious effects of heat and the proper methods of refrigeration; and they had to keep the costs of all these procedures within reason. There was no official procedure for testing or approving serums or vaccines and little awareness of the need for their regulation.[5]

In the midst of this confusion, catastrophe struck. The city of St. Louis, following the example of New York, had employed a bacteriologist to produce diphtheria antitoxin. In the fall of 1901 a number of children who had been given injections from one particular batch of this antitoxin died with symptoms of tetanus. Investigation showed that the antitoxin was contaminated with living tetanus bacilli. Pressure from public health officials and doctors for Congress to regulate the production of serums and vaccines resulted in the passage of the Biologics Control Act in 1902. This law, which prohibited the sale of such products in interstate commerce unless the producer had been licensed and the process of production approved by the Public Health Service, became the basis for all future regulation of biological products in the United States.[6]

Soon after the first patients were treated with antitoxin, doctors were surprised to find that some patients, after an initial improvement, became sick again. They developed a fever and often a rash, swelling in various parts of the body, and pains in the joints. These symptoms usually began about a week after the first injection of antitoxin and lasted from a day to a week or more. Although these illnesses could be mild, many patients were extremely uncomfortable. More serious were the immediate reactions to antitoxin. Within seconds or minutes after an injection the patient had a feeling of tightness or suffocation in the chest, followed by shortness of breath or wheezing. Although these symptoms usually subsided in a few minutes to about an hour, occasionally the patient became unconscious and died.

Harmful or potentially harmful reactions following the injection of animal serums into humans had been observed as early as 1667. In 1839 it was reported that dogs given two injections of egg white several days apart had an immediate adverse reaction to the second injection, and similar reactions were seen in other animals after a second injection of various proteins. In 1902 this phenomenon of increased susceptibility or hypersensitivity to a foreign substance in animals was named anaphylaxis, meaning literally "against protection." The resemblance of anaphylaxis to the reactions in humans following the injection of various therapeutic serums stimulated further research, which revealed that the forms of serum sickness were manifestations of an interaction between a foreign protein in the injected serum and an antibody present in the patient's tissues. In delayed serum sickness antibodies appeared several days after the serum was injected; they then interacted with the proteins that were still present in the body. In the accelerated or immediate types, including the one that resembled anaphylaxis in animals, a previous contact with the same or a similar protein was presumed to have caused the formation of antibodies. Sometimes there was a history of previous injection of serum from the same animal; in other cases the patients, particularly those with allergic constitutions, had been sensitized by inhaling tiny particles shed from the skin or hair of the animal, which also made them hypersensitive to the animal's serum.[7]

To avoid serum reactions, doctors learned to ask about previous symptoms of hypersensitivity and to test the patient's reaction by injecting a small amount of serum under the skin before giving a full dose. Reactions were prevented by withholding serum from patients found to be hypersensitive, or by giving it more cautiously, and by processing serums to remove many of the proteins that were not antibodies. This last procedure also increased the potency so that each dose could be given in a smaller volume of fluid. In 1905 bacteriologists in Park's laboratory led the way by finding a practical method to precipitate out the antibodies with ammonium sulphate. Through refinements of this procedure, serums were prepared that contained increasingly higher quantities of antibodies and caused correspondingly fewer adverse reactions. Yet occasional serum reactions have continued to occur, because there is no way of removing all extraneous proteins. A survey of patients in one hospital from 1928 to 1932 showed that some form of serum reaction occurred in 14 percent of patients receiving diphtheria antitoxin. Some of the reactions were severe, and the death rate of just under one for every thousand patients receiving the antitoxin was still an appreciable hazard.[8]

Before the development of antitoxin physicians, distressed by the agony and the tragedy of diphtheria, had tried many methods to check its course. In the late nineteenth century bleeding was still used, although a London physician, John Fothergill, had concluded over a hundred years before that it was "prejudicial" and "seldom fails to aggravate the symptoms; and in some cases it appears to have produced very fatal consequences." Death was so frequently brought on by obstruction of the larynx that desperate measures were justified to relieve it. Bretonneau first successfully opened the trachea below the obstruction, but the operation took time, especially if performed under sterile conditions, when a few minutes could mean the difference between life and death. Also it required great skill. John M. T. Finney, a Baltimore surgeon, told of the "struggling choking children" on whom he had to perform a tracheotomy in the late nineteenth century, "usually by lamp light, with poor assistants . . . or none at all. . . . Many were the times when I began the operation not knowing whether or not I should have a living patient at the end of it."[9]

Several doctors had tried to introduce a tube through the throat into the larynx to carry air past the obstruction, but none were successful until 1885 when Joseph O'Dwyer of New York City, using the right combination of mechanical ingenuity and manual dexterity, was able to devise suitable tubes and insert them so that they stayed in position until the swelling had subsided. His modest claim that this procedure would "be recognized by the profession as a legitimate and valuable method of overcoming obstruction in the upper air passages with a rapidity by no other means obtainable" turned out to be an understatement. The technique of intubation, as the method came to be called, was rapidly adopted by doctors all over the world and was the only feasible method of saving the lives of some diphtheria victims before the advent of serum.[10]

Yet intubation merely allowed patients to breathe while the disease ran its course. Other procedures had been tried with the object of combating the infection itself. At the end of the century doctors were fruitlessly painting, spraying, or irrigating the site of inflammation with anything from mild boric acid solutions to caustics such as bichloride of mercury and phenol, while they "supported the circulation" with wine, cognac, camphor, or strychnine, the first two of which were useless and the latter two dangerous. Meanwhile the disease went its inexorable way, leaving behind its toll of 40 percent dead.[11]

Since the death rate was so high and doctors almost helpless to affect it, it would seem that they would have welcomed diphtheria antitoxin with open arms; yet many did not, for a number of reasons. From a theoretical standpoint, giving serum to cure a disease was a totally new idea, based on the "germ theory," a concept that most doctors did not have the training to understand. From a practical viewpoint, if a doctor did use the antitoxin, the result was unpredictable because the serums varied so widely in potency. Yet some of the early results were impressive. For instance, in one hospital in Berlin 163 patients were given antitoxin between March and July 1894. Only 14 percent of these patients died, compared with 42 percent of those treated before antitoxin was available. One of the most convincing reports was made by Ehrlich and his associates in 1894. They observed an increasing percentage of deaths the later the patient was treated; from zero and 3 percent for those whose treatment was begun on the first and second days of the disease respectively, to 43 percent for those who received their first injection on the fifth day.[12]

Such reports, and the observations of physicians who had visited Germany and had seen the antitoxin used, stimulated American doctors to use it when they could get their hands on it. But still some were skeptical. Two interns were so afraid of dire results when they treated the first patient to receive diphtheria antitoxin in the city of Cincinnati that they decided to push the piston of the syringe together so as to share the responsibility. Other doctors were confused by the finding of diphtheria bacilli in the throats of well persons and wondered whether the patients who had seemed to improve dramatically following antitoxin administration had really had the disease. Still others thought they had been pressured unwisely into using antitoxin by the weight of the "great names" who advocated it and the resulting lay publicity.[13]

But most doctors probably hesitated to use diphtheria antitoxin because of an unfortunate first experience with it on the part of the doctor or a close colleague. Just as a doctor would take up the treatment enthusiastically when his first patient recovered promptly, so a doctor whose first patient died would tend to shun the new remedy from that time on. Still more disturbing to a doctor would have been a serious or fatal reaction to the serum itself, especially before the reasons for these reactions were understood. Even today doctors who have witnessed a fatal or near-fatal reaction to penicillin in a patient under their care confess that they have since avoided using penicillin if they can possibly substitute another drug. They are unconsciously responding to a maxim that every good doctor observes: *primum non nocere,* first of all, do no harm.

Better antitoxins and their use in large numbers of patients eventually proved beyond doubt the value of diphtheria antitoxin. Public health officials aided the movement toward acceptance by providing facilities for diagnosis and free antitoxin. Sometimes there were problems, as when a family physician in Richmond, Virginia, became incensed when a health officer called on one of his patients to confirm his report of diphtheria without first getting his permission. But such attitudes gradually disappeared. In Chicago the Health Department responded to an epidemic of diphtheria in 1896 by establishing an Auxiliary Medical Corps to advise physicians on the diagnosis and the method of administering the antitoxin. The statistics gathered at the end of the epidemic showed that of 805 patients treated with antitoxin, only 6.5 percent had died. Over three-fourths of the patients had received their first injection during the first three days of the disease, and of these only 2.1 percent had died. With examples such as this, cooperation between health departments and practicing physicians became the rule.[14]

Although tetanus antitoxin was introduced at the same time as antitoxin for diphtheria, the results of its use were not so dramatic. In fact fifteen years after the first report by Behring and Kitasato, Osler wrote that the use of antitoxin in the treatment of human tetanus had been disappointing. Early antitoxins were weak in potency and, because they were given under the skin, were absorbed so slowly that the full dose did not reach the area of infection for hours. Doctors gradually learned that large doses of antitoxin given intravenously produced better results and placed their reliance upon this therapy, plus heavy sedation to diminish the muscular spasm. But sedation carried other hazards, particularly congestion of the lungs and pneumonia. A drug that would counteract only the muscular spasm was needed.[15]

For many years doctors had known of a drug called curare, which was used by the South American Indians in hunting game. When applied to arrows, it killed the animals by paralyzing their muscles. This drug had been considered a tool for laboratory research until a British physician, Ranyard West, reported in 1932 that he had used minute doses to treat patients with tetanus and other muscular disorders. Soon afterward, chemists isolated the substance responsible for the drug's relaxing action and determined the chemical formula. Later other drugs with a similar action were synthesized. Today muscle-relaxing drugs are routinely used in tetanus to diminish the intensity of the attacks of spasm. When used in conjunction with antitoxin they apparently lower the fatality rate further than could be accomplished with antitoxin alone. With the proper use of the two, doctors at the Cincinnati General Hospital obtained a case fatality rate between 1947 and 1958 of only 21 percent. Yet in rapidly advancing cases the fatality rate was still 37.5 percent. Thus, prevention is still the sheet anchor in the control of tetanus.[16]

Doctors experimented with serums in other infections prevalent in the United States. In one of these, typhoid fever, many other measures were tried before and after serum therapy. In the eighteenth and early nineteenth centuries, when it was widely believed that fevers were caused by a general overreactivity of the body, bleeding was popular. By the late nineteenth century this procedure had been discarded by leading medical men, although as late as 1914 a professor of therapeutics at the University of Toronto was still urging that six to fourteen ounces of blood be removed in severe cases.[17]

After pathologists pointed out the significance of ulcers in the intestines of typhoid patients, some doctors concluded that the main objective of treatment should be to heal these ulcers. They gave a variety of drugs for this purpose, including hydrochloric and sulphurous acids, silver nitrate, creosote, phenol, and calomel, but these so-called intestinal antiseptics had no effect at all on the ulcers. Other therapy around 1900 was directed at specific features of the disease. Some of these procedures made the patient feel better, such as sponging and wet packs to bring down the temperature and bismuth and opium for diarrhea, while others were useless, for example, alcohol to combat the extreme prostration, which was incorrectly attributed to a weakened heart.[18]

Another therapeutic practice in the nineteenth century was to lessen the intake of food so as to decrease the work of the intestines. Avoidance of huge meals may have promoted healing, yet some diets were so meager that patients were literally starved to death. A diet recommended in 1835 of broth, toast, and jelly furnished only 300 calories a day or less. In 1870 the milk diet was introduced and soon became popular; it consisted of two quarts of milk given over the course of twenty-four hours and furnished about 1,400 calories. But even the slim diets in vogue at the turn of the century were being questioned as possibly giving too much food, and it was argued that some patients did well on cold water alone. Since a person with high fever consumes 50 percent more calories than one whose temperature is normal, it is not surprising that typhoid patients emerged from several weeks in the sickbed as gaunt and haggard shadows of their former selves.[19]

The knowledge that made diet a help rather than a hindrance to recovery from typhoid came from outside the field of infectious diseases. In the early years of the twentieth century a group of physiologists and clinicians at Cornell Medical School and Bellevue Hospital were intensively studying the metabolism of humans in health and disease. Eugene DuBois and Warren Coleman decided to study patients with typhoid fever because diet seemed to be so important in this disease and because it was easy to find many such patients in Bellevue Hospital. They found that large quantities of proteins were destroyed during the course of the fever. To replace these and to maintain the body weight, they recommended diets containing larger amounts of proteins and carbohydrates. Doctors who tried the more liberal diets were gratified to find that hemorrhage or perforation of the intestine occurred no more frequently than with the scanty diets of the past while patients retained their weight and recovered their strength much more rapidly. Thus another theoretically conceived regimen was displaced by one based on scientific experiments.[20]

But the laboratory was unable to produce similar results in serum therapy. Since 85–90 percent of patients ordinarily recovered from typhoid, even though it took several weeks, and they developed protective antibodies during that time, it seemed reasonable to assume that if the serum of convalescents were given to patients, it would shorten the disease. This concept led to the immunization of animals with killed bacilli or extracts from them and to the use of serums of these animals to treat the disease. Beginning in 1893, favorable reports emanated from Germany and France purporting to show that the fatality rate was substantially decreased in patients who received these serums. But the investigators committed the same fallacy as had many others before them, and as some still do today. They compared the results they had obtained with the observations of others on patients who had not received the serum. When such comparisons are made, the cases in the different groups may vary in several other particulars than treatment or lack of treatment with the new remedy. Also in such trials there is often an unconscious selection of the more favorable cases to receive the newer treatment. Eventually, even without the benefit of controlled studies, it became apparent that the serum was having no curative effect and it was gradually abandoned.[21]

Thus, until the 1940s typhoid was recognized as one of the most frustrating diseases to manage. The patients were highly uncomfortable and often miserably ill for many weeks, and the doctor could do little but stand by, watching for complications, hoping that they would not occur, and often treating them unsuccessfully when they did. The disease was a harrowing experience for patients, family, and medical attendants, as well as an expensive one for the patient if he could pay, and for society if he could not.

Although serums failed in typhoid, better results were obtained in two other serious diseases. Serums effective in pneumococcal pneumonia and meningococcal meningitis were prepared by immunizing animals with whole bacteria, rather than with toxins as was done in producing antitoxins for the treatment of diphtheria and tetanus. The more successful of the serums prepared with whole bacteria was the one used in pneumonia.

In the 1901 edition of the leading textbook of medicine, pneumonia was called the "most widespread and fatal of all acute diseases . . . the 'Captain of the Men of Death.' " There was much evidence to support this designation, for pneumonia had been a persistent killer as long as man had observed diseases, involving persons of all age groups, in all stations of life, and in all occupations. It has been detected in the lungs of mummies dating from the tenth century B.C. and is still common in present-day America, attacking each year eight in every 1,000 persons. At the turn of the century one death occurred each year from pneumonia for every 500 persons in the United States.[22]

Lobar pneumonia, the most characteristic form of this disease, usually begins dramatically with a chill or sharp pain in the chest. Cough is an early symptom, and the sputum is likely to be pink, red, or otherwise discolored by the presence of blood. Following the chill, the temperature rises as high as 104–105°F and, in the absence of specific therapy, remains high for several days. During this time the patient is miserable. He breathes rapidly and coughs frequently; each breath and each cough may cause a sharp pain in the chest, resulting from inflammation of the pleural covering of the lung. With the high fever goes a flushed face, hot, dry skin, and mental dullness that sometimes proceeds to delirium. After five to ten days, sometimes more, the fever subsides, either rapidly or over the course of two or three days. Complications are frequent; especially serious are meningitis, endocarditis, or inflammation of the inner lining of the heart, and empyema, or pus in the pleural cavity. The other variety of pneumonia, bronchopneumonia, affects one or more localized areas in the lungs rather than entire lobes and occurs particularly in elderly persons or as a complication of other diseases. Although its onset is less dramatic and its symptoms less pronounced, it is just as likely to be fatal as lobar pneumonia.

Descriptions of pneumonia have been recorded since the time of Hippocrates. For centuries the diagnosis depended upon the symptoms that the patient could report and upon simple observation. An important advance was made in 1761 by Leopold Auenbrugger of Vienna, who found that by tapping or percussing the chest wall, he could distinguish the normal portions of the lungs that contained air from the solid areas of the pneumonia. In 1819 a French physician, Rene-Theophile-Hyancinthe Laënnec, added the technique of auscultation: listening through a stethoscope to the sounds made by air as it moved through the bronchial tubes and into the lungs. He was thereby able to diagnose pneumonia with accuracy and to describe its several stages. A quarter of a century later Carl Rokitansky of Vienna completed the pathological picture by differentiating lobar from bronchopneumonia. Diagnosis was further sharpened in the last decade of the nineteenth century by use of the X-ray. By the end of the century doctors were proficient at detecting the presence of pneumonia, although there was little they could offer in the way of treatment.[23]

Meanwhile, in the widespread search for the causative agents of infectious diseases, the lungs of pneumonia victims were being carefully studied by bacteriologists. Their quest was made more difficult because several different microorganisms can cause pneumonia, including streptococci, staphylococci, and even smaller microorganisms, although the most common cause is the pneumococcus, and most of the progress in understanding the pneumonias as well as the advances in diagnosis and treatment have been made in connection with the pneumococcal variety. This bacterium is seldom round like other cocci; more often it is oval with pointed ends. It usually grows in pairs, called diplococci, as indicated by the technical name by which it has long been known, *Diplococcus pneumoniae.* The pneumococcus was first identified in 1881, not in the lungs but in the saliva of healthy persons, by Pasteur and independently by George Sternberg of the United States Army Medical Corps. At that time neither investigator related these cocci to pneumonia.[24]

A year later a German physician, Carl Friedlaender, reported seeing spherical and elliptical cocci, usually in pairs, which stained violet with Gram's stain, in the fluid drawn from the affected lobes of patients with pneumonia or from their lungs at autopsy. He was able to infect mice, guinea pigs, and dogs with these microorganisms, to culture them again from the sick animals, and with these to infect other animals. He thus satisfied Koch's postulates. His description both of the colonies produced when these cocci were grown on artificial media and of the capsule surrounding the cocci leaves no doubt that he was dealing with the pneumococcus. Confusion arose, however, when he later reported finding a rodlike bacterium in the lungs of other pneumonia patients. Although he made the sensible suggestion that more than one microorganism can cause pneumonia, the apparent ambiguity was seized upon by another investigator to initiate a controversy over the priority of the discovery. Albert Fraenkel, the rival professor who claimed the credit, insisted that Friedlaender had never seen the pneumococcus in cases of pneumonia but only the rod-shaped bacillus, later called Friedlaender's bacillus or *Klebsiella pneumoniae*, which is responsible for only 1 or 2 percent of pneumonias. The significance of pneumococci was verified when others found them not only in the lungs but also in highly invasive cases in the blood and in the pleural and cerebrospinal fluids.[25]

Early in the next century America came to the forefront in research on the pneumococcus, largely because of the founding of a new type of institution, the research institute. Frederick T. Gates, a member of the staff of John D. Rockefeller, president of the Standard Oil Company, learned from reading Osler's textbook of medicine that specific cures existed for only a few diseases. Realizing that remedies for the others were likely to come only if adequate facilities were provided for intensive research, he recommended that Rockefeller establish an institute for this purpose along the lines of the Robert Koch Institute and the Pasteur Institute, both of which had recently opened. Accordingly, the Rockefeller Institute was organized in 1901, opened its laboratories in 1904 and its hospital in 1910. The first director of the hospital was Rufus Cole, who decided to concentrate his research on pneumonia as "the most serious acute disease with which physicians have to deal."[26]

Remedies for pneumonia in the past had been numerous, varying from bloodletting and leeching to cold compresses and hot mustard foot baths; from inhalations of chloroform to injections of gold, silver, and platinum solutions; from doses of mercury, quinine, or digitalis to "two or three ounces of good whisky." It is an axiom in medicine that when many different remedies are used to treat a disease, none is likely to be very effective. Because this was true of pneumonia, Cole asked, "Is it possible that nothing can be done to decrease the frightful mortality due to this malady?"[27]

Fraenkel had reported in 1886 that when rabbits were injected with pneumococci, they became immune to subsequent infections with the same microorganism. Others showed that dead as well as living pneumococci could be used to induce this immunity and that the serums of immune animals would protect others from infection. Serums of animals previously injected with pneumococci were given to patients with pneumonia, with apparent benefit in some cases but not in others. Meanwhile, in 1897 investigators in the Pasteur Institute demonstrated that there were different strains of pneumococci and that serums made by immunization of animals against one strain would cause agglutination of some pneumococci but not of others. Thirteen years passed before two German bacteriologists, Fred Neufeld and Ludwig Haendel, announced that this principle could be applied to therapeutic serums. They produced a serum with a certain strain of pneumococci, which they labeled Type I. With this serum they were able to protect mice against other Type I pneumococci, but not against different strains.[28]

In 1913 two members of Cole's group announced that all pneumococci could be divided into three types—I, II, and III—and a miscellaneous Group IV. This finding gave impetus to the further classification of pneumococci and to the development of serums. Georgia Cooper, a bacteriologist in Park's New York laboratory, set to work to untangle Group IV. She and her associates expanded the number of identifiable types to thirteen by 1929 and to thirty-two by 1932. Eighty types were eventually identified, and there are probably still more. Clinicians determined the frequency and nature of the pneumonias caused by each type of pneumococcus and how these pneumonias differed from the ones caused by other types. In general, pneumonias caused by the "higher types," those above Type III, were less typical in their symptoms and more benign than those caused by Types I to III, although some of the "higher type" pneumonias were severe and fatal.[29]

The identification of pneumonias by type opened the way for effective therapy with type-specific antipneumococcal serums. In 1913 Cole and Alphonse R. Dochez reported that of eleven patients with Type I pneumonia to whom they gave specific serum only one died. They continued to improve their serums, to use them in a growing number of patients, and to impress upon doctors the value of serum therapy of pneumonia. A burgeoning social institution also contributed to the development of effective serums. By the early part of the twentieth century large municipally owned hospitals had grown up in Boston, New York, Philadelphia, Chicago, and other major cities in the United States for the care of the sick poor. Because of the great numbers of patients and the wide spectrum of diseases in these hospitals, they attracted a superior group of house officers, and medical schools used them for teaching and research. In 1919 a study of antipneumococcal serums was begun at the Boston City Hospital, the results of which were inconclusive, probably because these serums were not very potent. In addition, they, like other serums produced at the time, caused adverse reactions in a high percentage of persons, such as fever, rash, pains in the joints, asthma, and sometimes death, as had the early diphtheria antitoxins. Attempts were made to remove the substances that caused these reactions, and in 1924 Lloyd Felton finally devised a workable method of concentrating the serums so as to preserve antibodies while eliminating unnecessary proteins. When these serums were used in the city hospital in Boston and in Bellevue and Harlem Hospitals in New York, adverse reactions were far fewer than with unconcentrated serums and fatality rates were lower than in untreated patients.[30] Further progress occurred when investigators at the Rockefeller Institute found that serums produced by immunizing rabbits with pneumococci had a higher level of antibodies and caused fewer reactions than those produced in horses. When serums from rabbits were concentrated, large doses of antibodies could be administered with a small syringe instead of requiring a cumbersome reservoir suspended by the patient's bed.[31]

The time needed to determine the type of the infecting pneumococcus, once a problem to doctors, was reduced sharply when Neufeld showed that serum produced by one type of pneumococcus would cause the capsule of a pneumococcus of that type to swell, while leaving the capsules of other types unchanged. The type of pneumococcus could be determined by mixing one drop of a culture or of a body fluid from a patient suffering from pneumonia with each of several different antipneumococcal serums and observing which one caused swelling of the capsules. By this procedure the type could sometimes be determined in as short a time as fifteen minutes.[32]

Other problems slowed the acceptance by doctors of antipneumococcal serums. Typing required expertise and equipment that were not always available, many doctors were unfamiliar with the techniques of administering the serum, and the serum was expensive. To overcome these obstacles, the Massachusetts Department of Public Health in 1931 initiated a program which included typing pneumococci in the department's laboratory, teaching technicians how to perform this procedure in the laboratories of smaller hospitals, appointing consultants to verify the diagnosis of pneumonia and to demonstrate how to administer the serum, and providing free serums. Other states followed with similar programs, especially after the federal government helped by providing funds. A survey in the early 1940s of over 9,000 patients treated with antipneumococcal serum in various state programs and municipal hospitals showed that less than 15 percent had died.[33]

During the years when antipneumococcal serum was being developed, another serum made by immunizing animals with whole bacteria was used with some success in meningococcal meningitis. This infection is not widely prevalent at all times, as is pneumococcal pneumonia. Instead, in most years only sporadic cases occur; then suddenly an epidemic appears, in other words, large numbers of persons in a particular locality are affected at about the same time.[34]

In the typical case meningococcal meningitis begins with a chill or chilly sensations, vomiting, headache, a rash, and stiffness of the neck muscles. The course is so rapid that the full picture may be present within a few hours after the first symptom has appeared. The chills are followed by an irregular fever, the headache worsens, the neck stiffens further until in severe cases it cannot be bent forward even an inch, and sometimes the whole spine becomes rigid. Convulsions may appear, especially in young children, and mental dullness may deepen into coma or the patient may become delirious. The rash may look like flea bites or may consist of blotchy red or violet hemorrhages under or into the skin. In the absence of specific treatment one-half to three-fourths of the patients die. Those who survive usually improve in about a week. The most frequent complications are deafness and paralysis of the muscles of the eyes, face, or limbs. Arthritis, affecting one or more of the larger joints, is also common. The varied manifestations of meningococcal meningitis have given rise to a number of names, including cerebrospinal fever, spotted fever, epidemic meningitis, and cerebrospinal or spinal meningitis.

Although meningococcal meningitis has probably plagued man for centuries, its occurrence is not known for sure prior to 1805, when an epidemic in Geneva was so well described that it can be readily identified. In the next few years epidemics were reported from various parts of Europe, North America, and elsewhere, although the presence of cases between epidemics was not recognized. As late as 1864 most authorities believed that the disease was not contagious, but in 1887 an Austrian pathologist, Anton Weichselbaum, reported finding in the meninges of six patients who had died during an epidemic of meningitis gram-negative cocci that were usually arranged in pairs. The species is today called *Neisseria meningitidis*, or meningococcus for short. In a series of epidemics that occurred at the turn of the century, several bacteriologists confirmed Weichselbaum's findings. By 1910 it was generally accepted that only meningococci caused epidemics of meningitis, while pneumococci, streptococci, and a variety of other microorganisms might cause sporadic cases. The diagnosis in the individual case was facilitated in 1891 when a German physician, Heinrich Quincke, popularized lumbar puncture. By this technique a small quantity of the cerebrospinal fluid, which surrounds the brain and spinal cord, was drawn off through a needle inserted between the vertebrae so that it could be examined under the microscope for pus cells and cultured for microorganisms. Beginning in 1894 meningococci were cultured from the noses and throats of patients and well persons during epidemics, and in 1898 they were first cultured from the blood of patients with meningitis. Doctors correctly concluded that meningococci entered the body by way of the nose and in most persons they grew for a time in the upper part of the throat without producing an illness. In a small proportion of cases they entered the bloodstream, were carried to the meninges, and caused meningitis.[35]

Two syndromes in addition to meningitis were described soon afterward. In 1902 a case was reported in which the patient had fever for several months, during which time meningococci were recovered from the blood. In later epidemics this syndrome, called chronic meningococcemia, was thoroughly delineated: it consists of continuous or intermittent fever lasting for weeks or months, usually accompanied by a rash and often by pain and swelling of the joints. Meningitis occurs in less than a third of these patients.[36]

In 1906 a case was reported of a 50-year-old English physician who had died in less than two days of a meningococcal infection. At autopsy many hemorrhages were found under the skin and in various organs, including the adrenal glands. Five years later another English physician, Rupert Waterhouse, observed similar hemorrhages at the autopsy of an infant who had died within twelve hours of the onset of the illness. This syndrome, now generally known as the Waterhouse-Friderichsen syndrome, is attributed to widespread injury to the smaller blood vessels by meningococci, producing numerous hemorrhages, shock, and rapid death.[37]

Meningococcal meningitis occurs most frequently and certainly most spectacularly in the form of epidemics. The death rate rises several-fold during these periods. In Baltimore, Maryland, an average of 4.8 persons per 100,000 population died of meningococcal meningitis each year during three epidemic periods, compared with 0.6 persons per 100,000 during three nonepidemic periods. Nationwide epidemics have occurred in the United States during the twentieth century, about once every decade. Since the meningococcus is a notorious camp-follower, the most extensive and serious epidemics have appeared in times of war. Outbreaks occurred among troops in the War of 1812 and the Mexican and Civil Wars. In

World War I there were 5,839 cases in the American army; in World War II, 13,922.[38]

Although the discovery of different types of pneumococci was reported in 1897, no one apparently suspected that meningococci could also be divided into types until Charles Dopter of the Pasteur Institute reported in 1909 that he had found in the throats of well persons microorganisms that were similar in appearance and cultural characteristics to the known meningococci but which were differentiated from them by the agglutination test. Others were also finding strains that did not agglutinate with serums that were active against the commonly encountered strains, and in 1915 Mervyn H. Gordon and E. G. Murray of the British Army Medical Corps were the first to divide meningococci into types. They classified half of their strains as Type I, a third as Type II, and the remainder as Types III and IV.[39]

Over the next decades other groupings were devised for meningococci in an attempt to fit in all the variants, but this merely created a jumble of classifications. In 1950 a uniform system was finally agreed upon by an international committee, listing four groups— A, B, C, and D—to which several other groups have since been added. In spite of the deficiencies of all classifications, the ability to distinguish the different meningococci has helped materially in tracing epidemics. It is generally agreed that the large outbreaks have been caused almost entirely by meningococci of group A. Meningococci of other groups have caused infections in interepidemic periods and in about 5 percent of the cases during epidemics. But more important, differentiation of meningococci, as of pneumococci, was valuable in the production of therapeutic serums.[40]

Before serums became available, the seriousness of meningococcal meningitis and the likelihood of death for two or three of every four patients pushed doctors to extreme measures. Bloodletting, blistering, and large doses of mercury had their day and were discarded. At the end of the nineteenth century the only helpful measures were sedatives and opiates, which did not affect the course of the disease but did relieve spasm and headache. After the technique of lumbar puncture came into use, draining off quantities of spinal fluid often gave relief from headache, although it did not shorten the course nor reduce the fatality rate. In 1906 Georg Jochmann of Breslau immunized horses with meningococci and produced a serum that protected mice and guinea pigs against meningococcal infections. Later he treated patients with encouraging results. In 1907 Simon Flexner of the well-funded Rockefeller Institute, who was able to work with more expensive animals than many investigators, reported producing meningitis in monkeys with strains obtained during an epidemic. He passed the infection from monkey to monkey, and then protected monkeys from experimental meningococcal meningitis by injecting a serum produced by immunizing horses with meningococci. He too found that the overall fatality rate dropped when serum was given to patients.[41]

Hailed as a major advance in therapy, antimeningococcal serum was used extensively in the military and civilian populations during the epidemic that accompanied World War I. But deaths continued to occur after therapy which could not be blamed on either the lateness of treatment or the severity of the attack. When another epidemic occurred in 1927, antimeningococcal serum fell short of its expectations. In Indianapolis, Indiana, for instance, 65 percent of 141 patients treated with serum died, about the proportion one would expect in patients who received no serum. The reasons were complex.[42]

The ideal way to treat meningococcal infections would have been to follow the pattern used in pneumonia: find the type of meningococcus causing the patient's illness and give a serum prepared with strains of that type. Meningococci grew poorly on the artificial media available at the time, and when they did grow, several days had to elapse before they could be identified. Consequently, the pioneer laboratories in this country, such as the Rockefeller Institute and the New York City Health Department, used several types of meningococci, obtained during different epidemics, to prepare their serums. These were effective against some meningococci that caused epidemics and some sporadic cases, but not against all of them. Some clinicians advocated that, before an antimeningococcal serum was administered, several serums be tested to see which one best agglutinated the patient's microorganisms, but this took a long time and did not always indicate that the serum would work. Other doctors recommended shifting from one serum to another if the patient did not improve. This also meant a delay in starting effective treatment and was only possible in medical centers where a variety of serums was at hand. Still others believed that the solution lay in producing a serum that would be effective against a wider variety of meningococci; yet serums made by immunizing with many strains were not always sufficiently potent in combating any one strain. The best serums were probably those made by obtaining a high level of antibodies against about six strains.[43]

The problems relating to antimeningococcal serums were compounded by the fact that their effectiveness was never conclusively demonstrated in patients. The fatality rate in meningococcal meningitis varies widely, depending upon whether an epidemic is in progress, the stage of the epidemic, and the age groups affected. In one Detroit hospital where all patients were treated with serum, the fatality rate in the years between epidemics fell to as low as 14 percent, whereas it rose to 65 percent during the peak of an epidemic and declined to 42 percent toward the end. Fatality rates for groups of patients treated with serum in other places varied as widely, because each report usually included only those patients treated in a particular locality during a limited period of time. No adequately controlled study of the value of antimeningococcal serum was ever done. In the late 1930s, therefore, when other treatment became available, clinicians turned away from serum. The most accurate estimate today is that the best serums, when given early in the disease and administered properly, reduced the fatality rate by as much as one-half.[44]

By the time the serum era came to a close, a vast amount of research had been conducted in an attempt to develop effective serums. Some of the information obtained was utilized in diagnosing and tracing the spread of infections, and much of it helped explain the reaction of the human host to infections. Although serums had failed to affect the course of typhoid fever and certain other infections, they had saved the lives of many persons with tetanus and meningococcal meningitis and had scored major triumphs in the treatment of diphtheria and pneumococcal pneumonia, triumphs that raised the hopes of doctors and the public for additional successes in the future.

5.   Streptococcal Infections

In the early decades of the twentieth century a few diseases were brought under partial control by public health measures, by immunization, or by treatment with serums. Yet many other infections were not controlled by any of these measures. Among the most recalcitrant were diseases caused by streptococci. The process of defining these infections and explaining how streptococci brought them about took a longer time than it had taken to comprehend typhoid fever, diphtheria, and pneumonia, mainly because streptococcal disease had so many faces.

The diversity of streptococcal infections could have been seen by accompanying a doctor as he visited a municipal hospital on a typical day at the turn of the century. On the isolation wards he would have seen a number of children and a few adults with high fever, a sore throat, and a red rash extending from the cheek bones down over the rest of the body. These cases he diagnosed as scarlet fever. In other wards and the outpatient clinic he would have seen many patients suffering from sore throat without a rash; these illnesses he labeled septic sore throat. Off in a corner of a medical ward, isolated from other patients, might have been an old man with a sharply outlined area of inflammation extending from the nose outward across the face. A nurse might have been applying a purple liquid to the edges of the inflamed area in an attempt to halt its spread. The doctor called this erysipelas. In the obstetrical ward there would have been at least one patient who, a day or two after delivering a normal baby, sustained a sudden chill followed by high fever, pain in the lower abdomen, and increased vaginal discharge. Such a patient would have been fortunate if the infection remained localized in her uterus because she would then have had a fair chance of recovery, for if it had spread to the peritoneum, the membrane lining the abdomen, or to the bloodstream, she would almost certainly have died. Her disease was called childbed fever or puerperal sepsis. The doctor would have examined other patients who suffered from pain and swelling in several joints, some of whom were also short of breath; his diagnosis was rheumatic fever. All of these conditions could have been caused by streptococci because streptococci produce different diseases, depending upon the organ they attack and whether they cause the disease by direct infection, by the action of a toxin, or in some other way. Yet such was the state of knowledge at the time that the possibility of streptococcal infection was not even suspected in some cases, and in other cases where it was considered likely by one group of doctors it was categorically denied by another group.

The most dramatic episode in the history of streptococcal diseases was the discovery that puerperal sepsis was contagious. The fever that sometimes follows childbirth had been known and dreaded for centuries. Passages in the writings of Hippocrates and of Galen, a Greek physician of the second century A.D., clearly described puerperal infection, and other references appeared from time to time. Astute clinicians suspected that it was transmitted by persons attending the delivery, but their voices went unheeded until the middle of the nineteenth century when two men drove home the point in different ways. At that time, it was customary for doctors attending patients in labor to make vaginal examinations and for nurses to prepare patients for labor with little or no heed to cleanliness. Doctors thought nothing of attending a patient with an infectious disease, such as erysipelas, or performing an autopsy, and then proceeding directly to the obstetrical ward where, without washing their hands, they made a vaginal examination.

Oliver Wendell Holmes, the essayist and poet, was also professor of anatomy at Harvard University. Impressed by the observations of an English obstetrician as well as by cases that had come to his attention, he published in 1843 and 1855 two eloquent and impassioned articles in which he urged doctors to reform their practices. Holmes insisted, *"The Disease known as Puerperal Fever is so far contagious as to be frequently carried from patient to patient by physicians and nurses."* Holmes persuaded many physicians and laymen, but some physicians strongly disagreed, refusing to admit that a doctor or nurse could ever "carry disease and death instead of health and safety," and thus could become "a minister of evil" to women. Besides, Holmes had brought forward no proof of his own, having relied entirely on the observations of others.[1]

Fortunately, as Holmes and his critics disputed, the facts were being marshaled more scientifically by Ignaz Philip Semmelweis, a physician on the obstetrical wards of a hospital in Vienna. Concerned because so many women died of infection after childbirth, he studied the conditions under which these deaths occurred and found that, while 10 percent of the women delivered on the doctors' wards died, only 3 percent died on wards where midwives made the deliveries. None of the reasons given for the high fatality rate on the doctors' wards, such as the supposed existence of an epidemic or unnamed "cosmic influences," satisfied Semmelweis. When a close friend died of a generalized infection, the idea flashed through his mind that the features of the illness resembled those of puerperal sepsis. He deduced that doctors and medical students, coming as they did from attending patients with infections, or from autopsies or dissecting rooms, were transmitting the infections to the pregnant women, while the midwives, whose work was confined to the obstetrical wards, seldom did so. Beginning in 1847, Semmelweis required doctors and students on his wards to wash their hands thoroughly before performing a vaginal examination. The fatality rate promptly dropped to 3 percent, and in the following year, after examiners began using a chlorine solution on their hands, it fell to slightly over 1 percent. Despite the soundness of his data, Semmelweis' colleagues objected violently to his conclusions. So great was the opposition that he was finally forced to leave Vienna.[2]

Complete confirmation of Semmelweis' conclusions came several decades later, when the search for the bacterial causes of disease entered its most productive period. In 1874 the Viennese surgeon Theodor Billroth observed streptococci in pus, and a few years later Alexander Ogston of Aberdeen demonstrated two different varieties of cocci, that is, bacteria approximately spherical in shape. After producing infections with them in animals, he correctly concluded that both varieties caused not only abscesses, localized collections of pus, but also more widespread infections. Billroth had named the cocci that grew in chains streptococci; Ogston named the others, which grew in masses, staphylococci. Most of the streptococci that cause human infections produce hemolysin, which breaks up red blood cells and liberates the red pigment, hemoglobin. Although several different bacteria were found to cause puerperal infections, the epidemic forms and most of the fatal cases were caused by hemolytic streptococci. With the help of the pathologists, doctors were able to understand the genesis of puerperal fever. The inside of the uterus after childbirth resembles the raw surface of an open wound, providing an easy path of ingress for any bacteria that happen to reach it. If clothing, dressings, instruments, or the fingers of the patient or an attendant introduce virulent streptococci into the vagina, infection can easily enter the uterus and from there reach the peritoneum and the bloodstream.[3]

Another disease eventually shown to be caused by hemolytic streptococci is scarlet fever, which has for its main features a sore throat and a bright red rash which is frequently spread so evenly that it appears to be painted on the skin. Complications are most frequent in areas adjacent to the throat: abscesses around the throat and tonsils, sinusitis, and inflammation of lymph nodes and the middle ear. The combination of sore throat and rash was often described in the sixteenth and seventeenth century, and in the latter century the terms scarlet fever and its synonym, scarlatina, were in common use by doctors and laymen. Yet the throat symptoms were often confused with diphtheria and the rash with measles; not till 1861 was scarlet fever clearly differentiated from other diseases by the French clinician Armand Trousseau. In 1884 Loeffler, the codiscoverer of the diphtheria bacillus, found streptococci in patients with scarlet fever, and two years later an outbreak of scarlet fever was traced to a cow from whose udder streptococci were cultured.[4]

But doubts arose about the precise connection between streptococci and scarlet fever for several reasons: seemingly identical streptococci could be cultured from the throats of healthy persons, and streptococci could not be cultivated from the skin that peeled off as the rash subsided, nor could they be found in the blood of scarlet fever patients except in a few of the most severe cases. As a consequence, when Jochmann, a leading authority on infectious diseases, claimed in 1905 that streptococci were responsible only for the complications of scarlet fever and that something else caused the disease itself, this erroneous concept prevailed for almost two decades.[5]

The explanation of the relationship of streptococci to scarlet fever that eventually turned out to be correct was proposed as early as 1893 by André Bergé of Paris. He argued that "scarlet fever is a local infection" caused by a streptococcus which "grows in the crypts of the tonsils where it secretes an 'erythremogenic' toxin, the diffusion of which produces the exanthem," that is, an eruption on the skin. Yet for the next thirty years, bacteriologists searched for this toxin in vain.

The breakthrough came in 1924 with the investigations of George and Gladys Dick of the University of Chicago. The Dicks reported that when a culture of a hemolytic streptococcus obtained from the throat of a patient with scarlet fever was swabbed on the throats of two volunteers, one of them developed typical scarlet fever. Also, when the filtrate, or fluid remaining after the streptococci in a culture had been removed, was injected into the skin of patients early in the course of scarlet fever, it produced a round area of redness at the site of injection which resembled the rash of scarlet fever. This procedure is now known as the Dick test and the substance injected as the erythrogenic, or red-producing, toxin. When they injected the filtrate before a bout of induced scarlet fever, it produced redness, while after the illness it produced none. They rightly concluded that this shift from a positive to a negative test during scarlet fever indicated the development of immunity to the erythrogenic toxin. Within a few years the erythrogenic toxin was being widely employed as a skin test to determine whether people were susceptible to the disease.[6]

The Dicks claimed that the important features of the disease and most of the complications were caused by the erythrogenic toxin, but in 1926 Francis Blake and James D. Trask, Jr., of Yale University disproved this concept. They demonstrated that the level of erythrogenic toxin in the blood of patients paralleled only the extent of the rash and not the duration of illness nor the presence of complications. Thus the streptococcal infection itself was the important factor in scarlet fever and caused the complications and death, whereas the rash was an incidental, although spectacular, feature.[7]

The Dicks also hypothesized that scarlet fever was caused by a specific strain of streptococci. This theory, too, was refuted when hemolytic streptococci were divided into groups and these into types. In 1933 Rebecca Lancefield of the Rockefeller Institute reported the differentiation of hemolytic streptococci into broad groups, and in the following year an English bacteriologist, Fred Griffith, announced a method of distinguishing a number of types within the groups. Several different types from Group A were obtained from patients with scarlet fever. Final proof that scarlet fever was not caused by a single variety of streptococci came when it was shown that strains belonging to a number of types within Group A could produce erythrogenic toxin. Lancefield and her associates worked out a more exact method of distinguishing types according to a protein on the surface of the streptococcus called the M protein. There are more than 40 types of Group A streptococci, the main disease-producing group in humans, and each type has a different M protein. Immunity develops against this protein and is therefore against the individual type. A second streptococcal infection caused by the same type is rare. The recognition that a number of types of hemolytic streptococci caused scarlet fever explained why the early serums prepared by immunizing animals with whole streptococci were successful in some cases, where the type of infecting and immunizing streptococcus was the same, and unsuccessful in others, where the types differed.[8]

According to the present concept of the genesis of scarlet fever, if a person develops a sore throat caused by a strain of Group A hemolytic streptococci that produces erythrogenic toxin and he has no immunity to the toxin, he will have scarlet fever. If he is immune to the toxin or if the particular strain of streptococci does not produce the toxin, he will have a sore throat without a rash, or streptococcal sore throat. If a streptococcus to which he is immune lodges in his throat, he will most likely not develop any illness. This theory explains why in a single epidemic some persons will have scarlet fever and others only streptococcal sore throat. For instance, of 82 milkborne epidemics reported between 1870 and 1929, scarlet fever and streptococcal sore throat were present together in at least 17 epidemics. As further evidence, simultaneous cases of scarlet fever and sore throat were often observed in families. Sometimes the patient drew the right conclusion even though doctors failed to do so. For instance, midway in Abraham Lincoln's campaign for the presidency, his son Tad developed a bad case of scarlet fever. At the same time Lincoln had a sore throat and a headache, which convinced him that he too had the infection.[9]

Many therapeutic measures had been tried in scarlet fever, ranging from relatively sane advice, such as a bland diet, sponging to reduce the temperature, and applying "disinfecting" agents to the throat to prevent the spread of infection, to such drastic recommendations as cupping, bleeding, and purging, which had been advocated for many other diseases and which often made the patient worse instead of better. Investigators in the late nineteenth century had also succeeded in producing serums by immunizing animals with whole streptococci, but the results were not consistent.

Because they believed that the erythrogenic toxin was the important factor in the disease, the Dicks advocated therapy with antitoxin made by immunizing animals with this toxin and claimed that the antitoxin produced by their method diminished the fever, cleared the rash more promptly, and decreased the complications in comparison with untreated patients. Yet even in the large series of cases reported from the hospital associated with their laboratory the case fatality rate was the same, namely 1.6 percent, for the 882 patients treated with antitoxin, as for the 1,421 patients who received none.[10] The rapid subsidence of fever and the decrease in complications that followed the use of some serums may have occurred because these serums contained antibodies against products or chemical fractions of streptococci in addition to the erythrogenic toxin. In this connection Dochez's investigations were particularly helpful.

Three weeks after the Dicks' announcement of the rash-producing effect of toxin, Dochez and Lillian Sherman of Columbia University reported an ingenious method for serum production. They introduced under the skin of an animal a mass of agar, a semisolid substance in which bacteria grow readily, and injected into the mass streptococci obtained from a patient with scarlet fever. They hypothesized that if scarlet fever was caused by a toxin, the streptococci growing in the agar would diffuse the toxin into the blood, enabling the animal to produce an antitoxin. It was also possible that portions of the streptococci or products of their metabolism, in addition to the erythrogenic toxin, escaped from the agar and that they, in addition to the erythrogenic toxin, caused the development of antibodies. Augustus Wadsworth contended that immunizing animals with toxin alone did not produce the most effective serums and claimed that serums made in his laboratory by Dochez's method were much more potent in combating both animal and human streptococcal infections than those produced by the Dicks' procedure. Unfortunately, the Dicks obtained a patent on their method of production, by means of which they required licensees to produce antiscarlatinal serums according to their procedure. Thus opportunities for research on serums prepared by other methods were restricted, and a valid comparison of different serums was never made. In the late thirties, when serums were displaced by other therapy, no consensus had been reached regarding their value, although most doctors used them only in seriously ill patients.[11]

While these attempts were being made to understand and control scarlet fever, information was being accumulated regarding another streptococcal infection. Erysipelas is characterized by a sharply defined, shiny, red, raised area of inflamed skin. In nine out of ten patients the inflammation is on the face, usually extending from the bridge of the nose symmetrically over both cheeks, in severe cases reaching as far as the ears, the scalp, and even the neck. High fever and bouts of chills or chilliness accompany the inflammation. On the average, about 5 percent of patients die, but among young children and the aged the percentage is much higher. The distinctive rash makes the disease easily diagnosable on appearance alone.

Consequently, descriptions of erysipelas are recognizable in medical literature from the time of Hippocrates and other Greek and Roman authors. The disease became a special problem in the hospitals that burgeoned in the cities of Europe in the eighteenth century. It was part of a tidal wave of infection and suppuration that resulted both from the overcrowding of hospitals by thousands of sick poor who had flocked to the cities during the industrial revolution and from the ignorance of measures that would have prevented the spread of infection.[12]

After streptococci were cultivated from the skin of patients with erysipelas by Friedrich Fehleisen of Würzburg their role as causative agents was readily accepted. But a dispute arose as to whether the microorganism labeled *Streptococcus erysipelatis* was a separate entity or was identical with other disease-producing streptococci. Many clinical observations appeared to indicate that infection could be transferred from a patient with erysipelas to the vagina of a woman in childbirth, and it was not uncommon for nurses to develop streptococcal infections on their fingers when caring for patients with erysipelas. Also, investigators were able to produce erysipelas by injecting into the skin streptococci obtained from infections other than erysipelas. Yet in the twenties, at the time the Dicks were claiming that scarlet fever was caused by a distinct strain of streptococcus, Konrad Birkhaug of Rochester, New York, contended that another strain was responsible for erysipelas, and his viewpoint tended to prevail for a while. Once Griffith divided hemolytic streptococci into types, however, it was shown that erysipelas, like scarlet fever, was caused by a number of different types of hemolytic streptococci.[13]

These bacteriological findings fitted in well with clinical observations, because the course and complications of scarlet fever, streptococcal sore throat, erysipelas, puerperal sepsis, and infections in various areas from which hemolytic streptococci were cultured, such as the middle ear, the meninges, and the lungs, were much alike. Doctors finally understood that a number of different types of streptococci produce these infections and that the organ that becomes involved depends mainly on how the streptococcus enters the body, by way of the upper respiratory passages, through the skin, or into the uterus of the recently delivered woman. Furthermore, if the number of streptococci is too large for the host's immunity to counteract, they invade other areas. They may travel directly to adjacent tissues: upward from the nose and throat to the middle ear, to the mastoid cells and even to the meninges, or downward to the bronchi and lungs; from the uterus to the ovaries and the peritoneum; from a wound in the skin, even a tiny puncture, through the lymphatic channels into the bloodstream—which they may also reach eventually through any of the other routes. Once in the blood, they may be carried to the meninges, the joints, and the pericardium or endocardium, the membranes covering the outside and inside of the heart.

Many different treatments were used for erysipelas in the past. They included purging with calomel or the induction of sweating to get rid of the "toxemia" and applying various caustics or antiseptics to the skin or injecting them under the skin at the margin of the inflammation to prevent it from spreading. By the twentieth century the heroic general measures had been discarded, although various drugs taken orally and local applications to limit the advance of the infection were still being tried; by the late twenties these too had been found to be useless. When erysipelas was recognized as a streptococcal disease, antistreptococcal serums were tried, but the results were not convincing, and serum fell into disuse.[14]

Since doctors could not treat streptococcal infections successfully, there was all the more reason to try to prevent them. Yet there were many obstacles, as illustrated by attempts to prevent puerperal sepsis. Although the hazard of introducing infection during childbirth had been amply demonstrated and despite improvements in technique, such as the adoption of sterile gloves for vaginal examinations and deliveries, the maternal mortality from childbirth in the United States did not change appreciably between 1915, when data were first collected, and 1930. Since nearly two-fifths of these deaths were caused by infections, it was obvious that the principles of asepsis were not being universally applied.[15]

One reason that death rates remained high was that many deliveries were managed by midwives who, unlike the ones at Semmelweis' hospital, were usually uneducated and poorly trained and therefore not aware of the need for aseptic techniques. In 1910, 50 percent of all births in the United States were reported by midwives, and many more such deliveries were unreported. A movement to regulate and educate midwives spread gradually throughout the country. By 1921 many states had passed laws requiring the registration of midwives, and a few states also required them to be supervised. Eventually, as women came to know what constituted ideal medical care in childbirth, and as more and more of them had their babies in hospitals, midwives largely disappeared from the American scene.[16]

In some rural areas conditions were too primitive even for midwives, and childbirth often occurred under the most difficult conditions. In a section of rural Montana in 1917 40 percent of women in childbirth were attended by neighbor women, who were for the most part untrained in obstetrics, 10 percent were delivered by their husbands, and a few were even alone at childbirth. The mortality rate reflected the lack of skilled care: it was 12.7 deaths per 1,000 live births, double the rate for rural Wisconsin, where physicians attended 68 percent of births, and much higher than the 2.9 percent for rural Kansas, where 95 percent of the babies were delivered by physicians. Even in the cities the majority of births at the turn of the century took place outside the hospital, many under the most unsanitary conditions.[17]

In the final analysis improvements in maternal mortality and morbidity depended upon the ability of doctors to manage obstetrical cases properly, and at the beginning of the century many were still deficient in the necessary knowledge and skills. A questionnaire sent out by John Whitridge Williams in 1911 was answered by 43 professors of obstetrics. More than half had had no training in a maternity hospital, and 15 had not even had an internship! Six of the schools had no connection with a maternity hospital, and nine others had facilities for delivering less than 100 patients per year. If trying to study disease without patients, as Osler said, is like learning navigation without ever going to sea, neither the professors nor the students in many schools should have been allowed to leave the harbor. By the 1920s medical schools had improved to the extent that students were required both to participate in the care of obstetrical patients and to perform deliveries themselves under supervision. Strict asepsis was taught and observed, the average medical student being almost convinced that lightning would strike him dead if he performed a vaginal examination on a woman in labor.[18]

The newer generation of medical students could be taught better techniques, but something was needed to correct the methods of doctors already trained. Too many were like the doctor in rural Kansas who, according to Arthur E. Hertzler, "divided his time between practicing medicine and raising hogs . . . [and who] sometimes washed his hands after the completion of labor but never before."[19] Gradually the picture improved. Pressure from colleagues and from an increasingly better informed public persuaded older doctors to use aseptic techniques. Other changes affecting maternal mortality were the influx of doctors and public health nurses into rural areas and the establishment of prenatal clinics and programs for home visits by nurses in the cities. Since these changes took time and since they took place more rapidly in some localities than in others, their impact was not felt for some time. The mortality rate from puerperal infections began to decline in the United States after 1930. In that year it was 67 per 10,000 live births. By 1937 it had fallen to 49 per 10,000, after which the advent of drugs effective against streptococcal infections accelerated the decline.[20]

In erysipelas the spread of pathogenic streptococci could be prevented by proper isolation. But the situation was far different when the infection was in the respiratory tract. Although patients with scarlet fever might be isolated as soon as the rash was recognized, virulent streptococci could already have been present in the throat for some time. Patients with streptococcal sore throat were not routinely isolated, for without a rash to flag it, this disease was often undiagnosed. Sometimes the inflammation was so mild that it went unnoticed. Also, streptococcal infections could not be differentiated from those caused by other microorganisms except by culturing, a procedure that was seldom performed because the facilities were not usually available. Finally, as in diphtheria, too many persons were carriers of streptococci to isolate them all.

The Dicks advocated immunization with erythrogenic toxin to develop antibodies and thus prevent scarlet fever, and some clinicians and public health officials followed their suggestion. But after Blake and Trask demonstrated the limited action of this toxin, doctors realized that such immunization was of little value because it did not prevent the streptococcal infection of the throat. In fact, from a public health standpoint it was harmful, because by preventing the rash from developing, it removed a diagnostic sign and lessened the chances of the patient's being isolated. Consequently, immunization with erythrogenic toxin was never widely practiced.[21]

Two other illnesses, rheumatic fever and nephritis, are caused by hemolytic streptococci, but they differ from those already discussed in that the organs principally affected are not directly invaded by the streptococci. These syndromes begin several days or weeks after the onset of a streptococcal infection. In a typical case of rheumatic fever one or several joints become red, swollen, and painful. As these improve, other joints are similarly affected, the entire illness often lasting for weeks or even months. Involvement of the brain causes uncontrollable, rhythmic movements of the arms, legs, and face, a syndrome called chorea or St. Vitus' Dance. When rheumatic fever attacks the heart, it may weaken the muscle so that it cannot pump out the blood; it may deform the valves so that they do not close properly or partially block the onward flow of blood; or if it involves the pericardium, it may form an envelope of scar tissue around the heart which keeps it from expanding properly between contractions. It has been said that rheumatic fever "licks the joints but bites the heart," and these bites often lead to years of invalidism or early death.

It took centuries for this disparate picture to be recognized as a single disease. The acute arthritis was described by Hippocrates; the particular type of chorea associated with rheumatic fever was differentiated from other choreas by the English physician Thomas Sydenham in 1686, while the relationship of these to heart disease was not established until the end of the eighteenth century. Around the middle of the following century perceptive clinicians put together the broad syndrome of rheumatic fever and recognized that, although in a particular patient the manifestations might occur in any combination, they were all somehow related. Strangely, the presence of sore throat in association with the other manifestations of rheumatic fever was not emphasized until the 1880s. An important discovery was made in 1904 by a German pathologist, Ludwig Aschoff, who saw under the microscope small nodules, or round masses of cells, in the heart muscle of a patient who had died of rheumatic fever. The presence of "Aschoff's nodules" in various tissues is still accepted as positive evidence of rheumatic fever.[22]

As clinicians and pathologists became able to diagnose the disease with certainty, the importance of rheumatic fever became evident. It was common: in the late twenties, Alvin Coburn found that in one hospital in New York City one-fourth of the beds were occupied by patients with active rheumatic fever or its aftermath during the spring, the peak season for the disease. And it was deadly: of the 3,445 persons with rheumatic heart disease treated in Philadelphia hospitals from 1930 to 1934, over one-fifth died.[23]

Although doctors could diagnose rheumatic fever, they still did not know its cause. The continued search for a bacterial incitant had gotten nowhere. Every time someone claimed a particular microorganism as the cause, others were unable to culture it from their cases although they used the same techniques. In 1929 Homer Swift of the Rockefeller Institute clearly enunciated a theory that had been gaining ground in recent years. He made a number of deductions: since children under the age of two rarely have rheumatic fever, a particular kind of change must develop in the tissues before the disease can occur; since all children do not develop the disease after respiratory illnesses, something must differ in the reaction of children who do develop it; and since the symptoms of rheumatic fever usually appear several days after the beginning of a sore throat or of scarlet fever, some time must elapse before the tissue change occurs. In his opinion, these observations indicated that rheumatic fever occurred as a result of an allergic reaction, that is, an altered capacity of the tissues to respond to a subsequent exposure of a particular substance, in this case to streptococci or their products.[24]

Yet, throughout this excellent review Swift speaks equally of hemolytic and other streptococci. It was Coburn who first clearly implicated the hemolytic streptococci as the sole incitants of the tissue reactions in rheumatic fever. He noted that just before a return bout of rheumatic fever they appeared in throat cultures taken from patients who had had previous attacks. He cultured the tonsils of patients with rheumatic fever and found hemolytic streptococci in 90 percent of them. Noting that recurrences seldom appeared in patients who moved to warmer climates, he sent a group of such patients to Puerto Rico for the winter months. In this climate the hemolytic streptococci disappeared from the throats of all except one, and in this child's throat they were greatly reduced in number. Except for some fever in the first few weeks, all of the children remained free of rheumatic fever for the six months of their stay. Shortly after their return to New York streptococci appeared in their throats, and several had recurrences of rheumatic fever.[25]

These impressive observations were promptly supported by several British clinicians, who had reached the same conclusions independently. Koch's postulates could not be used to prove that hemolytic streptococci caused rheumatic fever, because the disease cannot be reproduced in animals. Further evidence, however, came from tests for antibodies against products of hemolytic streptococci. A number of substances produced by streptococci had been isolated in the search for toxins. One of them, hemolysin, was shown to produce antibodies, called antihemolysins, when injected into humans. The observation that the blood of patients with rheumatic fever was likely to contain an increased quantity of these antibodies, which tended to decrease after the acute attack subsided, helped to verify the association of Group A streptococci with this disease. Because streptococci were not present in the tissues affected in rheumatic fever, such as the joints, the heart, or the brain, Coburn's concept of the relationship of hemolytic streptococci was joined with the concept of allergy expressed by Swift. The general viewpoint on rheumatic fever today is that products of hemolytic streptococci produce an allergic reaction in certain tissues, thereby causing malfunction and, in the case of the heart, lasting damage.[26]

While progress was being made in understanding the cause of rheumatic fever, little was learned about how to prevent or treat it. A century ago doctors had found that the salicylates would cut short acute attacks of rheumatic fever, and the cheap synthesis of these drugs, particularly aspirin, in the late nineteenth century led to their extensive use. But despite hopes and claims to the contrary, no convincing proof was ever brought forward that salicylates would prevent recurrences or protect the heart from damage. Treatment of rheumatic fever was limited to lessening discomfort by means of salicylates, judicious use of rest, avoidance of respiratory infections, and the treatment of heart failure when it occurred.[27]

Discoveries of the injurious effects of streptococcal infections on remote organs did not end with their incrimination in the development of rheumatic fever. For years doctors had observed that acute nephritis often followed acute infections of the respiratory tract, particularly scarlet fever. After hemolytic streptococci were shown to be responsible for many cases of sore throat and for scarlet fever, their relationship to nephritis was also suspected. Once rheumatic fever was accepted as an allergic response to hemolytic streptococci, nephritis was suggested as belonging to the same category, and investigators began to look for a product of streptococci that would cause nephritis in animals.

In studying both the spread of streptococcal infections in military installations and the methods of preventing rheumatic fever in the 1950s, Charles Rammelkamp and his associates at Case Western Reserve University made a surprising discovery. They found that, whereas rheumatic fever followed infection by all types of Group A streptococci prevalent in the throats of the men examined, nephritis occurred only after infections with certain types, particularly Type 12. Although these observations did not supply the missing link between the streptococcus and nephritis, they simplified the task of investigators, who now experiment on the assumption that nephritis results from an allergic reaction to these types or from the direct action of fractions of these streptococci on the kidneys.[28]

The history of the streptococcal diseases in this century epitomizes what has happened in the broader field of infectious diseases in general. Diverse clinical syndromes were brought together into their proper relationship by means of clinical, pathological, and bacteriological observations. Discoveries in immunology enabled investigators to understand the role of toxins, as in scarlet fever, and of allergy, as in rheumatic fever. Before the mid-thirties no therapeutic agents were found to modify the more serious effects of streptococcal infections except for antistreptococcal serums, and these were not consistently effective. Yet the extensive observations on the symptoms and pathological changes in these diseases, on the methods by which hemolytic streptococci spread from person to person, and on the mechanisms that people develop to protect themselves formed a reservoir of knowledge that could be drawn upon to make prevention and treatment more effective when specific therapeutic agents finally appeared.

6. Tuberculosis

"That drop of blood is my death warrant, I must die." Thus the poet John Keats announced that tuberculosis was claiming another victim. Having been trained as a doctor, he recognized that the weakness, fever, and cough of the past year and the blood he had just spit up came from a tuberculous infection of the lungs. A year later he was dead, at the age of twenty-five. Tuberculosis was no respecter of persons, attacking alike the ruler and the ruled, the athlete and the aesthete, the learned and the ignorant, the genius and the average man. Keats died in 1821; at that time and for another century and more the disease was widely prevalent throughout Europe and America. In the United States at the beginning of the twentieth century it vied with pneumonia as the chief cause of death. In the year 1900 of every 1,000 Americans, 2 died of tuberculosis and 20 were ill with the disease.[1]

Tuberculosis is a widespread disease of man and many animals, sometimes acute, more often chronic, caused by the *Mycobacterium tuberculosis* or tubercle bacillus. The bacterium usually enters the human body through the air passages, although it may find its way in through the alimentary tract when food or drink is infected. From either portal of entry the bacteria may invade adjacent tissues or the lymph nodes draining the area, and they frequently enter the blood to be carried to organs in any part of the body. In the site where they enter or in other places they may be walled off by the body's defenses and remain quiescent for months, years, or a lifetime; they may produce an acute, rapidly spreading inflammation accompanied by a high fever and often fatal, or they may cause a chronic, slowly progressive disease with low fever, weakness, and loss of much weight. The acute form occurs particularly in the meninges and the lungs. Walled-off bacilli may become activated, whereupon the acute or the chronic form of the disease will develop. Until recent years people tended to picture the tuberculous patient as wasted, pale, and listless, coughing incessantly and expectorating chunks of

70

purulent sputum. Yet for every patient who fitted that concept, a dozen or more others infected with tubercle bacilli would have milder symptoms and some no symptoms at all. The popular image of the tuberculous patient was responsible for several of its former names. The Greeks called it phthisis, a derivative of the word for wasting away. In the past few centuries and up to the 1930s it was called consumption because it seemed to eat up the patient's flesh, and the acute pulmonary form was known as galloping consumption. It was also spoken of as the white plague because the skin was pale and unspotted.

Bones characteristic of tuberculosis can be traced as far back as 3000 B.C. in Egypt and possibly to the Neolithic period in Europe. Writings from ancient India and China tell of the presence of the disease. It was common in Greece in the days of Hippocrates and in Rome at the height of the empire.[2]

The frequency of tuberculosis may have diminished somewhat after the dissolution of the Roman Empire, because communities were smaller and travel less frequent. As people congregated within walled cities in the Middle Ages, it probably became common again, and the industrial revolution, bringing still more people together to work long hours in close quarters, was responsible for an even greater toll from the disease. In 1680 John Bunyan called it the captain of the men of death, a phrase that Osler later borrowed to describe pneumonia. The peak of mortality was probably reached in England around 1780, when records show that more than 11 persons of every 1,000 population died of the disease each year. In the United States the industrial revolution and urbanization came later than in Europe, and so also the peak of mortality from tuberculosis. Combined figures from Boston, New York, and Philadelphia show the death rate from the disease to have been between 400 and 450 per 100,000 population from 1812 to 1830 and approximately 350 per 100,000 between 1830 and 1880, after which it began to decline.[3]

The increase in frequency of tuberculosis helped doctors learn more about the disease. The Dutch physician Franciscus Sylvius in the mid-seventeenth century first clearly realized the relationship of the tubercles, the small, rounded masses of cells found throughout tuberculous tissues, to the various forms of the disease. He described the formation in the lungs of the cavities that occur in tuberculosis as a result of destruction of tissues, and he recognized tuberculosis of the lymph nodes. Little was added to his findings until the first quarter of the nineteenth century when Laënnec examined hundreds of patients by the techniques of auscultation and percussion and performed autopsies on those who died. He concluded that tuberculosis was a single disease, no matter in what organ it occurred and no matter whether the manifestations were cavities in the lungs, destructive changes in the bones, a painless swelling of the lymph nodes, or a rapidly fatal meningitis.[4]

The concept that tuberculosis was one disease was not generally accepted, however, until the agent that caused it was found. One reason for the lack of acceptance was that the contagiousness of tuberculosis was not so obvious as in diseases with a more rapid course, such as smallpox, diphtheria, and venereal diseases. Only a few people had glimpsed the truth about the contagion of tuberculosis. Italy and Spain had regulations to prevent its spread as early as 1699. Patients so afflicted were strictly isolated, and when they died, their bedding and the doors to their rooms were burned and their rooms were replastered. In general, northern Europeans tended to be either amused or annoyed at such meticulousness. When Budd, of typhoid fever fame, wrote to the *Lancet* in 1867 setting forth reasons why tuberculosis was contagious, the idea attracted little attention. Instead, doctors and laymen alike tended to assume that the disease was inherited, because it frequently occurred in several members of a family and in more than one generation.[5]

But the concept that tuberculosis was a single disease stimulated people to look for a single cause, and in 1865 a Frenchman, Jean Antoine Villemin, proved that the cause was a living agent. He infected animals with tuberculous material obtained from humans and cattle and was able to pass the infection from one animal to another. Koch then stained tissues from tuberculosis patients and discovered in them the curved rod of the tubercle bacillus. After growing these rods on artificial media, he announced in 1882 that he had found the cause of tuberculosis. Because the rods were difficult to stain by the usual methods, that versatile genius Paul Ehrlich improvised a new technique in which bacteria from suspected tuberculous tissues were first stained a pink color with one dye, subjected to treatment with acid, and then counterstained with a dye of another color, usually blue. The acid washed the stain from other bacteria, which then took up the blue counterstain, while the tubercle bacilli were unaffected by the acid and retained the original pink color. A modification of Ehrlich's technique is routinely used today to identify tubercle bacilli, and bacteria that retain the original stain are designated as "acid-fast."[6]

Prominent among the early contributors to knowledge concerning the tubercle bacillus was the American bacteriologist Theobald

Smith, who worked mostly with infections in animals. By comparing tubercle bacilli from the sputum of patients with those from tuberculous cattle and other animals, Smith clearly distinguished human from bovine tubercle bacilli. He showed that the former could infect cattle, although large doses were needed, and that bovine tubercle bacilli could infect humans, particularly children who ingested large quantities in infected milk. But the great Koch disagreed. In an address given in 1901, he maintained that "human tuberculosis differs from bovine, and cannot be transmitted to cattle," and what was more significant, concluded that the evidence "does not speak for the assumption that bovine tuberculosis occurs in man." Leading investigators in France and England supported Smith, as did the editors of the *British Medical Journal* and the *Journal of the American Medical Association.* As more studies were done, Smith was proved to be right, and his conclusions became the basis for an active campaign to stamp out bovine tuberculosis by requiring both the pasteurization of milk and the use of tuberculosis-free cows. As a result, bovine tuberculosis is today virtually absent in the United States.[7]

Once the causative agent of tuberculosis was established, doctors began to search energetically for measures that would contain the disease and, if possible, cure it. Remedies for tuberculosis in the past had reflected the philosophy of their day. Hippocrates emphasized diet and hygiene. Roman authors advocated that the patient eat wolf's liver boiled in wine or drink elephant's blood or, if necessary, bathe in the urine of a person who had eaten cabbage. Weasel's blood and pigeon's dung were prescribed in the fourteenth century. If these failed, the patient was advised to "betake himself to the king," because the mere touch of a duly anointed king was thought to have power to heal scrofula, or tuberculosis of the lymph nodes in the neck. English monarchs intermittently followed the practice of "touching" to cure disease from the time of Edward the Confessor in the eleventh century until the death of Queen Anne in 1714. Samuel Johnson was touched by Queen Anne for scrofula when he was two years old, and he wore the "touch-piece" given to him for the rest of his life. The monarchs were able to keep their celebrated reputations as healers for so many centuries because tuberculous lymph nodes of childhood usually heal spontaneously.[8]

Drugs used in other illnesses were also tried in tuberculosis, without success. Cod liver oil, prescribed for rheumatism in the late eighteenth century, was later given for tuberculosis. At about the same time the English physician William Withering, who had shown the value of digitalis in the treatment of heart failure, advocated its use also in tuberculosis. In 1821 Thomas De Quincey claimed that opium was practically a specific for the disease, and for several decades afterward many a patient with chronic tuberculosis became an opium addict in addition.[9]

But the age of bacteriology was not without its spurious remedies also, the most spectacular of which was offered by the father of bacteriology himself. In 1890 Koch announced that he had developed a substance named tuberculin, which would cure tuberculosis. Because of his reputation and because of the secrecy surrounding his method of preparation, the announcement created pandemonium. Doctors and patients alike flocked to Berlin. Even Lister, who took his niece to Koch for treatment, was fooled so completely that he wrote, "the effects of this treatment upon tubercular disease are simply astounding."[10]

But this time two great men were wrong. Tuberculin, which was later revealed by Koch to be a glycerin extract of cultures of tubercle bacilli, has no curative effect in tuberculosis and under certain circumstances may make the disease worse. The injection of tubercle bacilli or extracts made from them into patients with tuberculosis can produce an allergic reaction which, if the doses are large enough, can result in a flare-up of the patient's infection. Because tuberculosis was usually chronic and often stationary for long periods of time, or so slowly progressive that changes could not be detected for months with the diagnostic procedures available at the turn of the century, it took a long time before tuberculin was completely discredited as a remedy. The physician Edward Livingston Trudeau made a study in 1906 which was praiseworthy for that period. He compared the results in 185 patients treated with tuberculin and 864 not so treated, all of whom had been in his sanitarium in the past fifteen years. Among patients with advanced tuberculosis, 27 percent of those treated with tuberculin were apparently cured, compared with only 6 percent of those who had not received tuberculin. Although these figures were suggestive and Trudeau had no training in statistics, he recognized the possible fallacies in his investigation. The patients were not randomly selected for the treatment and control groups, and there had been a tendency to give tuberculin to patients who showed evidence of better nutrition, which could have influenced the results. He also remarked that patients treated with tuberculin often remained in the sanitarium longer in order to complete the treatment; there they would have been held to a strict regimen and thus the disease would have been less likely to relapse.[11]

Despite the fact that all clinicians were not as objective as Trudeau in their observations, doctors eventually realized that tuberculin was of no therapeutic value. In the 1914 edition of his textbook of medicine, Osler spoke of tuberculin as already in disfavor although it was still "lauded by fanatics as the one and only means of cure." If one can judge by subsequent editions, the "fanatics" continued to use it through 1938. In fact, it was still spoken of as useful in the treatment of tuberculosis of the eye in the 1968 edition of a popular American textbook of ophthalmology.[12]

Some investigators tried to apply to tuberculosis the method of therapy that had been so successful in diphtheria. They injected animals with tubercle bacilli, collected the serum, and injected it into patients with tuberculosis. But after clinicians had given serum therapy a thorough trial, it too proved to have no value.[13]

In the absence of effective drugs, doctors tried other measures. Some urged their patients to move to a warm climate: English physicians sent their patients to the shores of the Mediterranean, and American doctors recommended the southwestern states. Warmer climates may have been helpful because they were conducive to rest; yet many hopeless derelicts with scant funds and no friends simply wore themselves out wandering from one cheap boarding house to another. One victim, after seven years of such wandering, concluded: "It is not a medical problem but a sordid problem; a mere matter of food and air—and dollars."[14]

A movement toward high altitudes, where the air was presumed to be more beneficial, began in 1859 when a German physician, Herman Brehmer, established a sanatarium for patients with pulmonary tuberculosis at Görbersdorf in the Waldenburg Mountains. Since his patients seemed to do well, other doctors followed his example. The most famous sanitarium in the United States was founded by Trudeau at Saranac Lake in the Adirondack Mountains of New York in 1884. Among the measures that had been in vogue for over a century was exercise in the open air, especially horseback riding, and at first exercise was an important part of the treatment in these sanitariums. Trudeau, who was himself a victim of the disease, began to treat himself by vigorous hiking, riding, and hunting. After he nearly killed himself with strenuous exercise and was forced to rest for several months, his condition improved remarkably. Consequently, he made rest the mainstay of treatment for himself and his patients. Other doctors also observed that exercise often made their patients worse. It is now understood that the lungs heal best when they expand and contract the least and that exercise is harmful because it increases the activity of the lungs.[15]

Skeptics also began to question the need for a high altitude or a warm climate, and in 1891 Vincent Bowditch established a sanitarium 18 miles from Boston, only 250 feet above sea level. When patients in this and similarly located institutions improved as rapidly as those in higher altitudes or warmer climates, doctors gradually became convinced that the chief factor in improvement was rest.[16]

In the nineteenth century the diet of patients with tuberculosis was severely restricted. Keats was nearly starved to death, according to the practice of the time. Gradually doctors recognized that the "consumption" of the flesh which was taking place should be combated by giving the patient plenty of nutritious food. This led to excesses in the opposite direction. Around 1910 one doctor wrote in a popular magazine that patients should drink a quart of milk, rich in cream, twice a day, immediately after it had been obtained from the cow. Esmond Long, who later became an expert in the pathology of tuberculosis, was given four quarts of milk and sometimes as many as a dozen eggs a day while he was in a sanitarium in 1915. Years later he shuddered to think of the cholesterol bath in which his arteries had been soaked.[17]

Valid statistics on the effect of a nutritious diet were hard to accumulate, because other measures were used at the same time. The value of good nutrition in tuberculosis was demonstrated unintentionally during World War I in Denmark, where the mortality rate from tuberculosis rose during the early years of the war but began to fall in 1917. The decline coincided with the imposition of a submarine blockade by the Germans, which prevented the export of meat and dairy products to England, with the result that the Danes consumed these nutritious foods themselves. It took a while longer for doctors to learn that a diet should contain no more food than was necessary to replace the tissue consumed by the increased metabolism that accompanied the infection. Although the practice of "stuffing" patients with high calorie diets was condemned in 1916, the practice apparently continued for some time, because doctors were warned against overfeeding as late as 1947.[18]

When doctors correctly deduced that rest was helpful in tuberculosis, a natural sequel was to provide for still more rest of the lungs. As early as 1882 Carlo Forlanini of Italy had advocated the injection of air between the two pleural membranes, one of which covers the lungs and the other the inside of the chest wall, with the object of compressing the tuberculous area of the lung and putting it at rest until it healed. In 1895 he published a paper showing that this treatment was effective. The same idea occurred independently to a Chicago surgeon, John B. Murphy, who reported his results in

1898. By 1902 more than 400 persons in the United States had been treated by pneumothorax, as the method was called, although it did not gain general favor until well into the 1930s. The success of pneumothorax led surgeons to use other procedures to collapse the lungs when the injection of air was not successful. By 1942 about 70 percent of patients discharged from sanitariums had had some form of collapse therapy, which along with rest and a nutritious diet was the principal method of treatment.[19]

Pneumothorax and operations to collapse part or all of a lung were painful, time-consuming, expensive, susceptible to complications, and not always successful. Doctors therefore continued to look for drugs that would be effective. Creosote was in vogue in the early 1900s. When this drug was taken by mouth, some of it was excreted in the bronchial tubes and gave the breath a tarlike odor, which was enough to persuade many doctors and patients that it attacked tubercle bacilli. When doctors finally saw that patients who received creosote grew worse and died in the same numbers as those who did not, they discarded it. Nor was it any more effective when given by inhalation. Various chemicals, including the salts of copper, mercury, and iodine, and several dyes were also tried as drugs but found to be useless.[20]

The investigation in 1924 of one of the gold salts, Sanochrysin, as a specific for tuberculosis marked an advance in clinical research. Because differing reports had been issued as to its efficacy, doctors in Detroit conducted a clinical test which was one of the first trials of a drug in which matched patients were given a harmless, ineffective substance as a control. These investigators gave intravenous injections of the gold salt to 12 patients and distilled water to 12 others who had tuberculosis of equal severity. Patients who had received the drug became worse more often than patients who had received distilled water, and one patient who had received gold died of a disease of the liver, which was attributed to the drug.[21] This experiment was noteworthy not only because it successfully contradicted false claims for gold salts but more importantly because the value of the drug was determined by an adequately controlled therapeutic trial. The lesson was long overdue. If every therapeutic agent advocated for an infectious disease since 1900 could have been studied as rigorously, the medical profession would have had fewer remedies, but the patients would have been exposed to less discomfort and danger, the community would have had less expense, and fewer patients would have died.

A disease which during the first two decades of this century killed over 100,000 people and caused illness in another million in the United States each year, which fell heavily on young fathers and mothers, and which often spread within families, schools, and places of work—such a disease was not only a hardship to patient and family but also a heavy burden to society. Thus at the turn of the century a few began to recognize the duty of society as a whole to develop a program to control tuberculosis. The first of many local organizations formed in this country for the purpose of preventing tuberculosis was founded in Philadelphia in 1892, and in 1904 the National Association for the Study and Prevention of Tuberculosis, later renamed the National Tuberculosis Association, was organized. Meanwhile local, state, and federal authorities were establishing clinics, sanitariums, and diagnostic laboratories. In the beginning voluntary societies were at times in conflict with public health authorities, but gradually they worked out methods of collaboration which became a model for public-private cooperation in other countries and other diseases. In general, tuberculosis associations developed new programs, lobbied for laws to implement antituberculosis measures, and supported research, whereas local and state health authorities enforced regulations, maintained records, and operated facilities. Until a tuberculosis control division was set up in the Public Health Service in 1944, the National Tuberculosis Association functioned also as the chief coordinator for the control of the disease throughout the country. After that, planning on a national basis became a joint effort.[22]

Once the disease was accepted as infectious, the method of prevention that seemed most likely to yield to a large-scale effort was to break the chain of infection by which the tubercle bacilli reached their victims. The first need was to identify those who had the disease. Lawrence Flick asked the Philadelphia College of Physicians to approve the registration of cases by the city health department in 1893, but his efforts were defeated. Even though Biggs approached the matter more cautiously in New York, he still ran into difficulties. In 1894 the health department ordered institutions to report all cases of tuberculosis. Three years later, when the requirement was extended to physicians, the action met with a storm of opposition. One medical journal called compulsory reporting "mistaken, untimely, irrational and unwise" and "offensively dictatorial."[23]

Cooler heads realized that reporting and registration of cases were going to be necessary if tuberculosis was to be brought under control. The adversary could not be defeated until his whereabouts were known. Yet for some years after reporting of cases was required, valid data on tuberculosis were practically nonexistent, largely because doctors wanted to protect patients from the stigma attached to the disease. Figures for deaths were likewise inaccurate, for families often pressured doctors into writing another diagnosis on the death certificate because insurance companies would not pay benefits when tuberculosis was the cause. The fear of disgrace also operated to keep families from taking proper precautions to prevent the infection from spreading, because to do so would have revealed its presence. In 1915 in Chicago, for example, attempts to enforce the segregation of patients brought about "lamentations, yells, fist fights and knife play." By the time patients did seek medical aid, it was often late in the disease, when the chances of recovery were slim and the infection had been communicated to many others. Thus, a campaign to educate the public about tuberculosis became a major part of its prevention.[24]

Tracing contacts of patients was another early method of control, but it cost too much for extensive use. Some simple method had to be found to diagnose tuberculosis of the lungs, the most frequent form and the one most likely to spread the infection. X-rays provided that method. Less than a year after William Conrad Roentgen of Würzburg had discovered these rays in 1895, Francis H. Williams of the Massachusetts General Hospital demonstrated x-ray films of patients' chests. Improvements in technique over the next two decades developed x-ray examination into a practical means for detecting small areas of inflammation that could not be found on physical examination. Also, because the procedure took little time, it was ideal for examining large numbers of persons. In 1916 the National Tuberculosis Association launched a program in Framingham, Massachusetts, to determine the presence or absence of tuberculosis in every person in town by means of x-ray examinations. The number of identified cases increased sevenfold within a single year, and as a result it became clear that the ratio of active cases to deaths was not three to one, as formerly thought, but closer to nine to one. The Framingham demonstration did much to convince doctors and public health officials of the diagnostic value of x-rays in mass surveys.[25]

Another technique adapted to mass surveys was the tuberculin test. When doctors had first tried this substance for the treatment of tuberculosis, they had found that patients developed an inflammatory reaction at the site of the injection. Clemens Peter von Pirquet, an Austrian pediatrician, turned this observation to practical use in 1907 by showing that when a drop of tuberculin was placed on a spot of skin that had been scarified, patients with tuberculosis developed a redness and thickening of the surrounding skin. In 1910 the test was modified by a French physician, Charles Mantoux, so that a small quantity of tuberculin was injected intracutaneously, between the layers of the skin. This technique made it possible to test large numbers of persons in a short period of time. Testing with tuberculin was especially valuable where only a small percentage of the persons tested were likely to have the disease. Since only those with positive tests would have to be examined by x-rays, that more expensive test could be kept to a minimum. By the 1920s campaigns to detect tuberculosis in schoolchildren and in industrial groups were well under way.[26]

During these years hope persisted that some way would be found to increase the resistance of the individual so that he would not acquire a tuberculous infection in the first place. Killed tubercle bacilli, in contrast to typhoid bacilli and many other bacteria, did not produce immunity. In 1906 two Frenchmen, Leon Charles Albert Calmette and Camille Guérin, announced that live tubercle bacilli, after being transferred many times on artificial media so that they were no longer infectious, still retained the property of immunizing animals. During the 1920s Calmette showed by immunizing over 50,000 infants that a vaccine made from these attenuated bacilli, called BCG, could be successfully used in humans. Although BCG was widely used in several countries, it found little favor in the United States. One reason was the Lübeck disaster, in which 271 infants in that German city were vaccinated and 77 died shortly afterward. An investigation showed that BCG vaccine had been mixed accidentally with a virulent strain of tubercle bacilli which was being cultured in the same laboratory. This episode raised doubts about the safety of the vaccine, and confidence was further eroded when some investigators claimed to have restored the virulence of the bacilli by rapidly passing them from animal to animal, even though most investigators believe that this cannot be done. A number of studies have shown that BCG vaccination can confer some protection against tuberculosis. One disadvantage, however, is that the tuberculin test becomes positive after vaccination with BCG, thus making it impossible to use that test for the detection of tuberculosis. Since the tuberculin test is an important tool in the control of the disease where the incidence is low, BCG vaccination in the United States has been confined mainly to persons in susceptible age groups with a high risk of developing tuberculosis, such as medical and nursing students and children of tuberculous mothers. In countries where tuberculosis is a major problem the vaccine is used extensively.[27]

The decline in the death rate from tuberculosis had begun before tuberculosis control programs were put into effect. In Massachusetts, for instance, deaths from tuberculosis of the respiratory system fell from 365 per 100,000 population in 1861 to 308 in 1880, 190 in 1900, and finally to 37 in 1945. The decline in mortality before 1900 has been attributed to improvement in the general health of the population following the rise in the standard of living, which brought with it improvements in nutrition, housing, sanitation, and education. From 1900 to the mid-1940s measures aimed directly at the treatment and control of tuberculosis contributed further to the decline in the death rate. The number of new cases detected fell also, although more irregularly, since this figure depended partly on the zeal with which cases were tracked down and on other factors, such as the rate of immigration. In New York City, for instance, 23,000 new cases of pulmonary tuberculosis were reported in 1908. The number increased to 32,000 in 1910 and then declined, reaching 14,000 in 1920, 11,000 in 1930, and 6,000 in 1945. Yet in spite of these gains, deaths from tuberculosis continued to be unacceptably high in many places, so that much still remained to be done.[28]

Between 1900 and the 1940s the control of tuberculosis in the United States had come a long way. Resistance of the population to the disease continued to improve as the standard of living climbed and as more people learned the principles of healthful living. The white plague nevertheless remained a serious problem. In the decade from 1940 to 1950 it was the seventh leading cause of death.[29] The army of workers who had enlisted in the battle were knowledgeable and dedicated; all they needed were better weapons. And in the 1940s such weapons appeared.

7.   The Venereal Diseases

"It was a pestilence . . . evil, sordid, cruel to excess and very contagious, it was terrible and impossible to conquer. Those it attacked were prey to atrocious pains, their feebleness was extreme, their complexion lost its freshness and became livid. A very villanous evil which has begun on the most villanous part of the body." Thus Francisco Villalobos, a sixteenth-century Spanish physician, described syphilis, the most serious of the venereal diseases.[1]

Two centuries later James Boswell, Samuel Johnson's biographer, wrote of another venereal disease, gonorrhea: "I have had two visitations of this calamity. The first lasted ten weeks. The second four months. How severe a reflection is it! And, O, how severe a prospect!" He was thinking of the several weeks in which he would be incapacitated and the painful manipulations he would have to endure at the surgeon's hands. But the "prospect" was much more severe than that. When he was fifty-five, after several more gonorrheal infections, the disease finally presented all its IOUs for payment. He was attacked with chills, fever, severe pain, and swelling in the region of the bladder, and one month later he died of kidney failure as a result of abscesses or obstruction in the urinary tract.[2]

The venereal diseases are those that are almost always transmitted by sexual intercourse. The two important ones from the standpoint of severity, complications, and social consequences are syphilis and gonorrhea. The microorganism that causes gonorrhea is similar to the meningococcus. In the distant past all the gram-negative cocci that grew in pairs were probably harmless inhabitants of man's nose and throat and genito-urinary tract. At some point a few of these cocci broke through the host's defenses in the throat, entered the bloodstream, and invaded the meninges, causing an infection. Their descendants are called meningococci. Similarly, others invaded the deeper tissues of the genito-urinary tract, where they caused an infection. Their descendants are named gonococci, because the disease that they cause in the urethra, the tube that leads from the bladder to the outside, has for a long time been called gonorrhea, meaning literally a flowing of seeds or sperm.

In addition to attacking genital and urinary organs, gonorrhea causes complications elsewhere, some of them fatal and others producing sterility and blindness. Three to five days after contact with an infected woman the male begins to have a purulent discharge from his penis and burning on urination. Hence the name *chaude-pisse* or "hot urine" was used for the affliction in the fourteenth to sixteenth centuries, replaced by "clap" in the sixteenth century.[3] If the victim is fortunate, the infection will remain in the front end of the urethra and will subside in four to six weeks. But frequently the back part of the urethra is invaded also, in which case the infection involves the prostate gland and sometimes invades the epididymes, the tubes that carry sperm from the testicles. In these structures a chronic infection may smolder for months. Sterility probably results occasionally, but the main disabilities are the pain and discomfort and the interference with urination and sexual function. The general symptoms are more serious as far as life is concerned.

When the female is infected by a male, she may have no symptoms or may have a transient inflammation in the urethra. But often the infection is more serious and extends to the cervix, or neck of the uterus, and even into the Fallopian tubes, where it may last for a long time. Since the function of these tubes is to convey eggs from the ovaries into the uterus, gonococcal infection frequently produces sterility by matting the walls of the tubes together. In the female the rectum may be involved by transmission of the gonococcal infection from the urethra; in both the male and the female it may become involved as a result of anal intercourse.

In both sexes the gonococci may overwhelm the local defenses and enter the bloodstream, in which case they have a special affinity for the joints, the eyes, and the endocardium. One of the most tragic consequences of gonorrhea is blindness resulting from infection of the eyes of an infant during childbirth.

Gonorrhea is as old as the history of man. The Assyrians described it around 1000 B.C., and there is an account of the disease in a Japanese manuscript dating from 900 B.C. The "discharge from the body" which is labeled unclean in Leviticus 15:2 is thought to be gonorrhea. Greek and Roman authors described symptoms that were almost certainly gonorrhea. The disease was identified by physicians of the Middle Ages, and in 1162 an ordinance of the bishop of Winchester directed "That no Stewholder keep noo woman wythin his hous, that hath any sycknesse of Brenning [burning].'"[4]

The term syphilis comes from a sixteenth-century poem about a shepherd of the same name who was a victim of the disease. It is caused by the *Treponema pallidum*, which means pale treponeme, a thin, spiral-shaped bacterium. It differs from gonorrhea in nearly every way except its mode of transmission. Both gonorrhea and syphilis usually start as a local infection in the external part of the genito-urinary tract, but gonorrhea spreads mostly to adjacent structures, while syphilis becomes a generalized infection and invades distant organs. The primary stage of syphilis consists of an ulcer, called a chancre, at the point where the treponemes enter. The primary ulcer heals in a few weeks but is often followed by a secondary stage in which the infection is carried throughout the body and shows itself as a rash on the skin and the membranes of the mouth and genitals. The rash, too, disappears, but unfortunately in many patients the disease remains and becomes chronic. The treponemes localize around small arteries in one or more sites. The inflammation they produce gradually chokes off the blood supply, killing the functional tissue and resulting in replacement by scar tissue. So slow is the process that impairment of the function of the organ is not apparent for many years. Late foci of infection, formerly called tertiary and now designated late syphilis, may occur anywhere but most frequently involve the heart, the great blood vessels, the brain, and the spinal cord. When the disease occurs in either of the latter two organs, it is called neurosyphilis. Worse still, syphilis may be congenital. An infected mother may pass the disease to her child, which may be stillborn or die soon after birth. Those who survive may have evidences of syphilis at birth or may show them later.[5]

The early history of syphilis is clouded by controversy over whether the disease was present at the dawn of civilization in the Mediterranean area or whether it was imported there from the New World by Columbus' sailors when they returned from their first voyage in 1493. The Columbian theory is based on finding syphilitic lesions in bones of American Indians that predated his voyages and on the occurrence of a virulent epidemic of syphilis in Europe at the beginning of the sixteenth century. Other scholars believe that the venereal diseases and some cases of leprosy referred to in the Bible and other ancient literature were actually syphilis and that the lesions found in ancient bones in Europe and the Near East could have been syphilitic. They argue that the characteristics of syphilis could not have been so widely known and the method of treating it with mercury so well understood in 1495 if the disease had been unheard of a scant two years before. The problem cannot be solved on the evidence available, but a new theory about the historical relationship of man to the treponemes has brought better insight.[6]

According to this theory, treponemes were once free-living microorganisms, which eventually attached themselves to man or his ancestors and at some point began to cause disease. In warmer climates and in more primitive societies where general body contact was common, they were usually transmitted to young children, in whom they produced sores on the skin which healed after months or years as immunity developed. As men became more civilized and wore more clothing, the disease did not spread in children, and the intimate contact by which it was transmitted was sexual intercourse. Persons infected in this way developed the disease known as syphilis, although in some periods it could have been a very mild disease. The appearance of a more virulent form around 1500 resulted from the loosening of morals following the Black Death, the increase of travel during the Renaissance, and the papal bulls abolishing leper houses in 1490 and 1505. These societal changes speeded up the transmission of treponemes from person to person, a situation that characteristically raises the virulence of a microorganism. There is much to be said for this theory. Treponemal infections closely resembling venereal syphilis and predominantly transmitted by nonsexual contact are found today in many parts of the world under a variety of names, and the treponemes causing these diseases are almost indistinguishable from those causing venereal syphilis.[7]

Regardless of how it originated, around 1495 a virulent, highly infectious, often fatal form of syphilis began to spread rapidly across Europe. Within the next fifty years the course of the disease changed until it resembled the slowly progressive form seen today. During the sixteenth century doctors learned the external features of the disease, recognized that it was usually spread by sexual intercourse, and observed that children could be born with it. With the rise of the science of pathology they came to understand also how the internal organs were affected.[8]

Up to the sixteenth century gonorrhea seems to have been clearly differentiated from syphilis, but at that time doctors, impressed by the virulent form of syphilis sweeping across Europe, decided that gonorrhea was merely one of its manifestations. The confusion lasted for over two hundred years, until discerning clinicians began to suggest that syphilis and gonorrhea were distinct diseases. In

1767 John Hunter, a Scottish surgeon and pioneer experimenter, decided to find out. He pierced his penis and swabbed the area with pus from a patient whom he had diagnosed as having gonorrhea. Within a few days there appeared at each site an ulcer which developed the characteristic hard margin of a syphilitic chancre. Not content with the damage already done, Hunter gave himself only partial treatment so that he might better observe the course of his illness. When he died twenty-six years later, the autopsy showed that he still had signs of syphilis. In the light of today's knowledge these events can be explained only by assuming that the pus for the experiment had been taken from a patient who had both gonorrhea and syphilis. Hunter, on the contrary, drew the false conclusion that the two diseases were the same, and so great was his reputation that his viewpoint predominated for another seventy years.[9]

Because investigators were unable to produce gonococcal infection in animals, they continued to use humans for their experiments. Between 1767 and 1903 many studies were reported in which pus or secretions from patients with gonorrhea were inoculated into the urethras of males. Owing to the crude methods used, some experiments were unsuccessful, but a typical gonorrhea infection was produced in at least 21 persons. Finally, the experiments of the French venereologist Philippe Ricord, who performed 667 inoculations between 1831 and 1837, convinced the doubters that gonorrhea was a separate disease.[10]

The gonococcus was one of the first bacteria to be identified with the disease it caused. In 1879 Albert Neisser, a German dermatologist and bacteriologist, described small, beadlike bodies, occurring mostly in pairs and flattened on the opposing sides, which he found in pus and secretions from patients with gonorrhea. Their relationship to the disease was verified six years later when the German gynecologist Ernst Bumm inoculated cultures into the urethras of two women and observed the development of gonorrhea in both.[11]

Although the gonococcus was firmly established as the cause of gonorrhea, an enigma surrounded the transmission of some cases. When a man's sexual partner had an obvious discharge and he subsequently developed gonorrhea, it was not hard to conclude that the infection came from her, especially if gonococci were found in the discharge from her vagina. But there were many cases through history in which a male contracted gonorrhea though he had had no recent intercourse with anyone except a woman who had neither a discharge nor any other symptoms of the disease. Such events were particularly tragic when they led to the husband's accusing his wife of infidelity.[12]

To explain this perplexing situation, doctors and laymen alike at first assumed that some cases of gonorrhea were "strains" which followed excessive intercourse. After the gonococcus was discovered, doctors were still puzzled because they could not find it in the vaginal secretions of many of the sexual partners of male victims. To the concept of "strain" they added the idea that the secretions of some females contained irritating substances which could produce gonorrhea in males. Meanwhile the intensive study of pathology that was taking place in the nineteenth century uncovered the answer. Autopsies performed on women who had had gonorrhea years before and no symptoms in the interval, or in some cases no history of any infection, often revealed a dormant gonococcal infection of the Fallopian tubes and surrounding structures. The concept of latent gonorrhea in women was first clearly set forth by an American physician, Emil Noeggerath, in 1872. Further proof came when gonococci were found in the diseased organs of women who had had no recent symptoms. Even then, however, not everyone was convinced that a latent infection could exist. Whether from Victorian modesty or a tender regard for the patients' feelings, many doctors clung to the belief that gonorrhea did not always originate from another gonococcal infection.[13]

The proper understanding of how transmission occurred was helped by the study of latency in the male. As doctors adopted the techniques of staining and culturing urethral secretions, they began to find cases in which gonococci disappeared from the secretions of males, only to return after months or years. The explanation for a wife's seeming unfaithfulness thus became evident: thinking that he had been cured of his gonorrhea, a husband would resume intercourse with his wife. She would become infected but have only inconsequential or transient symptoms. Nevertheless, a focus of infection would remain, and at some later date she would discharge a few gonococci and reinfect her husband. In the same way women who had various sexual partners could infect many men without knowing they were doing so.

Sadder still were the numerous cases in which women unknowingly infected the eyes of their offspring during childbirth. The connection between gonorrhea and eye infection was recognized early in the eighteenth century by Charles de Saint-Yves, a Parisian ophthalmologist, who believed that the infection traveled from the urethra through the bloodstream to the eyes. Later it was found that the more common methods of infection were by fingers, by articles such as towels and clothing, or by direct contact as the fetus passed through the infected birth canal. While the infection was usually confined to the conjunctiva, the membrane that lines the inner surface of the eyelid and the front of the eye, sometimes it became more than a conjuctivitis, invading the deeper structures and causing partial or complete blindness.[14]

Arthritis is a frequent complication of gonorrheal urethritis. It involves especially the knees, ankles, hips, shoulders, elbows, and wrists, tending to move from joint to joint at first and finally settling in one joint. The occurrence of gonorrhea and arthritis at the same time had been noted as early as 1507, although for a long time the relationship was thought to be coincidental. In the last half of the nineteenth century Jean-Alfred Fournier of Paris pointed out arthritis' distinct characteristics and its close connection with gonorrhea. In 1893 the association was confirmed when gonococci were cultured from the fluids of inflamed joints.[15]

Whereas the vaginas of adult women are resistant to gonococci, the vaginas of girls up to the age of puberty easily become infected. Gonococcal infection in the vaginas of young girls was recognized in the early part of the nineteenth century, and the mode of transmission became evident when doctors observed case after case of children whose only source of infection was a mother or older sister with gonorrhea. In 1885 Eugene Fraenkel of Hamburg added to the evidence by showing that the gonococcus could be consistently found in the vaginal pus of these unfortunate children.[16]

The treatment of gonorrhea went through various vicissitudes. Doctors tried two kinds of remedies: drugs taken by mouth, on the supposition that they would have either a soothing or an antiseptic effect on the urethra when excreted in the urine, and chemicals injected into the urethra, with which they hoped to kill the causative organisms without injuring the tissues of the host. The remedies taken by mouth did no good and rarely any harm, while those injected locally did much harm and seldom any good. Oils, balsams, and mucilagenous substances, such as slippery elm tea, were prescribed on the assumption that, since they flowed smoothly down the throat, they would do the same as they passed through the urethra. The various concoctions devised by doctors and swallowed by trusting patients had one virtue: because the patients drank large quantities of liquids, they passed large quantities of urine, which washed out the pus more effectively and made it less irritating through dilution.

Beginning in 1891, when doctors were starting to use synthetic chemicals as drugs, synthetic dyes were tried in gonorrhea. The first was methylene blue, which was such a weak antiseptic when excreted in the urine that its only value was to make the patient feel he was being helped because his urine turned bluish green. Another synthetic used was methenamine, which releases formaldehyde in acid solutions, as urine usually is, and was therefore expected to kill bacteria, but again the concentrations in the urine were ineffectual. Until the 1940s the only useful drugs taken by mouth were sedatives and opiates, which allayed anxiety and pain but did not alter the course of the disease.

Injections into the urethra were used as early as the eleventh century, and in the fourteenth century John Arderne, an English surgeon, injected lead lotion. By 1756 Jean Astruc, physician to the king of France, was recommending the injection of cow's milk, frog's spawn water, or a decoction of marshmallow roots. The medical profession soon discarded such substances, which did neither harm nor good, but unfortunately retained others with a great potential for harm. Mercury was extensively prescribed to be swallowed, inhaled, or rubbed into the skin for syphilis and gonorrhea when the two were thought of as the same disease. After gonorrhea was differentiated, mercury was injected into the urethra. By the end of the nineteenth century mercury had been generally rejected as more toxic than beneficial. For a while zinc and bismuth were used instead but were also finally discarded as worthless.[17]

Other chemicals injected with the intent of killing gonococci were highly colored solutions of potassium permanganate and acriflavine and the silver salts. In chronic gonorrhea, the prostate gland was massaged through the rectum, so as to push secretions into the urethra and thus to drain pus from the recesses of the gland. These treatments, though tedious, uncomfortable, sometimes painful, and always expensive, were probably helpful in some cases. Nevertheless, the doctor's main contribution was in warning the patient that strenuous exercise, sexual stimulation, and the use of alcohol would make the disease worse and in keeping him busy with pills and injections until nature cured the gonorrhea. In fact, in 1929 so little progress had been made in the treatment of the disease that doctors still subscribed to Ricord's statement made a century before: "a gonorrhea begins and God alone knows when it will end."[18]

When inflammation caused a narrowing or stricture of the ure-thra, the skills of the surgeon were essential to enable the patient to void properly. Chronic strictures called for a long course of treatments designed to stretch the narrowed walls. Benjamin Bell of Edinburgh was inserting solid rubber instruments for this purpose as early as 1793, but later curved metal rods, called sounds, were substituted. Eventually the standard treatment required a number of months, during which the patient submitted to the painful process of the insertion of sounds of progressively increasing diameter.[19]

In some cases of arthritis the inflammation in the joints subsided after days or weeks; in others it settled in one joint for months and sometimes caused so much destruction of the adjacent cartilage and bone that the joint never again functioned normally. Doctors tried everything to halt the disease, including bloodletting and the application of leeches, blisters, liniments, and hot solutions, without success. Drugs that were of value in other forms of arthritis, such as aspirin in rheumatic fever and colchicum in gout, did nothing to shorten the course of the gonococcal form. Around 1880 the application of splints and casts to the joints was the fashion, until it became evident that these served only to fix the joints so that they remained rigid for life.[20]

Eventually one measure was found that would arrest the arthritis in most cases. After it was discovered that gonococci could be killed in the test tube by subjecting them to a temperature of 41.5° C. (about 105.5°F.) for several hours, doctors began to treat gonococcal arthritis by inducing fever. Injections of killed bacteria or proteins to produce an artificial fever proved unreliable, but then in the late 1920s a cabinet was designed which enclosed all of a person's body except his head and was heated by light bulbs, and within which a patient could tolerate high temperatures for several hours without too much discomfort. This technique brought about improvement in all patients with acute gonococcal arthritis and in 80 percent of those in the chronic stage.[21]

Women were spared the heroic treatments to which men were subjected because it was not recognized that the disease usually involved internal organs. After doctors learned this fact, they first prescribed douches, as they had for simple vaginal discharge, but these had no effect on the complications, no matter how brightly colored with various chemicals. Eventually doctors resorted to prolonged douching with hot water, realizing that the application of local heat was the best method of aiding the body's defense mechanisms. Unfortunately, many women had so much abdominal pain that the Fallopian tubes and sometimes the ovaries had to be removed. When the Fallopian tubes were not removed, they often became so completely obstructed that these women were sterile for life. It was estimated that in 1900 in this country 12 percent of all marriages were sterile, of which at least two-thirds were the result of gonococcal infection.[22]

For a long time doctors were just as helpless in treating young girls with gonorrhea, in whom the primary gonococcal infection occurred mostly in the vagina. In 1928 it was found that when hormones derived from ovaries were injected into immature monkeys, their vaginas matured into adult form. The next logical step was to use these hormones in girls with gonorrhea. When such treatments were tried, beginning in 1933, the infections consistently improved as the vaginal membranes became like those of adults and thus resistant to gonococci.[23]

By the end of the nineteenth century the management of syphilis was less advanced than that of gonorrhea. Doctors were able to diagnose syphilis fairly reliably and to understand how it was transmitted and where it spread within the body, but progress was hampered because they had not found the causative agent. In 1905 two Germans, Fritz Schaudinn, a protozoologist, and Eric Hoffmann, a dermatologist, observed in various syphilitic tissues a pale, spiral or corkscrewlike microorganism, first called *Spirochaeta pallida* but subsequently renamed *Treponema pallidum*. This treponeme was not immediately accepted as the cause of syphilis, because twenty-five other so-called causes had been "discovered" in the preceding twenty-five years. But skeptics were convinced when treponemes were consistently found in the primary and secondary foci and in the internal organs of syphilitics.[24]

In 1906 another German, August Wassermann, and his colleagues announced that they had devised a test for the diagnosis of syphilis which became known as the Wassermann test. This development made the diagnosis of syphilis much more certain than when it had depended upon clinical examination alone and gave an impetus both to the treatment of patients and to the public health control of the disease. Other tests for syphilis have since been devised, popularly but incorrectly called "Wassermann tests." The proper designation for all tests for syphilitic antibodies is "serological tests for syphilis."[25]

Hideyo Noguchi of the Rockefeller Institute then achieved notoriety by claiming that he had cultivated the *Treponema pallidum*. Although investigators could not reproduce his results, his claim was long honored, probably because he worked in such a prestigious place. Yet investigators anywhere can be wrong at times, and it is now generally agreed that no one has succeeded in cultivating the treponeme that causes syphilis.[26]

Noguchi, however, did make one significant contribution to the knowledge of syphilis. The coincidental occurrence of paresis, a form of rapidly progressive and highly disabling insanity, had long been noted. Some believed that paresis was actually syphilis of the brain, whereas others thought that it was a parasyphilis, meaning something that accompanied syphilis elsewhere in the body. In 1913 Noguchi and J. W. Moore showed that syphilis was present in the brain itself by finding *Treponema pallidum* in the brains of patients who had died of paresis, and the disease came to be understood as a chronic, progressive infection of the meninges and brain by *Treponema pallidum*, the first manifestations appearing many years after the original infection.[27]

In the late fifteenth century European doctors began to treat syphilis with mercury, taken internally. It was probably introduced by the Arabs, who had used it for centuries for diseases of the skin. Since mercury was toxic to the treponemes, it was sometimes curative, occasionally spectacularly so, in the early stages. When it was given in sufficient doses in late syphilis, the disease sometimes retreated slowly, although complete cure was probably infrequent. Unfortunately, it was also toxic to human tissues, particularly if given in large doses, and it was often impossible to find a dose that was high enough to bring about improvement but would not also produce one or more toxic effects, which included profuse salivation, loosening of the teeth, diarrhea, anemia, mental changes, and damage to the kidney with consequent renal failure. Mercury therapy was an excruciating experience for many patients, and for some it was fatal. Other drugs were tried and eventually discarded as worthless, except for the iodides of potassium and sodium. These were introduced in 1834 by William Wallace of Dublin to treat the late stages of syphilis. They apparently caused some improvement in a few cases, although the frequency with which they were prescribed probably represented hope more than evidence of efficacy. Fortunately, other forces were at work to change this discouraging situation.[28]

The synthetic dye industry had its beginning in 1856 when an Englishman, William Henry Perkin, trying to synthesize quinine, produced instead a purple dye. Four years later salicylic acid was synthesized, the first drug to be developed as a by-product of research on dyes. Although its use was limited to the skin, where it was valuable in the treatment of fungal infections, one of its derivatives, aspirin, became the most widely used drug for minor pains and mild fever. About this time Ehrlich began to study the affinity of dyes for bacteria. While staining bacteria in brilliant reds and blues, he conceived the idea that he might find a chemical which, by combining with a microorganism, would destroy it. In 1891 he and an associate reported the first successful application of this principle in the treatment of malaria with methylene blue. Later investigators found that an arsenical compound, atoxyl, would cure infections caused by trypanosomes, which are protozoans or single-celled animals. Unfortunately, this compound sometimes caused atrophy of the optic nerve. Although treponemes, being bacteria, are single-celled members of the vegetable rather than the animal kingdom, Ehrlich believed he could find a less toxic arsenical compound which would kill treponemes. In 1907 he patented his 606th preparation, which produced dramatic results in first rabbits and then patients infected with syphilis. As one clinician recalled, when the drug was first given, "the chancre, the rash, the Wassermann reaction, all cleared up like magic. People . . . were astounded. . . . It was a miracle."[29]

Ehrlich was immediately besieged by letters, cablegrams, and crowds of doctors, all begging for some of the precious drug. He provided sixty-five thousand doses to a selected group of investigators for clinical studies between June and December 1910, and by the end of that year the drug was available commercially under the trade name of Salvarsan. Because Ehrlich had patented the drug, because he insisted upon rigidly controlled animal tests and extensive studies in humans before it was distributed generally, because he personally searched out the causes of toxic reactions and looked for ways to avoiding them, and because a large pharmaceutical company began promptly to synthesize the drug under his direction, the testing and introduction of this drug stand out as a model for its time.[30]

From the beginning, however, a perplexing problem was the high price of the drug. By World War I, after the Allies had blockaded Germany, the price rose to prohibitive levels in the United States. Although no one outside Germany had been made privy to the method of synthesis, American chemists finally succeeded in producing the compound and gave it the generic name of arsphenamine. By 1917 a dose in this country cost $1.00 to the army and $1.50 to physicians, in contrast to the price of $4.00 before the war. Later

Ehrlich and others developed improved analogues of arsphenamine, compounds that differed slightly from it in chemical structure while having similar therapeutic effects. One of these, neoarsphenamine (Neosalvarsan in Germany), which was less toxic and could be given more conveniently than arsphenamine, became the most widely used antisyphilitic remedy up to the 1940s.[31]

Another early problem with the drug was its toxicity. Ehrlich's insistence that the optimal therapy consisted of a single dose that would kill all the treponemes at once also caused doctors to use excessively high doses and to expect recovery from one or a few injections. When adverse reactions developed, especially if they were fatal, or the patient relapsed after a short course of therapy, some doctors began to doubt that this was such a worthwhile drug after all. As late as 1914 the editor of the *Journal of the American Medical Association* warned that arsphenamine might not be as effective as mercury in the treatment of syphilis. But further studies showed that the doubters were wrong. Doctors found that smaller amounts of the drug should be given in courses extending over a period of time. Toxic reactions could also be minimized by carefully neutralizing the acid drug with an alkali immediately before injection and by giving it intravenously instead of into the muscles, where it produced large, painful lumps that lasted several days.[32]

In 1917 Johan Almkvist of Sweden and Albert Keidel of Johns Hopkins independently proposed that patients with syphilis be treated for at least a year to ensure complete elimination of the treponemes. To lessen the toxic effects of prolonged therapy, they recommended that courses of arsphenamine and mercury be alternated. In 1921 doctors began to substitute bismuth for mercury because it was less toxic. In the twenties the various regimens of treatment used and the difficulty of following up patients to determine the results of therapy produced wide differences of opinion as to the best treatment. In 1929 the Cooperative Clinical Group, composed of the chiefs of five leading clinics in the United States and Thomas Parran, then assistant surgeon general of the Public Health Service, began using uniform regimens in their clinics and pooling the data on the results. The information incorporated in fifteen extensive reports published by this group between 1931 and 1940, which represented the most complete study of the therapy of a disease carried out to that time, determined that the best method of treatment consisted of courses of weekly injections of arsenicals alternating with courses of weekly injections of bismuth without a rest period, the total duration to extend over several months or years, depending upon the severity and amount of disease.[33]

The arsenical drugs, though effective in other forms of the disease, proved of little value in most forms of neurosyphilis. A totally different kind of therapy gave more promising results. In 1887 Wagner von Jauregg of Vienna began using fever therapy in paresis. The fever produced by tuberculin or bacterial vaccines was so variable that he tried inducing fever by injecting malarial parasites, which produced a chill and fever every second or third day, depending upon the type of parasite. By 1926 62 percent of patients at the Mayo Clinic who had received this treatment improved, 25 percent of them to such an extent that they could return to work. Later, doctors found a better way to induce fever by the same fever cabinet used in treating gonococcal arthritis. A comparative study reported in 1941 that 69 percent of patients with paresis improved with this method, compared with 58 percent of those who had been given malaria.[34]

Ehrlich had discarded as too toxic another arsenical that was slightly different from arsphenamine in chemical structure. Years later pharmacologists at the University of Wisconsin reported that smaller doses of this drug than of neoarsphenamine cured experimental syphilis in rabbits. In 1935 it was tried in patients with gratifying results. Under the name of arsenoxide, it became widely used, especially for intensive courses of arsenical treatment.[35]

By the 1930s syphilis could be diagnosed with certainty, several remedies were available to treat it, and clinical investigators had worked out optimal regimens. If these were properly followed, a high percentage of patients with early syphilis could be cured and the late stages of the disease could often be ameliorated and sometimes cured. Yet here too, as with gonorrhea, the main focus was on trying to control the spread of the disease.

Society's attempts to prevent the spread of infections are always hindered by ignorance, carelessness, unconcern, and unwillingness to forsake accustomed patterns of action. But in the case of the venereal diseases there is another barrier to control. Sexual practices are so intimately bound up with the moral code of a social group that this code and the way people interpret it often interfere with the treatment and prevention of these diseases. The prevalent attitude toward syphilis in the sixteenth and seventeenth centuries was one of hopelessness and in the next two centuries was one of indifference. Toward the end of the nineteenth century syphilis came to be looked upon as a disgrace. This attitude placed it outside the concern of respectable citizens; in fact, the word could not even be mentioned in polite conversation. When Samuel Gross of Philadelphia spoke of syphilis before the American Medical Association in 1874, he apologized for raising such an unpleasant subject and hoped that it would not give offense to his colleagues. Around 1900, when Osler was enlisting the help of a civic leader in the fight against syphilis, he cautioned her against using the word in public lest people think she was not "nice." The attitudes of the public were reflected in Upton Sinclair's novel *Sylvia*, published in 1913, in which the women of the family "now and then . . . spoke in awe-stricken whispers of this mysterious taint, using the phrase 'a bad disease.'" Such was the "conspiracy of silence" that surrounded syphilis. Those who were religious considered it God's punishment for sin; those who were not were just as certain that it was a punishment for violating the laws of nature.[36]

Because the late complications of gonorrhea were often not identified with the original disease and because it was so common and frequently so mild, there was a tendency in the eighteenth century to regard it mainly as a nuisance or, as Boswell put it, "merely a chance of war." How trivial it was considered to be is reflected in an old jingle:

> A clap in spring
> Is physic for a King.[37]

Yet as Victorian morality took over, gonorrhea came to share in the opprobrium of syphilis. This attitude blocked efforts to control either disease.

The winds that eventually shifted the weather vane of public opinion arose first in Europe in the late nineteenth century. The French clinicians who advanced medical knowledge of the venereal diseases spoke out also for disseminating that information and pushed for control by public health authorities. European physicians and public health officials sponsored a series of international conferences on syphilis, the first in 1899. In 1905 Prince A. Morrow of New York founded the first voluntary society whose purpose was to fight venereal diseases. European writers, such as Ibsen in *Ghosts* and *A Doll's House*, did not hesitate to show the disastrous effects of venereal diseases on family life. One play in particular revolved completely around the subject of syphilis. *Damaged Goods*, by a Frenchman, Eugene Brieux, was the story of a young man who was diagnosed as having syphilis and warned by his physician not to marry for several years. Thinking that he was cured, he married earlier, only to give the disease to his wife. The play struck hard against secrecy. One of the characters cries, "No, no! Not a word about that without blushing; but as many dirty jokes as you like. Pornography, as much as you please; science, never!" The writer counseled men to tell their sons "to love only one woman, to be her first lover, and to love her so well that she will never be unfaithful."[38]

When the United States was faced with a world war, the ubiquity of venereal disease forced the problem to the attention of authorities. Examinations of servicemen at the time of induction showed that 5.6 percent of a total of 3.5 million men were infected with a venereal disease. Prior to the war the Army Medical Corps had undertaken to combat venereal diseases by instruction in sex hygiene and the provision of prophylactic materials. These measures were rewarded by a drop in the frequency of new cases of syphilis from a peak of 35 per 1,000 average strength in 1911 to 15 per 1,000 in 1917. When the war started, however, the Allied armies in Europe had to fight a battle with venereal diseases that almost equaled the one against the Germans. At the time the United States became involved, 800,000 cases of gonorrhea and 200,000 cases of syphilis had been treated in the French army. General John J. Pershing issued an order requiring prophylaxis by the application of a mercury ointment following exposure, treatment if syphilis appeared, and punishment if soldiers did not follow prescribed measures. The Americans also declared French houses of prostitution "off limits." As a result, the rates of venereal disease remained lower in the American than in the British and French armies, although still high by peacetime standards. In the United States, military and civilian authorities collaborated in suppressing prostitution and other sources of promiscuity in the vicinity of army camps, a Venereal Disease Division was set up in the Public Health Service, and venereal disease clinics were opened in the vicinity of 25 cities.[39]

Whatever slight gains in the education of the public had occurred during the war were mostly lost afterward as the lid of secrecy clamped shut again. Thus the matter remained until 1926 when a new champion entered the lists against the venereal diseases with the appointment of Thomas Parran as chief of the Division of Venereal Diseases of the Public Health Service. Despite his efforts to get the story of these diseases before the public, as late as 1934 he was told by the Columbia Broadcasting System that he could not use the word "syphilis" in a scheduled radio broadcast. This episode merely increased Parran's determination to break through the wall of silence. Finally in 1936 he persuaded the *Survey Graphic* to publish an article by him on syphilis, which was extracted at length in the *Reader's Digest*. In it Parran mentioned syphilis repeatedly by name, described its symptoms, pointed out the dangers of the prevalent attitude of sweeping it under the rug, and explained that early treatment could often bring about a cure. Despite the fact that he did not mince words, he carefully avoided, as he wrote to one of the editors, "the lurid description and the over-statement." The response of readers, estimated in the millions, was overwhelmingly favorable. This reversal of attitudes was formalized in 1937 when Franklin Roosevelt, after appointing Parran as surgeon general, became the first President to make "sexual conduct a matter for unembarrassed adult discussion" by endorsing federal intervention to reduce "the disastrous results of venereal disease." The way was opened for Parran's program.[40]

When World War II again made venereal disease in troops an urgent problem, the campaign of publicity was stepped up. Conservative groups decried the emphasis on the curability of syphilis, contending that health authorities should "promote morality and clean living rather than open and shameless discussion of the checking of disease contracted through sinful practices." This was an echo of a previous battle among the reformers, who had often disagreed on whether the spread of venereal diseases could best be stopped by moral suasion or by using scientific knowledge for prevention and treatment. Workers in the front lines saw the futility of relying on the former approach alone. From their observations they tended to agree with Montaigne that "a hundred students have caught syphilis before they come to Aristotle's lesson on temperance." Also they suspected what Alfred Kinsey later demonstrated, that premarital and extramarital intercourse was much more frequent than most people supposed. Parran, while agreeing that continence was an ideal to be striven for, insisted that syphilis could not be stamped out unless other measures were also used.[41]

One important breeder of venereal diseases was prostitution, which flourished in the nineteenth century with only occasional and ineffectual molestation. Police knew that the trade was booming and had the necessary legislation to stop it. Yet they made no serious effort to interfere, either because they realized that the majority of the public did not want suppression or because they were profiting handsomely from the status quo. The cities of continental Europe had long placed their faith in the registration of prostitutes and the requirement that they undergo periodic physical examinations, and in some American cities in the late nineteenth century, pressure was put on authorities to use this method. In 1870 St. Louis established a system of inspecting prostitutes and isolating those found to have a venereal disease. But opposition arose on moral grounds, and four years later the enabling legislation was repealed. The case against regulation was persuasively made by Abraham Flexner, who found in a 1914 study of prostitution in Europe that registration and inspection did not materially diminish venereal disease because the regulations did not reach the "clandestine prostitutes." He concluded that these measures actually increased the amount of prostitution by perpetuating the conditions out of which it arose: "an indulgent attitude towards the male sex, on the one hand, and a disregard of woman's dignity, on the other." Howard Woolston in a study conducted in the United States concluded that regulation was practically unenforceable because it depended upon the vigilance of the police, the docility of the women, and the support of the public, which factors were never present for long in any city.[42]

Yet if regulation would not work, something else had to be done, for in 1913 the Department of Justice estimated that 63,000 women in the United States were supporting men from their earnings as prostitutes, 26,000 of them in New York City alone. In *Jews without Money* Michael Gold told how, when he was a child in New York in the late 1890s, prostitutes "called their wares like pushcart peddlers." Public actions to diminish prostitution consisted mainly of periodic bursts of arresting when the police were sufficiently pressured by aroused citizens and by the attempts of public officials to improve the slums. The Progressive reforms of the early 1900s led to the passage of the Mann Act in 1910, which prohibited interstate traffic in women. The specter of prostitution was also one factor that brought about the passage of minimum wage laws beginning in 1913. In general the federal government did practically nothing to combat prostitution, and state governments were equally remiss. Such was the situation when Parran made the battle against prostitution part of his war against venereal diseases. He advocated strict enforcement of laws against prostitution and urged that the public be kept informed of the enormity of the problem. But he was realistic enough to see that progress would come mainly through prevention and treatment of the diseases themselves.[43]

The public health control of venereal diseases required first that the cases be reported. Since syphilis could not be detected by physical examination during much of its course and since, even when visible, it could be confused with other diseases, requirements that doctors report cases were not feasible until 1906 when the Wassermann test made accurate diagnosis possible. The availability of arsphenamine for treatment four years later provided another reason for reporting. Health authorities began to conceive of a comprehensive program which included registration of cases, provision of diagnostic facilities, tracing of contacts to uncover other cases, isolation of infectious patients, and clinics for treatment of all patients so that some would be cured and the treponemes in the remainder would be diminished and sequestered internally so that they could not infect others—or as the program was later summed up, "find and treat." In gonorrhea, since no specific treatment was available to break the chain of infection, control depended upon early and exact diagnosis, identifying the source, and preventing patients and contacts from having sexual intercourse until they were no longer infectious.[44]

In 1911 California became the first state to require the reporting of venereal diseases, and the next year health authorities in New York City ordered public institutions and requested doctors to report cases of syphilis. Although some hospitals established clinics for treatment, these were few and inadequate. A physician at the Pennsylvania Hospital in Philadelphia complained in 1911 that the most important cases of syphilis from the community point of view, the most contagious ones, were "treated in some out-of-the-way corner by the least thoughtful and most uncouth junior member of the hospital's surgical staff, with little or no supervision or enforced sense of responsibility."[45]

In 1921, of 83 cities surveyed, 82 required doctors to report all cases of venereal disease and 81 had free clinics for their treatment. Yet health department laboratories were equipped to diagnose gonorrhea in only 64 cities and to diagnose syphilis in 49, while only 38 cities provided hospital care for venereal disease patients. The report of the survey concluded: "Fewer city health departments are equipped to aid in the diagnosis and control of the venereal diseases than of diphtheria, typhoid fever or tuberculosis." It is not surprising that venereal diseases remained rampant in the United States. In 1922 death rates from syphilis were 18 per 100,000 population, approximately what they had been before the war, and in 1936 they were still 16.2 per 100,000. A survey in 1930 showed that nearly 4 persons out of every 100 were suffering from syphilis. In that same year the rate of new cases reported rose to 185 per 100,000 population, and increased gradually to a peak of 372 per 100,000 in 1938, partly as a result of more complete detection. Clearly this country was getting nowhere with controlling venereal diseases.[46]

Although the problem of the venereal diseases was most obvious and most urgent in the cities, they could not cope with it alone. Some states began to recognize their responsibilities. By 1923 Pennsylvania had developed the rudiments of a comprehensive plan. State-supported clinics were opened to demonstrate how "to find and treat." Approval was given for the sale of a preventive to be used after exposure. Educational programs were mounted for both the public and doctors. New York State's program went even further, requiring that doctors confirm their diagnosis of syphilis and gonorrhea by a laboratory examination and that they report every case. The state health department also attempted to coordinate the efforts of public and private groups. Since a husband, or sometimes a wife, often brought syphilis unknowingly to the marriage bed, laws requiring premarital examinations for venereal diseases had been advocated for many years, and in 1935 Connecticut passed the first law requiring that a blood test for syphilis be performed on both partners. By 1939 17 more states had enacted such laws, and by 1964 the number of states had reached 45.[47]

Because of the economic status of many victims of venereal disease, the cost of treatment was a major obstacle to control. In 1932 the actual cost of drugs and laboratory tests for the 76 visits required to treat a patient with early syphilis was $78, while the doctor in private practice paid $380 for drugs and laboratory tests, and his average fee was $650. According to one estimate, 80 percent of patients with syphilis could not afford to pay for treatment. State and municipal resources were inadequate to the task, while private contributions were difficult to obtain because of the prevalent attitude that donors to this cause were suffering from venereal disease. Finally in 1936 the federal government untied the purse strings by making Social Security funds available to the states and empowering the Public Health Service to administer the grants and conduct research. The results were soon evident. In the year ending June 30, 1939, the number of clinics for the treatment of venereal diseases in the United States increased more than 30 percent as compared with the previous year, the number of patients brought under treatment for the first time by 60 percent, and the number of laboratory tests performed by venereal disease clinics by 78 percent.[48]

Increased funds made it possible to mount special programs. In some areas where syphilis was particularly prevalent, congenital syphilis was appallingly frequent. A study of a black community in

Manhattan in the early 1920s showed that one pregnancy out of five in syphilitic mothers resulted in either a stillbirth or a miscarriage and that, of children born alive, 383 per 1,000 died during their first two years of life. In 1938 New York State became the first of many states to require that a serological test for syphilis be performed on every pregnant woman. In California, where such a law was passed in 1939, the number of reported cases of congenital syphilis fell from 1.6 per 1,000 live births in 1938 to 0.43 per 1,000 in 1943. Further progress was slow because the first examination was often not made until the latter months of pregnancy, whereas syphilis in the offspring could only be prevented if the mother received treatment early. In one study, when treatment was first given during the seventh month, half the children had syphilis, and over two-thirds had it when treatment was started during the eighth month. In contrast, when treatment was begun between the first and the fourth months, only 5 percent of the children were syphilitic, and all the children were free from the disease when the mothers were started on treatment before the pregnancy.[49]

Testing for syphilis showed particularly high rates in some of the southern states, particularly among blacks. In Savannah, Georgia, in 1945, among approximately 70,000 persons aged 15 years and over, 31.5 and 30.3 percent of black males and females, respectively, were found to have syphilis, compared with 3.4 and 2.3 percent of white males and females. Beginning in 1943, Alabama required every person between the ages of 14 and 50 to have an examination of his blood for syphilis. Once the facilities in Alabama for control and treatment had been expanded, the number of new cases dropped from a peak of 32,507 in 1948 to 2,124 in 1955.[50]

In rural areas, where patients were less mobile, they were more easily traced than in big cities; consequently they could be persuaded, and if necessary compelled, to finish a regimen of therapy so that they would not be able to transmit the infection. In cities the retention of patients in clinics until the completion of therapy was much more difficult. The major complaints by urban clinic patients involved the long waits for treatment, the lack of cleanliness and privacy, and the understaffing. As a result, many patients did not finish the prescribed therapy and some did not even take enough treatment to prevent relapse of the infection. To cope with this problem, another method of treatment was devised, known as intensive therapy. Previously, treatment with arsphenamine had lengthened from a single dose to a course spread out over a year or more. As less toxic drugs became available, particularly arsenoxide, and doctors became more adept at administering them and more knowledgeable about toxic manifestations, it became possible to shorten the course of therapy without raising the risk to the patient immoderately. Late in 1942 Chicago opened the first hospital for intensive treatment, where patients with early syphilis were required to remain until a short, intensive course of therapy had been completed. By 1947 there were 63 rapid treatment centers throughout the country.[51]

All these public health measures eventually had their effect on the frequency of syphilis. The death rate from syphilis and its sequels declined slowly, if irregularly, from a peak of 19.1 per 100,000 population in 1918 to 15.9 in 1938. For the next two decades it fell more rapidly, reaching 9.3 per 100,000 in 1946. Because opportunities for health education and facilities for treatment were sadly deficient among certain underprivileged groups, the United States did not succeed in controlling syphilis as completely with arsenicals and public health measures as did some other countries with a wider availability of services. The results were nevertheless impressive, and the experience gained by the cooperative endeavors of various groups—federal, state, and local; lay and professional; public and private—was utilized in other fields of public health.[52]

The gratifying progress made in the control of syphilis was not paralleled in gonorrhea except in gonococcal infection of the eyes of the newborn. In 1880 Carl Siegmund Franz Credé, a Leipzig obstetrician, initiated the practice of instilling silver nitrate into the eyes of every newborn child immediately after birth to prevent conjunctivitis. He found that when this technique was properly carried out, not a single child developed gonococcal infection of the eyes. But adoption came slowly, for doctors who could understand the need for preventive measures for babies born in charity wards could not understand that same need for children of private patients. The problem was not solved until laws were passed requiring the instillation of silver nitrate or another effective drug in the proper dilution into the eyes of every baby at birth. Silver nitrate was also found to be effective in the treatment of gonococcal infections of the eyes. As a result, gonococci, which in 1907 caused 28.2 percent of the cases of blindness in a sample group of children in schools for the blind, caused only 0.1 percent of such cases in 1954 and 1955.[53]

Given the unwillingness of patients with gonorrhea to seek treatment, the tendency of the disease to become chronic, and the almost complete inability of doctors to shorten the course of the disease, the possibility for further reducing the number of victims was small indeed. At the beginning of the twentieth century gonorrhea was probably the most frequent disease treated by practicing physicians, and estimates of the number of men who had had gonorrhea at least once varied from 48 to 99 percent. In World War I, among a million men drafted in this country, 4.5 percent had gonorrhea when they were examined. In 25 communities surveyed in 1926, between 5 and 6 of every 1,000 persons had a new gonorrheal infection each year, and 474,000 persons with the disease were under medical care at any one time. Obviously, education and hunting down contacts were not enough without an effective method of prevention or therapy. The urgent need for a curative drug was voiced by one public health worker, who pledged: "the world will build a monument of historical significance" to the scientist who discovers a specific for the treatment of gonorrhea.[54]

8. The Sulfonamides

"The darkest hour is that before the dawn." This old English proverb, though scientifically unfounded, is psychologically sound. Nothing illustrates it better than the attitude toward the treatment of infections during the first third of the twentieth century. At that time the prospect of specific remedies for most of the infectious diseases seemed dim. True, diphtheria antitoxin had been highly successful and tetanus antitoxin somewhat less so. True also, doctors were gradually being won over to the use of antipneumococcal serum, although its effectiveness was diminished by the problem of typing pneumococci beforehand and by the cost of the serum. Despite these few achievements, attempts to develop serums against other bacteria had been disappointing, and the partial success of antimeningococcal serum only demonstrated how hard it was to match the type of serum and the microorganism causing the disease in a particular patient. Also, adverse reactions still occurred, including fatal ones, even when refined serums were used.

The failure of serum therapy to fulfill its early promise except in a few diseases had driven those interested in public health to stress the prevention of disease by vaccines, where these were available, the reporting and isolation of contacts, and improvements in sanitation, housing, general hygiene, and nutrition—with one exception. In the therapy of syphilis synthetic chemicals had proven so effective that the identification and treatment of patients had become the cornerstone of public health measures for its control. The success of chemotherapy in this disease had stimulated a search for chemicals to cure other infections. In 1924 investigators in the Bayer pharmaceutical firm in Germany announced the synthesis of a remedy against malaria, pamaquine (Plasmoquin). In the next few years they synthesized other antimalarial drugs, including quinacrine (Atabrine), which was later used extensively by the Allies after the Japanese shut off their supplies of quinine in World War II. In a similar search for new compounds British scientists synthesized proguanil, which became the prototype for another series of anti-malarial drugs. Also during these years chemical compounds were synthesized that effectively combated some protozoal infections, such as amoebic dysentery and the type of encephalitis caused by trypanosomes.[1]

In contrast to these successes, attempts to synthesize drugs for the cure of bacterial diseases had been a failure. In 1911 German investigators announced that a synthetic derivative of quinine, ethylhydrocupreine, would kill pneumococci in the test tube and cure infections in animals. But when blindness sometimes followed its use in patients, it had to be abandoned. Gold salts were tried in the treatment of tuberculosis beginning in 1924; within a few years they too were discarded as ineffective. From time to time desperate clinicians tried using arsphenamine and related drugs in serious bacterial infections on the assumption that since they worked well against treponemes, they were bound to have some effect against other microorganisms. Failure followed these efforts also.[2]

After Lister had demonstrated that phenol prevented surgical wounds from being infected, unsuccessful attempts were made to cure local infections with this and other antiseptics. Koch even tried injecting mercuric chloride as a cure for generalized infections in guinea pigs. Eventually it became clear that chemicals like phenol and the salts of mercury, which killed the common bacteria by destroying their protoplasm, could not be used to treat infections because they were just as destructive to the cells of the host. Certain dyes with a selective affinity for particular bacteria sometimes prevented infections when spread on the skin prior to a surgical incision or applied to a wound before bacteria had a chance to multiply; yet they failed to check infections once they had occurred. In a desperate attempt to save the life of a seriously ill patient, one of these synthetic dyes would sometimes be tried internally. Although a doctor occasionally persuaded himself that the drug was responsible for a patient's recovery, examination of the results in a large number of cases showed that the toxic effect of the chemical far outweighed any possible benefit. An example was merbromin, known best as Mercurochrome, which commanded much attention in the 1920s. After it had been tried in a number of patients with different bacterial infections, doctors realized that it represented just one more failure.[3]

Thus, by the mid-thirties practically all investigators and clinicians working in the field of infectious diseases had given up hope that effective chemical agents would be found for the treatment of bacterial infections. The general impression was that protozoa and treponemes could be attacked by chemicals without seriously injuring the host, whereas for some reason bacteria other than treponemes could not. Or as Behring had put it long before, "inner disinfection is a vain dream."[4]

Among the few investigators who did not succumb to the prevalent attitude of despair were a small group in the research laboratories of the I. G. Farben Industrie in Elberfeld, Germany. In 1927, at the age of thirty-two, Gerhardt Domagk had been appointed director of research in experimental pathology and bacteriology and had immediately set to work to find a chemical that would cure bacterial infections. The major products of the company were dyes for textiles, and its chemists had found that the azo dyes, compounds in which two nitrogen atoms were linked by a double bond, had a particular chemical affinity for protein materials such as wool or silk. Domagk therefore began testing azo dyes by injecting them in mice that had been infected with streptococci. One of these dyes, later named Prontosil or Prontosil rubrum, which had been synthesized in the same laboratories by the chemists Fritz Mietzsch and Joseph Klarer, cured mice that had been given lethal doses of hemolytic strepococci. Domagk made this observation in 1932, although he did not announce it until February 15, 1935.[5]

French scientists, after reading Domagk's report, asked for samples for investigation. When these were not immediately forthcoming, they synthesized the drug themselves and verified Domagk's results. A team of investigators at the Pasteur Institute in Paris, Dr. and Mrs. Jacques Tréfouël, Frederic Nitti, and Daniel Bovet, confirmed another of Domagk's observations, that Prontosil had no effect on streptococci in the test tube, and tried to find out why it was effective only in animals. Knowing that azo compounds were easily split where the two nitrogen molecules were linked, they hypothesized that this occurred in the animal body, and by studying the products of such a cleavage, they found that one of them, para-aminobenzenesulphonamide, later named sulfanilamide, was as active as Prontosil against hemolytic streptococci or even more so. In their report, also published in 1935, they concluded that it was the sulfanilamide liberated from Prontosil that cured bacterial infections. This simpler compound had been synthesized by a German chemist, Paul Gelmo, in 1908 and could be sold cheaply because no patent was in effect. It had the further advantage of not turning the patient as red as a boiled lobster, as Prontosil did. It Domagk can be said to have opened the treasure house of bacterial chemotherapy, the workers at the Pasteur Institute were the first to reveal what was inside.[6]

The question arises as to why Domagk himself did not discover that sulfanilamide was the active agent. Some think that he did. The official explanation given for the delay of more than two years in the publication of Domagk's successful experiments was that the intervening period was devoted to confirming the results and proving that Prontosil was effective in human infections. Yet Domagk's report as finally published cited only one experiment in animals, and the clinical reports that appeared up to that time included only a few sketchy cases. Then, too, there was the enigma of the long delay in providing samples of Prontosil to reputable scientists in France and England. One leading British bacteriologist, Ronald Hare, concluded that Domagk and his colleagues discovered that sulfanilamide was the active principle quite promptly but realized that, since it was not protected by a patent, the firm could expect scant profit from its sale. Consequently, Domagk's group spent the time between the discovery and the announcement of Prontosil's effectiveness in a vain attempt to find an analogue of sulfanilamide that would be more effective and at the same time patentable. Prontosil had to be given to clinicians, however, so that they could assess its value in humans, and when they began to spread word of the striking improvement in streptococcal infections, publication of Domagk's original experiments became necessary. If Hare's hypothesis is correct, the desire for profits blocked for a time the path from a scientific discovery to its application for the benefit of mankind. Such a stain on an otherwise brilliant achievement is greatly deplored.[7]

The manner in which the discovery was announced explains why it created only a ripple of interest at first. The few case reports that preceded and accompanied Domagk's meager report were more like testimonials than careful scientific appraisals. They did not give enough facts about individual cases to enable the reader to reach a conclusion on the basis of his own judgment. Doctors were familiar with publicity from Germany, where new drugs rose like a rocket, lit up the sky for a while, then fizzled out and were never heard from again. Hare's reaction when he heard of Prontosil for the first time was typical: "Another of those damned compounds from Germany with a trade name and of unknown composition that are of no use anyway."[8]

But fortunately Hare's chief, Leonard Colebrook, decided to investigate the new drug. He and his research team at Queen

Charlotte's Hospital in London had been trying to reduce the high death rate from puerperal infections caused by hemolytic streptococci. Although these infections could usually be prevented by strict aseptic techniques, once an infection occurred, nothing could be done to help the victim. Colebrook's team, after trying many chemical agents without success, had finally decided that they could do nothing more for these unfortunate women than give them the best possible nursing care. Colebrook was unable to obtain samples of Prontosil until late 1935. He confirmed Domagk's results in animals and then began treating patients. In May 1936 Colebrook and Meave Kenny reported that of 38 patients treated with Prontosil for puerperal sepsis, only three had died. This fatality rate of 8 percent contrasted with rates of 26 percent for the 38 patients admitted to the hospital immediately before Prontosil became available and of 29 percent for all the patients treated during the preceding four years. Since no other measures used in the management of these patients had changed, Colebrook and Kenny cautiously concluded that "the very low death rate, taken together with the spectacular remission of fever and symptoms observed in so many of the cases, does suggest that the drug exerted a beneficial effect." In December of the same year they reported treating an additional 26 patients without a single death. Thus the fatality rate for the entire 64 cases was only 4.7 percent.[9]

Still, certain clinicians doubted, calling attention to the low fatality rates for puerperal sepsis in some other hospitals in 1936. But the editor of the *Lancet* immediately grasped the significance of Colebrook's experiments. He pointed out that Prontosil had cured peritoneal infections in mice, although it had been given several hours after the bacteria were injected and had been administered by a different route. He contrasted these results to tests with acriflavine, which had created a false hope a few years before; acriflavine had checked streptococcal infections in mice only when it was injected directly into the peritoneal cavity shortly after the bacteria had been placed there. In other words, not enough time had elapsed for an extensive streptococcal infection to develop; instead, the peritoneal cavity had merely served as a test tube allowing acriflavine to act directly on the streptococci. He called for more clinical cases, and when these were reported the following December, he told doctors, "it has now become an imperative duty to employ these remedies in puerperal fever and other severe forms of streptococcal infection."[10]

In July 1937 the value of Prontosil in streptococcal infections was further demonstrated by a report of 312 patients with erysipelas. The fatality rate of 4.8 percent in these patients compared favorably with a rate of about 10 percent usually observed in this disease. Also, sulfanilamide was found to be effective in meningococcal infections, initially in mice and then in animals.[11]

At first Domagk's report caused even less of a stir in the United States than in Great Britain, but when Perrin H. Long of Johns Hopkins heard of Prontosil's dramatic effects in 1936, he immediately began working on both Prontosil and sulfanilamide. In collaboration with Eleanor A. Bliss, he tested these drugs in a variety of animals and in humans. E. K. Marshall, Jr., a pharmacologist at Johns Hopkins, devised a simple method for determining the content of sulfanilamide in body fluids. Marshall and his associates then made an exhaustive study of the absorption, diffusion, metabolism, and excretion of sulfanilamide. They found that it was rapidly absorbed from the intestines when given by mouth, that it diffused well from the bloodstream into the various tissues and body fluids, and that it was gradually excreted in the urine. Armed with this information, clinicians worked out dosage regimens which would ensure that high concentrations were reached rapidly and maintained adequately in the blood.[12]

Despite sulfanilamide's successes, it was not as effective in pneumococcal as in streptococcal infections, even though these two bacteria were similar in many of their growth characteristics. This failure spurred investigators in the laboratories of various pharmaceutical firms to search for more effective analogues. The first to strike pay dirt was a group headed by A. J. Ewins of the British firm of May and Baker. The new compound, eventually called sulfapyridine, was tested in mice by Lionel Whitby, a bacteriologist at the University of London. In May 1938 he announced that it provided "the one striking success in the chemotherapy of pneumococcal infections in an assessment of no less than 64 related sulfanilamide compounds." Sulfapyridine was soon found to be as effective as sulfanilamide, and usually more so, in all of the infections in which both were tested. Consequently, it rapidly displaced sulfanilamide, and from 1938 on the important clinical tests were made first with sulfapyridine and later with other compounds of equal effectiveness.[13]

Sulfapyridine had two serious drawbacks: in therapeutic doses it frequently caused nausea and vomiting, severe enough in some cases to require stopping its use, and it tended to form crystals in the urine, aggregates of which were sometimes large enough to block urinary passages, and could even cause cessation of the kidney function. This provided the incentive to search for a less toxic analogue. Chemists responded with hundreds of new sulfonamide compounds, only a few of which survived the sieves of animal and human testing, while fewer still were actually marketed. One of them, sulfadiazine, synthesized in 1940 in the laboratories of the American Cyanamid Company, stood out as therapeutically superior and minimally toxic. Yet, sulfadiazine, which became for a time the standard drug for the treatment of infections susceptible to the sulfonamides, still produced renal stones in a few patients. Several years later a research team at the Hoffman-LaRoche laboratories showed that sulfisoxazole was much more soluble in urine. Extensive use established the therapeutic effectiveness of this newer analogue as similar to that of sulfadiazine and showed that it rarely produced renal stones. It largely displaced sulfadiazine for therapy, although the latter continued to be used for prevention because the small doses administered were not followed by significant crystal formation. These two uses of the sulfonamides, for treatment and for prevention, were developed in relation to a number of different disease-causing bacteria.[14]

The early successes in hemolytic streptococcal infections were followed by others. In addition to puerperal sepsis and erysipelas, other serious streptococcal infections such as pneumonia, mastoiditis, and meningitis usually responded well, and often dramatically, to sulfonamide therapy. Only in streptococcal sore throat, including scarlet fever, were the results disappointing. Sulfonamides had little effect on the fever or the symptoms, and opinion was divided on whether or not they diminished the complications. Not until years later was the inadequacy of sulfonamide therapy in streptococcal sore throat satisfactorily explained. In 1953 and 1954 at an air force base in Wyoming where streptococcal infections were particularly prevalent, throat cultures were performed at frequent intervals on 525 patients with sore throat caused by Group A streptococci. Half of them were treated for five days with sulfadiazine, and the other half were observed as controls. During the first week after therapy was stopped, fewer streptococci were detected in the throats of patients treated with sulfadiazine and local complications of the disease were somewhat less. Yet within another week, the number of patients harboring streptococci and the frequency of complications were the same in both groups. Sulfadiazine had not eradicated the streptococci but had merely suppressed the infection temporarily so that it returned after therapy was stopped. Worse still, the percentage of patients developing rheumatic fever after the infection was as great in the treated as in the control group.[15]

Although sulfonamides were of little value when given for a short time for the treatment of streptococcal infections of the throat, when they were given continuously they were highly effective in preventing these infections and the rheumatic fever that sometimes followed them. In 1939 clinicians at Columbia and Johns Hopkins independently reported that patients who had had an attack of rheumatic fever, and were therefore particularly susceptible to additional bouts, usually remained free from the disease while they were taking small, daily doses of sulfanilamide. Others quickly adopted the procedure. Experience over the next fifteen years showed that recurrences occurred in 2 percent of patients given sulfonamides continuously as a preventive, as compared with 13 percent of untreated control patients. Since the possibility that a patient would develop a permanent deformity of a heart valve was apparently directly related to the number of attacks of rheumatic fever that he had had, the use of sulfonamides was in part responsible for the diminishing frequency of rheumatic heart disease.[16]

Because streptococcal infections were common among recruits in World War II, sulfonamides were given as a preventive measure to large numbers of men in military installations. The results were often striking. For instance, when several thousand men took a small daily dose of sulfadiazine during an epidemic of scarlet fever, the number of new cases dropped practically to zero, while new cases continued to occur in the untreated controls at the previous rate. All went well with this prophylactic method until the nemesis of extensive chemotherapy made its appearance: strains of streptococci resistant to sulfonamides emerged and were able to spread widely in groups of people receiving sulfonamides because competing bacteria had been suppressed. The experience of the navy, where at least a million streptococcal infections occurred during World War II, is a good example. Mass prophylaxis with sulfonamides reduced the frequency of streptococcal infections from 6.7 cases per 1,000 persons in 1943 to 5.1 in 1944, yet in the following year the rate rose to 7.4 cases per 1,000 persons because of the spread of resistant strains. There was nothing to do but abandon prophylaxis and return to the former, relatively ineffective measures of control, such as segregation of patients and contacts and avoidance of crowding in barracks and elsewhere, measures that were difficult to achieve in civilian life and much more so among troops in wartime. Fortunately, resistant strains did not become a problem when sulfonamides were used in civilian life to prevent recurrences of rheumatic fever, probably because the conditions were not present that favored the rapid passage of streptococci from person to person, namely the crowding of large numbers of susceptible persons in almost continuous contact, all of whom were receiving a drug that suppressed sensitive streptococci and thus allowed resistant strains to grow.[17]

Compared with streptococcal diseases, infections with pneumococci yielded even more completely to therapy with potent sulfonamides. In 1938 English clinicians claimed highly successful results in the treatment of 100 pneumonia patients with sulfapyridine. In the next few months scattered reports of the pronounced effect of this drug on pneumonia appeared. Around and often ahead of these reports swirled rumors that the "ultimate cure" for pneumonia had been found. Some doctors tested new sulfonamides in patients before adequate studies of the action of the drug had been done in animals. Others based their favorable evaluations on only a few cases. In fact, one report on the use of sulfapyridine involved a mere eight patients with pneumonia, and the type of pneumococcus had been identified in only half of these cases. Such unreliable tests would be repeated many times in the next few decades as new sulfonamides appeared faster than investigators could be trained to test them or than doctors could be educated to judge whether testing had been proper. Because these drugs could be produced rapidly and cheaply and large profits could be expected from them, because doctors were willing to try them and patients anxious to take them, because there was no law requiring that their efficacy be proved before they were sold, and because they could be easily, even casually, administered since they were usually taken by mouth, it is not surprising that the clinical investigator was often by-passed.[18]

Doctors experienced in assessing new drugs counseled a wait-and-see attitude until large-scale trials then in process were finished and evaluated. Fortunately, these assessments were not long in coming. By the end of 1939 clinicians in large hospitals in Boston, Philadelphia, New York, Washington, and St. Louis had treated more than a thousand patients with pneumococcal pneumonia, of whom only 12 percent had died, as compared with an expected fatality rate of about 30 percent in patients who received no specific therapy. But in contrast to streptococcal infections, an effective remedy for pneumococcal pneumonia was already available in antipneumococcal serum, and some doctors were loath to give it up, especially those who had used it extensively. The fatality rates among patients treated early with sulfapyridine were slightly lower than those obtained in patients given a specific serum, although experienced observers recognized that this improvement may have been the result of unconscious selection of the more favorable cases for treatment, as usually happens when a new drug is first used. Sulfapyridine was much easier to administer, typing of the infecting pneumococcus was not required, and it could be given both orally and by injection. Yet the patient's fever and toxic symptoms often disappeared less rapidly than after serum therapy. A reasonable solution seemed to be to give both remedies. Tests showed that pneumococci were killed more quickly in the presence of serum and sulfapyridine together, as compared with either alone. Controlled clinical trials, however, showed that combined therapy gave no better results, or at most was only slightly more effective, than therapy with sulfapyridine alone.[19]

As it turned out, practical considerations strongly influenced doctors' decisions. Therapy with sulfapyridine and later with other sulfonamides took over completely because they were so much simpler to use. Furthermore, as experience increased, the results obtained with the superior sulfonamides proved to be better than they had appeared at first. Mark H. Lepper and I found that among 1,275 patients with pneumococcal pneumonia treated with sulfonamides, 12 percent died compared with 17 percent among 889 patients treated with serum. Of even greater significance, the fatality rate was 32 percent among patients over 59 years of age treated with sulfonamides, compared with 48 percent among those who received serum. Within a short time most doctors had abandoned the typing of pneumococci, pneumonia control programs had closed down, and pharmaceutical companies had stopped producing the serums because they were no longer profitable. As an example, the Lederle Laboratories, probably the world's largest producer and the only one producing serums for all the types of pneumococci then identifiable, wrote off their investment in 28,000 rabbits and the quarters, laboratories, and equipment used to produce antipneumococcal serum from them. Once more, obsolescence had been the price of scientific progress.[20]

Among the most serious of pneumococcal infections is meningitis, which before the appearance of the sulfonamides was almost always fatal. Treatment with sulfanilamide was followed by a few recoveries, and sulfapyridine appeared even more effective. For instance, clinicians in the New York City Health Department treated 53 patients with sulfanilamide, of whom 17 percent recovered, and 30 patients with sulfapyridine, of whom 33 percent recovered.[21]

More impressive still were the results obtained when sulfonamides were used in the variety of meningitis caused by meningococci. When serum was used as therapy, it was not always possible to match the type of meningococcus causing a particular patient's illness with one of the serums available. Even when patients received the proper serum, the fatality rate was reduced by only half, and many of the patients who recovered had residual paralyses, deafness, or impairment of vision. Consequently, when sulfanilamide was shown by British investigators to protect animals against meningococci, clinicians were quick to try it. The announcement early in 1937 by Long and his associates of eleven patients treated with sulfanilamide, of whom only one died, was followed by other favorable reports. As in pneumonia, the question was posed whether sulfanilamide gave better results than serum. In the contagious disease hospital of Baltimore the fatality rate among 368 patients treated with serum during 1935 and 1936 had been 27 percent. After sulfanilamide became available, among the first 106 patients to receive it, only 18 percent died, although some of them had also received serum. Skeptics contended that this apparent superiority of sulfanilamide may have occurred because all of the patients who received it were treated in a period between epidemics and that a new drug could only be tested properly against the more virulent strains prevalent during an epidemic. The opportunity came all too soon.[22]

As the Allied armies mobilized for World War II, a wave of meningococcal meningitis swept across England. The sulfonamides were tried and proved highly effective. Among 2,491 patients who were given one of these drugs, usually sulfapyridine, only 14.3 percent died. The epidemic reached the United States somewhat later, the peak year being 1943, when it caused nearly 3,000 deaths. The response to sulfonamide therapy at this time was as good as in interepidemic periods: the fatality rate in 1,232 patients treated with these drugs was only 9.9 percent. Among military personnel the fatality rates were much lower because they were in a favorable age group. During World War II, among 14,504 American soldiers with meningococcal infections, only 3.8 percent died—a far cry from the 39 percent fatality rate of World War I.[23]

As in pneumonia, when the sulfonamides were first introduced, some investigators contended that the best results could be obtained if a specific serum were also given, particularly in severe cases. The difficulty of matching the serum with the type of meningococcus infecting the patient, the cumbersomeness and cost of serum therapy, the risks of hypersensitivity reactions to serum, and the lack of proof that serum would increase the benefit obtainable from sulfonamide therapy militated against combined therapy, and antimeningococcal serum accompanied antipneumococcal serum to the scrap pile.[24]

It was predicted early that if sulfanilamide should cause meningococci to disappear from the throat as rapidly as from the meninges, "the carrier condition will for the first time be capable of ready control." During 1940 and 1941 doctors in Australia, England, South Africa, and the United States tried prophylaxis with sulfonamides during localized outbreaks of meningococcal meningitis and were uniformly successful in lowering the carrier rate. It appeared also that reducing the number of carriers had prevented the occurrence of new cases of meningitis, but this was not certain because only a few control patients were observed in one of these trials and none in the others.[25]

When the incidence of meningococcal meningitis began to rise in the United States in 1941, sulfadiazine was just coming into use. This drug was soon found to clear meningococci from the throat more consistently than sulfapyridine and to be more acceptable for use in large groups because nausea was less frequent. It had been noted that when healthy individuals took sulfonamides for some time, their throats usually remained free from meningococci for several weeks, but as soon as they came into contact with carriers who had not received sulfonamides, they were likely to become infected again. With this in mind, one investigator wrote, "The use of the sulphonamide drugs in the treatment of carriers is likely to have only a limited application since their wholesale administration is impossible." But he reckoned without the exigencies of war. In early 1943 army officials were confronted with several epidemics among troops in the southeastern states. Harry A. Feldman and his colleagues in the Third Service Command Laboratory initiated prophylaxis with sulfadiazine for every soldier barracked in a certain area in two large military training facilities where meningitis had been prevalent. Over 15,000 men received sulfadiazine, while nearly 19,000 living in separate areas in the two installations were observed as controls. There was no contact between the two groups. The carrier rates among the soldiers receiving prophylaxis fell from 36 to 2 percent in one installation and from 30 percent to zero in the other, while among the controls they rose from 38 to 57 percent and from

29 to 33 percent, respectively, during the same period. In the eight weeks after sulfadiazine was given, only two cases of meningitis occurred among the 15,000 treated soldiers, in contrast to 40 cases among the controls. At approximately the same time F. Sargent Cheever used sulfadiazine to stop epidemics in naval personnel with similar success.[26]

Prophylaxis with sulfadiazine was widely practiced in the presence of epidemics of meningitis for the next two decades, during which time it almost appeared that this dread disease had been brought under control. Resistant strains of meningococci did not appear under these circumstances as had resistant streptococci—or so it seemed until 1963. In May of that year investigators throughout the United States received a letter from Feldman informing them that sulfadiazine-resistant strains of meningococci had been found in military installations and requesting cultures of meningococci so that he could determine how widespread these strains were. Shortly thereafter the medical profession was informed that sulfadiazine-resistant strains had been causing meningitis at the Naval Training Center in San Diego and that mass prophylaxis with sulfadiazine had not stopped the epidemic. At nearby Fort Ord a similar situation produced such panic among the citizens of California that the base had to be closed as a training center. Sulfadiazine-resistant strains subsequently appeared in other military installations, and in 1974 the Center for Disease Control of the Public Health Service reported that 23 percent of all strains of meningococci submitted to them from civilian patients in the United States in the past three years were resistant. Most of these strains belonged to Groups B and C and only a rare one to Group A, the strain that has caused all the large epidemics since typing began.[27] Outside the United States sulfonamide-resistant Group A strains were found in many parts of the world and caused devastating epidemics in Africa.[28] As a result, the use of the sulfonamides to prevent meningococcal infections has been confined to localized epidemics where the meningococci have been found to be sensitive to these drugs. Emphasis shifted instead to vaccination.

Almost simultaneous with the report on the effectiveness of sulfanilamide in meningococcal infections, an announcement came from Johns Hopkins that sulfanilamide had cured 15 of 19 patients with gonorrhea, and similar results were reported by many other groups within a short time. Although there were some disappointments, probably as a result of inadequate doses or the failure of patients to continue treatment after symptoms disappeared, it soon became clear that for the first time doctors had a specific cure for this ubiquitous disease. With the use of other sulfonamides that were both more potent and less likely to cause adverse reactions, about 90 percent of patients with uncomplicated gonorrhea could be cured with one course of treatment and the remainder with further courses. Even the more serious complications, such as arthritis and endocarditis, sometimes responded.[29]

Public health officials were elated. It now seemed possible to control gonorrhea, the foundling of the antivenereal disease effort, the "wet-diapered and puny little bastard that would neither get well nor die." Venereal disease clinics were opened to patients with gonorrhea, and 3.5 million tablets of sulfanilamide were distributed to clinics and doctors for the treatment of this disease. But the very simplicity of the treatment carried the seeds of its eventual failure. Patients, given a box of tablets comprising a full course of treatment, would stop taking them when their symptoms disappeared, before their infections were eradicated. Worse still, they gave or sold them to others, who proceeded to treat their infections even more casually. Among the gonococci that remained behind, some were partially resistant to sulfonamides, and these became the sole survivors when subsequent courses were given. Also, the rapid disappearance of symptoms meant to many patients that they need not refrain from sexual intercourse.[30]

Just at this time the mobilization of troops for World War II was making the control of venereal diseases urgent. Doctors found that by giving sulfonamides to every man when he left a military post in the evening and again when he returned, they could prevent gonorrhea. At one post, among 1,400 men who were given one of the more potent sulfonamides, sulfathiazole, as a preventive, only four cases of gonorrhea developed in five months, while among 4,000 controls there were from 100 to over 200 cases per month during the same period. Since it was clear that sulfonamides could not only cure but also prevent gonorrhea, public health officials were urged by the commissioner of health of New York City to "Stamp Out Gonorrhea Now!"[31]

But it was not to be. As early as 1938 strains of gonococci resistant to sulfonamides had been detected in the laboratory, although little attention was paid to them at the time. By 1944 British doctors found that sulfonamides were curing less than 25 percent of soldiers in Italy, compared to 70 to 75 percent cures among the same troops in Africa a few months before. The same problem was soon encountered in American troops. Tests showed that a high proportion of the strains of gonococci now being recovered were resistant to sulfonamides. Obviously sulfonamides had not turned out to be the panacea for gonorrhea that some had hoped for.[32]

Up to the 1930s drugs had been of little or no value to the surgeon in combating infections. Sulfonamides were tried early and seemed to help recovery from infected wounds, appendicitis, and peritonitis. Surgeons thought that they were about to reach the millennium, when wounds would never become infected or could quickly be made sterile if they did. They sprinkled sulfonamides liberally into wounds and incisions, into the peritoneum and other cavities during operations. But before long the disadvantages of local application became apparent. When large quantities of sulfonamides were placed in the peritoneal cavity, they frequently damaged the liver, because the blood containing the high concentrations of these drugs was filtered through that organ before it entered the main circulation.[33]

Local administration of the sulfonamides had other drawbacks. It caused hypersensitivity to develop to these drugs more frequently than when they were given orally or by injection, and it delayed the healing of tissues. In 1947 the American Medical Association withdrew approval of local use as a routine. Given by mouth or by injection, however, the sulfonamides proved successful in altering the course of some local and general infections cared for by surgeons, especially those caused by hemolytic streptococci, staphylococci, and the bacteria that produce gas gangrene, an infection following injury to muscles characterized by bubbles of gas within the destroyed muscular tissue.[34]

One important concept arose from experience with the local use of sulfonamides. As early as 1938 it was noted that whenever any appreciable amount of dead tissue was present in a wound, the activity of sulfonamides was diminished. Out of this observation came the hypothesis that sulfonamides acted on bacteria by interfering with the assimilation of some important nutritional substance, and that these drugs failed to act in the presence of dead tissue because that tissue was an alternative source of the necessary substance. Two British investigators, D. D. Woods and P. Fildes, verified the concept by demonstrating that sulfonamides prevented bacteria from utilizing an essential growth requirement, para-aminobenzoic acid. Because the sulfonamides resembled this compound closely in chemical structure, they attached themselves to bacteria in its place and, since they could not be utilized metabolically, brought growth to a halt. This process, called competitive inhibition, has since been widely utilized in the search for drugs for the treatment of infections and of other diseases, particularly cancer.[35]

Dysentery, inflammation of the bowel wall with resulting diarrhea, may be caused either by amoebas or by bacteria. The causative agents of the bacterial type, which is usually more acute and is more frequently lethal, are gram-negative rods called *Shigellae*. Early attempts to use sulfonamides in this disease gave questionable results. In 1940 Marshall and his associates announced the synthesis of sulfaguanidine, a sulfanilamide analogue that was only slightly absorbed from the intestines, and the following year reported its use in shigella dysentery. Cases of this disease are usually much more severe in tropical and subtropical regions and in underprivileged countries than in the United States, as the Allied armies were finding out in the Mediterranean and South Pacific theaters of World War II. Sulfaguanidine was promptly used to treat dysentery among Allied troops and found to be dramatically effective. For instance, in an epidemic in New Guinea only two deaths occurred among 10,000 patients treated. Since a sulfonamide that was poorly absorbed from the intestines did not enter the blood in enough quantity to form precipitates in the urine, sulfaguanidine was thought to be the best sulfonamide to use under battle conditions in the hot, steamy jungles of the South Sea islands and in the North African deserts, where sweating would cause the urine to be concentrated. Yet it was subsequently shown that this theory was wrong and that the best results in dysentery were obtained with a sulfonamide that was readily absorbed, such as sulfadiazine. After all, dysentery was a disease of the wall of the intestines, not of the bacteria in the intestine itself, and diffusion into the blood and from there into the intestinal wall made for optimal therapy.[36]

With the use of the most effective sulfonamides, the death rate from shigella dysentery in the United States fell from 8.2 deaths per million population in the five-year period 1935–1939 to 5.9 in 1940–1944. It reached 2.7 deaths per million during 1945–1949, although this additional reduction can be accounted for in part by the introduction of antibiotics. Widespread use of the sulfonamides, however, was followed by the appearance of sulfonamide-resistant *Shigellae*. In 1953 American investigators reported that in a prisoner-of-war camp in Korea 55 percent of the patients who received sulfadiazine still had *Shigellae* in their stools after two days of therapy, only 18 percent less than the percentage of untreated patients with positive cultures on the same day.[37]

Although the poorly absorbable sulfonamides did not turn out to be the optimal drugs for dysentery, they were soon put to another use. Surgeons found that these compounds would reduce the number of bacteria in the intestines and thus could be given before operations to decrease the chances of contaminating surrounding areas when the intestines were opened. Other poorly absorbable sulfonamides were synthesized, which displaced sulfaguanidine because they were even more effective in diminishing the bacterial population of the intestines. The widespread use of these compounds helped to reduce the number of infections and deaths following operations on the intestines. The effect of preoperative sulfonamides plus the use of sulfonamides for therapy when an infection occurred was reflected in a drop in the death rate from appendicitis from 11.4 deaths per 100,000 in 1936 to 5.5 in 1943.[38]

Other diseases in which the sulfonamides were effective were trachoma, a serious, scarring infection of the conjunctiva; chancroid, a venereally acquired ulcer of the genitals which resembles the syphilitic chancre; and many of the common bacterial infections of the urinary tract. Some infections caused by staphylococci responded, although often the severe ones did not. Sulfonamides were tried in a number of other infections and found ineffective.

As early as 1940 the diseases in which the sulfonamides were useful and those in which they were not had been delineated quite accurately.[39] Many of the definitive clinical evaluations were carried out in the United States and Great Britain, not in Germany where the first laboratory studies had been performed. One reason was the development in the United States by the mid-thirties of a number of well-equipped university hospitals and the establishment of laboratories for clinical research at several large municipal hospitals, such as the Boston City Hospital and Bellevue Hospital. These institutions had well-organized medical staffs, including a core of full-time teachers and investigators, and the laboratories were equipped to perform the sophisticated tests required at the advancing edge of research. In contrast to the situation in Germany, where each professor controlled only a small group of patients and dictated to his subordinates what would be done with them, the chief of a medical service in the United States usually considered it his function to facilitate the research of his junior colleagues. As a consequence, an investigator working on pneumonia might have access to patients with this disease throughout the hospital. In Great Britain, where the hospitals were in general smaller and the compartmentalization between groups of physicians attending on different wards sometimes more rigid, the Medical Research Council frequently sponsored studies on particular diseases in which adequate clinical and laboratory facilities were made available in one or more hospitals. This method paid dividends in Colebrook's early demonstration of the value of sulfonamides in puerperal sepsis and would continue to be highly productive in obtaining rapid and accurate evaluations of newer treatments.

One additional factor apparently aided clinical research in Great Britain and the United States. Relationships between clinical and basic science departments were closer than in Germany, where science departments were more likely to be set up as separate institutes. A good example is the collaboration at Johns Hopkins of the Department of Pharmacology with the Departments of Medicine and Pediatrics and with surgical groups in evaluating sulfanilamide, sulfapyridine, and sulfaguanidine.

In contrast to the experts on infectious diseases, many other doctors were overenthusiastic in their use of sulfonamides. Some prescribed them for diseases in which there was no possibility of their being effective or, what was worse, for illnesses that were merely suspected of being infectious because the patient had fever. Sulfonamides were so easy to use—just a prescription for a few tablets was all that was needed in most cases—and some infections took so long to diagnose that the temptation was ever present to prescribe these drugs instead of making a diagnosis. Pressure on doctors to prescribe them also came from patients who were influenced by extravagant eulogies in magazines and newspapers. Even the staid *New York Times* called sulfanilamide "the drug which has astounded the medical profession" and spoke of the miracles that sulfapyridine performed in pneumonia. In the years 1935–1950 over 5,400 articles on sulfonamides appeared in medical journals, 586 in the peak year 1943. It was estimated that in 1941 in the United States 1,700 tons of sulfonamide compounds were administered to between 10 and 15 million persons. They continued to be prescribed extensively for several more years until replaced by even newer "miracle drugs."[40]

All drugs carry some risk of untoward reactions, and the sulfonamide drugs were by no means innocuous. In New York City in 1941, 28 persons were reported to have died as a result of adverse reactions from sulfonamides, and undoubtedly many more deaths were unrecognized as such. Because sulfonamides were being given to so many people, it was difficult to ascertain the exact frequency of adverse reactions, but it was estimated in 1941 that they caused one death among every 1,600 patients who received them as treatment for pneumonia. When small quantities were used as a prophylactic measure, adverse reactions were less frequent and usually less severe. As an example, when a single dose of two grams of sulfadiazine was given to 25,000 persons at an army air base, only 128 persons, or 0.51 percent, reported reactions, of which 13 were severe and three critical for a time, although there were no deaths.[41]

One catastrophe occurred that was only indirectly the result of sulfonamide administration. In 1937 a chemist in a small pharmaceutical firm, seeking a liquid in which sulfanilamide could be dissolved, so that it could be administered easily to children, found that a commercial solvent, diethylene glycol, would answer the purpose very well. Without testing the solvent or the final product for toxic effects, the firm sold the solution widely through the South as "Elixir of Sulfanilamide." Before its lethal effects were appreciated, it had killed over 100 persons, and subsequent tests showed that diethylene glycol was poisonous when ingested. The proprietor claimed that he had done nothing illegal, and as a matter of fact he could only be convicted on a technicality: that the word elixir means a medicine containing alcohol, and there was none in the so-called Elixir of Sulfanilamide. Fortunately, the tragedy shocked Congress into action. The following year it passed the Food, Drug and Cosmetic Act, which required, among other provisions, that before a new drug could be marketed, its sponsor had to satisfy the Food and Drug Administration as to its safety.[42]

Great as was the effect of the sulfonamides upon the infectious diseases themselves, their discovery had an additional impact upon the development of drugs for other diseases. During the period when new sulfonamides were being eagerly sought, the Germans were occupying France and the disruptions of war were producing many cases of typhoid fever. Marcel Janbon of the University of Montpelier tried to treat them with a new sulfonamide. Because the patients were poorly nourished before they became ill with typhoid, they were already predisposed to hypoglycemia, or a low concentration of glucose, the main sugar in the blood. After a number of them died of an undetermined cause and many who lived developed prolonged coma or suffered from convulsions, Janbon found that the glucose in their blood was indeed low. Auguste Loubatières then studied the action of the same sulfonamide in dogs and discovered that it caused hypoglycemia. German investigators, who learned of these results, began a search for analogues of the sulfonamides that would lower the blood sugar in diabetics, and they developed tolbutamide and carbutamide, which became widely used in the treatment of diabetes mellitus, a disease characterized by an abnormally high concentration of glucose in the blood.[43]

In another area, investigators noted incidentally that the sulfonamides caused diuresis, or an increased secretion of urine. British and American investigators found that this diuresis resulted from the inhibition of an enzyme, carbonic anhydrase. Acting upon this clue, chemists of the American Cyanamid Company produced a more effective inhibitor of the enzyme, which was named acetazoleamide. Later, another inhibitor, chlorothiazide, was produced in the laboratories of Merck Sharpe and Dohme. These became the prototypes for a number of other similarly acting drugs that have been found useful in a variety of conditions from kidney disease and high blood pressure to disturbances of the fluid system in the eyes.[44]

The sulfonamides did not prove to be a panacea; they did not cure every infection even when the causative agent was highly susceptible. But they did represent a big step forward in the control of bacterial diseases. In many infections they shortened the duration of the disease, lowered the fatality rate, and lessened the number of complications. Even though their use as prophylactics was hampered by the appearance of resistant strains, the sulfonamides were often effective in preventing individual cases of disease and in stopping epidemics, and they continued to be useful in preventing return bouts of rheumatic fever. Finally, the success of the sulfonamides showed that drugs could be found that would cure bacterial infections. Thus they stimulated a search for other antibacterial agents, a search that led within a very short time to penicillin.

9. *Penicillin*

Events that change profoundly the lives of many people tend sooner or later to become enveloped in myths. And, although the remarkable power of penicillin to cure infections has been known for only a third of a century, the myths have already begun to form. The popular idea of the introduction of penicillin is that its potential was apparent immediately: when in 1928 Alexander Fleming saw a mold on an agar plate, surrounded by a zone in which no bacteria were growing, he knew immediately that the mold was producing a substance that would cure human infections, and that nothing remained to be done except for a few minor technical maneuvers. Also, many people believe that if this event had not occurred, there would never have been any antibiotics. Both of these ideas are wrong, but like most myths, both contain a germ of truth. Actually, the discovery of the antibiotics, like most important advances in science, involved the work of many people extending over several decades.

The path to the discovery of penicillin had its beginning half a century before Fleming's important observation and did not reach its goal for several years afterward. In the 1870s a number of bacteriologists, including Pasteur, recognized that some microorganisms inhibited the growth of others. In 1885 Victor Babès, later professor of bacteriology at the University of Bucharest, showed that this inhibition of growth was caused by a substance manufactured by a microorganism which was liberated into the liquid or semisolid medium on which it was growing. In other words, the microorganism was producing an antibiotic. This term, which did not come into use until the 1940s, designates a substance produced by microorganisms which in dilute solutions will inhibit the growth of or destroy other microorganisms. Two years later a Swiss surgeon and bacteriologist, Carl Garré, employed a test for detecting such substances which in modified form is still in use today. He streaked cultures of two bacteria in alternating lines on solidified gelatin and observed how the substance from one microorganism diffused through the gelatin and inhibited the growth of the other. Between 1899 and 1913 a number of investigators tried without success to treat generalized infections with pyocyanase, an antibiotic produced by *Bacillus pyocyaneus* (now named *Pseudomonas aeruginosa*). Although pyocyanase was capable of lysing or dissolving several species of bacteria, including anthrax, diphtheria, and typhoid bacilli, it proved to be too toxic when injected into animals. Thus it could only be used for local application to surface infections—a fate that was to be the destiny of many antibiotics in the years to come. Other substances derived from bacteria and molds were studied, but all were either too weak or too toxic. Several investigators even observed that a mold, that is, a type of fungus that often grows on vegetable or animal matter, called *Penicillium* inhibited the growth of certain microorganisms, but this clue was not followed up.[1] Meanwhile, the success of vaccines and serums as preventive and therapeutic agents and the spectacular results achieved with the arsenical compounds in syphilis drew most investigators away from research on antibiotics. Then came Fleming's important contribution.

Alexander Fleming was born in Scotland and studied medicine at St. Mary's Hospital, London, receiving his degree in 1908. Afterward he worked there as a bacteriologist, investigating vaccines, the bacteriology of war wounds, and agents that gave promise of curing infections. He shared in the early success of chemotherapy, using some of the first arsphenamine to reach England, and in its frustration, as trial after trial demonstrated that antiseptics did not cure systemic infections. In 1922 Fleming thought he had found such a cure when he discovered a substance in tears and other body fluids which lysed various bacteria. This proved to be an enzyme, a substance produced by living cells that brings about or accelerates chemical changes in other substances while remaining unchanged itself. Unfortunately, extensive investigations by Fleming and others failed to show any way in which lysozyme could be used in the treatment of infections. Then in 1928 he stumbled upon penicillin.[2]

The story of this discovery is best told in his own words: "While working with staphylococcus variants a number of culture-plates were set aside on the laboratory bench and examined from time to time. In the examinations these plates were necessarily exposed to the air and they became contaminated with various microorganisms. It was noticed that around a large colony of a contaminating mould the staphylococcus colonies became transparent and were obviously undergoing lysis . . . Subcultures of this mould were made and experiments conducted with a view to ascertaining something of the properties of the bacteriolytic substance which had evidently been formed in the mould culture and which had diffused into the surrounding medium. It was found that broth in which the mould had been grown at room temperature for one or two weeks had acquired marked inhibitory, bactericidal and bacteriolytic properties to many of the more common pathogenic bacteria."[3]

The mold was later identified as a variety of *Penicillium notatum*. Fleming showed that the substance produced by the mold, which he named penicillin, had a powerful effect on bacteria. When diluted to 1/800 of its original strength, it inhibited the growth in the laboratory of staphylococci, pneumococci, and several varieties of streptococci. It had a lesser inhibitory effect on diphtheria and anthrax bacilli, but even in strong concentrations it had no effect on the growth of gram-negative rods, including typhoid bacilli. Fleming also found that penicillin was not injurious to white blood cells in the test tube and not toxic when injected into rabbits and mice or when used to irrigate surface wounds in humans. He recommended penicillin as an aid in isolating certain bacteria in the laboratory, because of its power to inhibit the growth of other bacteria, and he suggested that it might also be "an efficient antiseptic for application to, or injection into, areas infected with penicillin-sensitive microbes."[4]

Fleming asked two associates, Stuart Craddock and Frederick Ridley, to try to purify penicillin. They worked hard at the task, but when the end product of their manipulations always turned out to be a brown, syrupy mass from which they were never able to extract the active principle, they gave up. Fleming tried the crude "mould-juice" locally on the infection of an amputated stump without success, and on Craddock's staphylococcal sinusitis with no apparent effect. When it was dropped into the eye of a patient with pneumococcal conjunctivitis, however, it seemed to help. Fleming asked the surgeons in his hospital for cases in which wounds had become infected, but they never had a case when he had a supply of penicillin and he never had penicillin on hand when they came up with a case. "In this way," he explained, "therapeutic use [of penicillin] lapsed."[5]

In 1932 Harold Raistrick and his associates at the London School of Tropical Medicine and Hygiene, in the course of their studies on metabolic products of fungi, examined Fleming's penicillium mold. They made several important observations: the mold would grow better on a synthetic medium than on the one used by Fleming, the activity of penicillin could be preserved for at least three months if bacteria were filtered out and the filtrate was made mildly acid, and penicillin could be extracted into ether if the solution was acidified. But they could carry the process no further. Penicillin was so unstable that additional chemical manipulations caused it to "vanish away," as Raistrick later commented. Since their interest was in molds rather than in the treatment of human infections, they then moved on to other molds. No one interested in clinical research had the opportunity to use their extracts of penicillin.[6] Meanwhile the sulfonamides were discovered and began to absorb the efforts of investigators interested in infections, including Fleming. Then in 1938, nine years after Fleming's first report, the other major actors entered the scene.

Howard Walter Florey, an Australian, began his research career at Oxford and later returned there as professor of pathology. He gathered around him an impressive team of investigators, among them Ernst Boris Chain, a refugee biochemist from Hitler's Germany. Together they began an intensive study of the antibacterial substances produced by microorganisms.[7]

Meanwhile, evidence was accumulating on the existence of substances that were able to control bacterial infections. In 1931 René Dubos and Oswald Avery of the Rockefeller Institute announced the extraction of an enzyme from a soil bacillus that would break down the capsule of the Type III pneumococcus. Stripped of its capsule, this coccus was then an easy prey for the cells that ingest and destroy bacteria and other foreign bodies, called phagocytes. They demonstrated that this enzyme could protect animals against infections with virulent pneumococci and could cure pneumococcal pneumonia in monkeys. Although the enzyme proved too toxic to be used in patients with pneumonia, it stimulated a search for similar substances of less toxicity. Next came the sulfonamides, which were evidence of the existence of substances that could attack bacteria without injuring the cells of the host. How much these discoveries helped motivate Florey and Chain to pursue their search is hard to tell, but probably each had some influence.[8]

Although at first the main emphasis of Florey's group was on lysozyme, work on penicillin was also started in a small way. The early results were "not impressive," according to one participant, but Florey and Chain were not easily turned aside by small setbacks. Florey assigned Norman Heatley, a biochemist, to develop a procedure for assaying simultaneously several samples of penicillin. Such a method was needed before the group could know whether the procedures being tried for extracting penicillin were getting anywhere. Heatley placed glass cylinders in a plate of agar medium and filled them with the liquid containing the penicillin. As the penicillin diffused under the cylinders and through the agar, its concentration was measured by the size of the area within which a microorganism previously seeded on the agar would not grow. The method, which was simple and reliable, became the prototype for methods that are widely used today.[9]

Chain, Florey, and Heatley now focused on finding a way to obtain a sufficient amount of reasonably pure penicillin for animal experiments. They found that penicillin which had been extracted from an acid solution into ether could then be extracted from ether into a neutral aqueous solution, which was necessary so that it could be injected into animals. This procedure removed many impurities and could be carried out with little loss of antibacterial activity if the manipulations were made at a temperature of 2° to 5°C. Although each plate of penicillin mold yielded only a small amount of crude penicillin, the team kept steadily at it until in May 1940 enough penicillin had been obtained so that when eight mice were injected intraperitoneally with streptococci, four of them could be given penicillin subcutaneously. In seventeen hours all the control mice were dead, while all the penicillin group still lived and two survived completely. Here, after nearly seventy years of research on naturally occurring antimicrobial substances, was the first evidence that one of them could be an effective and nontoxic, systemic therapeutic agent. Penicillin was no longer a laboratory curiosity, no longer just another substance to apply to external wounds. Rather, in Florey's words, "penicillin was a chemotherapeutic drug of great power . . . [because] it could, without producing any toxic signs, be injected . . . in amounts which would stop the growth of bacteria in all parts of the body." After verifying these results in more animals and determining that they applied also to staphylococci and to a bacterium that caused gas gangrene, the group made its first report.[10]

Florey immediately moved to throw the resources of his laboratories behind the research on penicillin. He tried to interest British pharmaceutical firms in producing penicillin pure enough and in sufficient quantities to be given to humans, but when he received what was to be the first of many refusals, his team tackled the job themselves. In their desperate efforts to produce large amounts, they tried every conceivable receptacle for growing the mold and finally settled on rectangular porcelain vessels modeled after old-fashioned bedpans which afforded a large surface for the growth of the mold. Florey's group set their sights at 500 liters of broth culture of the mold a week, from which they planned to extract 100,000 to 200,000 units of penicillin, less than one-fourth of a single day's treatment of pneumococcal pneumonia today. A unit of penicillin is an arbitrary measure, established by Florey's group as the amount that produced a zone of inhibition of growth of staphylococci similar to that produced by a standard solution of penicillin. As soon as they had enough penicillin, they tested it in volunteers. When the injections caused chills and fever, Edward Abraham, who had recently joined the team, found a way to eliminate the substance that was responsible.[11]

During 1940 the Oxford group increased their production of penicillin a thousandfold, and on February 12, 1941, they began to treat their first patient. Unfortunately they undertook too rigorous a test, for they selected a policeman who was almost dead from a widespread staphylococcal infection of four months' duration complicated by a secondary streptococcal infection. Sulfonamides had had no effect on the hundreds of abscesses nor on the osteomyelitis, an infection in the bones. By the fourth day of penicillin therapy the abscesses were healing, the patient's temperature was normal, and his appetite had returned. But by the fifth day the scanty supply of the antibiotic had been exhausted; then the signs of infection began to return, and three weeks later the man was dead.[12]

The team went back to producing more penicillin. By May they had treated five more patients, four of them seriously ill. In August 1941 they published a second report which covered their laboratory investigations more fully, plus the treatment of ten patients, five of whom had received it intravenously, one by mouth, and four, who had eye infections, by local application. The results were highly favorable, in spite of the harrowing case of the first patient and another death from a complication which occurred after cure of the infection was almost complete. Spurred on by these results and convinced that he could not get the help he needed from the British pharmaceutical industry when their factories were being bombed and Britain was fighting for its life, Florey decided to go to the United States, and the penicillin story entered a new phase.[13]

Some Americans had read the first report of the Oxford group, and had sent for cultures of the mold to work on in their laboratories. One group at Columbia-Presbyterian Medical Center in New York, headed by Martin Dawson, had produced their own penicillin and even begun treating patients with it four months before Florey's group. Four patients with bacterial endocarditis improved temporarily, while eight patients with staphylococcal infections of the eyes responded well to local treatment.[14]

When Florey and Heatley arrived in the United States, the reaction to their story was mixed. Positive action came only when the Department of Agriculture referred them to the Northern Regional Research Laboratories in Peoria, Illinois, where Heatley remained for several weeks to work with experts in fermentation procedures. Out of this union of forces came two significant advances and the beginning of a third. First, it was found that the yield of penicillin could be increased tenfold by adding to the culture medium cornsteep liquor, a by-product of the extraction of cornstarch. Second, a world-wide search for variants of Fleming's mold resulted in strains that produced more penicillin than the original. Finally, the first attempts were made to grow the molds in deep tanks instead of on the surface of media in small vessels. Florey later cited these three advances as the important contributions of the United States to the development of penicillin.[15]

Meanwhile, Florey was making the rounds of the pharmaceutical companies. He got no cooperation in Canada, but in the United States the outcome was different, largely through the influence of Alfred N. Richards, chairman of the Committee on Medical Research of the Office of Scientific Research and Development, which had just been organized to help prepare for the war that was fast approaching. Within a year Richards had persuaded three pharmaceutical companies—Merck and Company, E. R. Squibb and Sons, and the Charles Pfizer Company—to share with each other and with the committee the information obtained in their research on penicillin.[16]

By early 1942 enough penicillin had been produced by American companies to treat one patient, and a year later 100 patients had been treated. The production of penicillin increased from 400 million units in the first five months of 1943, enough by today's standards to treat ten patients with pneumococcal pneumonia or two with staphylococcal bacteremia, to approximately 300 billion units a month by the end of 1944. At the time of the Allied invasion of Europe in that year enough penicillin was on hand to treat all severe casualties in British and American troops. At the same time the cost decreased. The Office of Scientific Research and Development paid commercial companies $200 per million units in early 1943, which was less than the cost of production, but in 1945 only $6 per million units.[17]

These achievements were accomplished so well and so rapidly in the midst of a war only by skillful planning. One crucial decision—

whether to perfect the fermentation method or to learn how to synthesize penicillin chemically—was made early by Richards in favor of fermentation, and he stuck to it in the face of considerable pressure from those who believed that the antibiotic could be made more quickly and cheaply by synthesis. Events proved his critics wrong, for when penicillin was later synthesized, it became evident that the mold aided by man could produce penicillin more efficiently than man alone in the chemical laboratory. Another decision, whether to continue the surface cultivation of the mold as perfected at Oxford or to experiment with large-scale production by deep fermentation, was decided more easily. Some companies were assigned to produce penicillin by the surface method and others, particularly those with experience in producing other substances by deep-tank fermentation, were assigned to try that procedure. When it became clear that the deep-tank method required only one-twentieth the space of the surface method to produce the same quantity of penicillin and that it increased productivity 72-fold in terms of man-hours, the surface method was abandoned.[18]

By the time the pilot plants of the commercial companies of America were beginning to produce penicillin, extensive clinical trials had become mandatory. The results in animal infections had been highly successful; the results in the 22 patients treated by the combined efforts of the Oxford and Columbia groups had been encouraging but not completely convincing, especially since only nine of them had received it by injection for serious infections and six of these had died. The challenge was taken up by Richards, who in 1942 asked the Committee on Chemotherapeutic and Other Agents of the National Research Council to direct the trials. Chester Keefer, who took charge of the job, was tough-minded enough to parcel out the scant supply of penicillin to cases where the illness was serious and other agents were not likely to be effective, while resisting all pleas from harassed doctors pursued by frantic families to use it in cases where it offered no hope of help. In 1943 the committee reported on 500 patients treated with penicillin, nearly half of whom had staphylococcal infections. Among 91 patients in whom bacteremia was present, only 40 percent failed to recover, as compared with an expected fatality rate of nearly 100 percent. Of 55 patients with osteomyelitis, 48 recovered or improved. Results were equally impressive in other staphylococcal infections, although harder to evaluate because in many cases spontaneous recovery would have been expected. Since sulfonamides were seldom effective in serious staphylococcal infections, these observations were especially important. Good results were also obtained in infections caused by gonococci and streptococci resistant to sulfonamides.[19]

In the same year Champ Lyons of the Massachusetts General Hospital began pilot studies in army hospitals on the treatment of surgical infections. Here, too, penicillin was beneficial in infections caused by staphylococci and hemolytic streptococci. To these were added infections caused by the *Clostridia,* anaerobic bacteria which were especially likely to grow in the torn and dirty tissues of wounds. Florey's group had reported that penicillin inhibited a clostridial infection in mice; Lyons found it to be highly effective even in the worst of war wounds.[20]

Another use was found for penicillin, also in 1943, when John F. Mahoney and his colleagues in the Public Health Service reported that it would cure experimental syphilis in rabbits and seemed to have cured a few patients with syphilis. Its effectiveness was fully confirmed as more patients were treated. In view of the need to determine quickly how penicillin would help in the war, comparisons with untreated controls were not made in any of these diseases. Yet only experts used penicillin at first, and they were able to draw upon their knowledge of the course of infections in patients given serum, sulfonamides, or no specific therapy to determine whether penicillin gave better results. Their judgments stood up well when larger numbers of patients were treated later.[21]

As the production of penicillin increased in the United States, the Oxford group, who had made this possible, should have been assured of an adequate supply, but external events worked against them. By then the United States had been catapulted into World War II. Material for building and equipping penicillin plants was given priority over certain other needs, and in return for this priority, once penicillin was produced, it was rationed according to this country's requirements. By late 1942 Florey had received only three samples of penicillin from America, all of them low in potency. He resumed his role of gadfly toward British pharmaceutical firms with some success, although his main supply still came from his Oxford laboratories. With their meager stock of penicillin, Florey and his wife, M. Ethel Florey, began a second series of clinical trials. Ten patients with staphylococcal infections improved rapidly, among them three who would certainly have died without it. In a study of over 300 battle casualties in North Africa, Florey and Hugh Cairns, professor of neurosurgery at Oxford, found that penicillin was effective in war wounds and from this experience were able to lay down methods for the use of penicillin by military surgeons. By 1943, with support from the government, enough penicillin was available in Britain so that some could be used in civilians, and studies similar to those in the United States were started. In 1945 supplies were placed in a number of hospitals, and in 1946 penicillin was freely available through commercial channels.[22]

Florey and his British colleagues were just beginning to solve the problems of production when another obstacle appeared. Andrew J. Moyer, an American bacteriologist, had been assigned to work with Heatley at the Peoria laboratory. Together they got the *Penicillium* mold growing and showed that corn-steep liquor would increase the yield of penicillin. But while Moyer was soaking up all the information that Heatley could give him about the processes used at Oxford, he was keeping his own laboratory notes to himself. In 1945 he applied for a patent in the United Kingdom for the use of corn-steep liquor and other substances in the growth media used in the production of penicillin. To some in the scientific community this seemed improper, although Moyer was legally entitled to obtain a foreign patent for work done while he was a civil servant. Florey, who had been advised by leaders of British science not to apply for a patent on the work of the Oxford group because "the people paid for this work and should have the benefits," was also criticized for allegedly giving away to America a project that might have earned millions of dollars for Britain. Yet while Moyer is said to have become rich from royalties, no one need doubt that Florey's name will be written in the angel's book alongside Leigh Hunt's Abou Ben Adhem as one who loved his fellow men.[23]

While the battle to produce penicillin was under way, another one was shaping up over who should get the credit for penicillin. The conflict was started by that arch disputant Almroth Wright, who was Fleming's chief. When a friend of Fleming was dying from streptococcal meningitis in 1942 at St. Mary's Hospital despite large doses of sulfapyridine, Fleming obtained penicillin from Florey to treat the patient. The London *Times* reported the patient's dramatic recovery and described the work on penicillin at Oxford, but without mentioning any names. Wright wrote a letter to the editor claiming the credit for Fleming. From that point on the legend began to take shape that "[o]n that summer day in 1928 occurred one of the great events in human history" when the lone scientist saw on an agar plate the effects of the mold that had dropped through the window.[24]

There is no doubt both that Fleming was an acute observer and that he had the good sense to follow up his observation by culturing the *Penicillium* and measuring in the test tube the effect of its product upon several bacteria. He also tried, though not very systematically, to use this product locally in infections. When the results in a few cases were equivocal, he seemed to lose interest. He later said that he had gone no further because there were no experienced chemists at St. Mary's, but he did not take the obvious step of injecting the crude filtrates into animals infected with staphylococci or other bacteria to see whether it would be effective systemically, that is, against a generalized infection. As late as 1939 he told Dubos to forget about penicillin as an antibiotic because it was so unstable. Thus, when Florey's group began to experiment with it, penicillin was no more than another one of a large number of antibiotic substances obtained from various microorganisms which had been demonstrated to inhibit the growth of bacteria in the test tube and which had been tried locally in infections. True, Fleming had injected penicillin into an animal and observed no toxic effects, but this hardly constituted an adequate study of its systemic toxicity.[25]

Fleming had made an important observation and followed it up a certain distance. Florey, however, took the scientific concept that some microorganisms produced substances inimical to other microorganisms and hypothesized that at least one of them would be effective in human infections when given systemically. He directed his efforts toward finding that substance, purifying it, and proving its value. Toward that end he begged for funds, directed the work of his team, and put in operation the strategy that produced a usable penicillin and proved its effectiveness. In attributing Florey's success to the presence of a team of workers in various disciplines, Fleming overlooked the fact that the concept came first, along with the drive to see it proved. The team was secondary to the concept. For their work on penicillin both Fleming and Florey were awarded the Nobel Prize, along with Chain, who saw the possibility of purifying and concentrating penicillin and pushed through the work on chemical extraction. And so, as Abraham remarked, "all three of those who played leading roles in the story of penicillin received appropriate recognition."[26]

Certainly the Nobel Laureates richly deserved that distinction, and there is enough glory in the accomplishment to include all of those who participated. Without the work of any of the three, penicillin would not have been developed when it was. As a matter of fact, the history of antibiotics and chemotherapy might have been entirely different. After the value of the sulfonamides had been established, the inclination of the pharmaceutical companies was to turn to the synthesis of chemical compounds for the control of infectious diseases. This was one of the reasons that Florey found it so hard to interest them in a substance that was biologically derived. Even though the antibiotics have dominated the field of infectious diseases for the past three decades, some valuable synthetic drugs have in fact been developed, particularly those for infections of the urinary tract. It is therefore conceivable that synthetics of greater value might have been discovered in the intervening years if penicillin had not already occupied the center of the stage.

One effect of Wright's letter to the *Times* was to stir up the press, who descended upon the quiet laboratories of St. Mary's and Oxford. The latter received them not, and Florey vanished out the back door. But at St. Mary's the press were welcomed, and Fleming, who gave lengthy interviews and posed for innumerable pictures, soon became a folk hero. No doubt people needed such a hero to help them absorb the enormous changes that the antibiotics were bringing. Yet in becoming their willing victim, Fleming virtually gave up his scientific career. Florey, on the contrary, by keeping the press and the public at arm's length, was able to push through the development of penicillin and later to return to other fields of research. Perhaps Fleming unwittingly did Florey and the Oxford group a favor by monopolizing the limelight.[27]

One of the first questions to answer regarding any new drug is the dose that will ensure a sufficient concentration for a sufficient length of time at the site of action and yet be as low as possible so as to cause the fewest adverse reactions. In the early days of penicillin its scarcity was an added reason for holding down the dose. The interplay of these factors in determining the dose of penicillin at various times in its history is illustrated by the treatment of pneumonia. Because patients with this disease usually responded so well to therapy with the sulfonamides, Florey's group did not want to expend their precious supply of penicillin to treat it. They did show, however, that pneumococci were highly susceptible in the test tube. When larger supplies became available in America, Keefer's committee arranged to have pneumonias studied by William S. Tillett of New York University. Most of his patients received injections of 10,000 units three or four times during the daytime and none at night, which meant that there was no detectable penicillin in the blood for eight to twelve hours of the twenty-four-hour period. Even with these small doses and discontinuous therapy, only three of 46 patients with pneumococcal pneumonia, or 7 percent, died. Just as striking was the observation that among the 14 patients whose blood contained pneumococci on admission to the hospital and who were therefore seriously ill, none had a positive blood culture after the first dose of penicillin.[28]

As the supplies of penicillin increased, progressively larger doses were used. By 1945 my colleagues and I were routinely giving patients with pneumococcal pneumonia 75,000 units intramuscularly every three hours, or 180,000 units in twenty-four hours. Although highly effective, these doses required many injections before a patient was cured. The Oxford group had shown that penicillin was absorbed poorly when given by mouth and somewhat better when instilled through a tube into the small intestine. Walsh McDermott and his associates at Cornell demonstrated that penicillin was inactivated slowly in the moderately acid solutions usually present in the human stomach. They found that penicillin was absorbed mostly from the duodenum, the first part of the intestine, where it remained for only a short time. They reasoned that, if large doses of penicillin were given by mouth, some would be inactivated in the stomach and some would be carried down into the lower intestines, but that a sufficient amount would be absorbed while the penicillin was passing through the duodenum to give therapeutic concentrations in the blood and in the tissues. Accordingly, they recommended doses approximately five times the dose that would have been given by injection. Patients with pneumonia and other highly susceptible infections responded as well to this regimen as to the smaller amount intramuscularly and were much more confortable.[29]

Sulfonamides had worked best when their concentration in the blood was kept constant for the duration of therapy. This had been easy to accomplish because the excretion of sulfonamides by the kidneys extended over many hours. Maintaining constant levels of penicillin was harder because its more rapid excretion made the attempt resemble trying to fill a bathtub when the stopper was missing. Florey's group had taken advantage of the rapid excretion of penicillin in the days when it was scarce by collecting urine from patients and reextracting the penicillin, but this was hardly feasible as a routine. The problem was solved in two ways, by delaying absorption from the muscles after injection or by delaying its excretion by the kidneys. Karl H. Beyer found that a synthetic drug called carinamide acted upon the minute excretory tubes in the kidneys to delay the excretion of penicillin but was itself excreted so rapidly that it had to be given in large doses. His second discovery, probenicid, worked better. One tablet four times a day would double the concentration of penicillin in the blood. By the time probenicid therapy was perfected, however, penicillin was being supplied in large quantities at a low cost. As a result, probenicid is seldom used for the reason for which it was discovered. Yet it turned out to have another, more significant use: it increased the excretion of uric acid, thus getting rid of the deposits of crystallized uric acid that cause the crippling deformities of chronic gout. This was one more example of the spillover of the results of research into fields other than that where it was first directed.[30]

Investigators also looked for ways to delay the absorption of penicillin from the muscles after injection, first by combining it with oil and beeswax and later by devising less soluble compounds of penicillin, such as procaine penicillin. The penicillin contained in these preparations entered the blood gradually over a period of several hours to several days, depending upon the total dose. The ultimate in delayed absorption was achieved in benzathine penicillin, which liberated small amounts of penicillin steadily for as long as a month. While some clinicians believed that consistent levels of penicillin were desirable, others argued that it was important to achieve high peaks at intervals because the concentration remained high in the tissues for some time after it fell in the blood. Eventually the question lost much of its significance because, when large quantities of the antibiotic became available and the price dropped, doctors almost always overtreated their patients to "make assurance double sure." This can be appreciated by comparing the dosage usually used in the treatment of pneumococcal pneumonia today, 600,000 to 1,200,000 units a day, with the dosage recommended by Tillett in 1944, 30,000 to 70,000 units a day.

Subsequent experience in the penicillin therapy of pneumococcal pneumonia justified the conclusions of the original report. By 1948 my colleagues and I had treated 400 patients with a case fatality rate of 6.3 percent, as compared with a rate of 11.8 percent among 1,250 patients treated with the sulfonamides. The superiority of penicillin was evident in all age groups. In patients over fifty years of age, for instance, the fatality rate was 18 percent with penicillin and 29 percent with sulfonamides. For a short time some doctors felt safe in using penicillin only if the patient was also given sulfonamides, but when the superiority of penicillin became evident, it replaced the sulfonamides, just as the latter had supplanted antipneumococcal serum.[31]

Another problem that faced clinicians in the early days of penicillin was whether to depend on penicillin being carried into a localized infection from the bloodstream or to instill penicillin directly into the infected area. The shift in attitude that took place was illustrated by the treatment of meningitis. In 1942 when Fleming treated the meningitis patient with penicillin supplied by Florey, he gave it first by intramuscular injection, so that it would enter promptly into the veins and, once in the bloodstream, travel throughout the body, but when this did not arrest the disease, he injected penicillin directly into the spinal fluid. When doctors found no penicillin in the spinal fluid after systemic, that is, intravenous or intramuscular administration, or at most only a small amount, they usually followed Fleming's lead, giving penicillin both systemically and into the spinal fluid in meningitis. Among 66 patients with pneumococcal meningitis treated in this way by 1946, my colleagues and I obtained recoveries in only 38 percent. Since nearly two-thirds of the patients died and since penicillin itself, particularly the impure preparations available in the early years, sometimes produced inflammation of the spinal cord when it was injected locally, a better method was sought. Our investigations showed that the concentration of penicillin in the spinal fluid rose with the concentration in the blood. At the same time others reported that penicillin was consistently present in the spinal fluid of patients who were given 20 million units or more intravenously during the course of twenty-four hours. Accordingly we began in 1946 to treat patients with pneumococcal meningitis with large doses of penicillin by the intramuscular route alone, and 62 percent of them recovered. Systemic administration of large doses has since become the usual method of treatment.[32]

In 1952 we also demonstrated that patients with meningococcal meningitis responded as well to the same large doses of penicillin systemically as control patients given sulfonamides. When sulfonamide-resistant meningococci began to appear a year later, systemic treatment with large doses of penicillin was quickly adopted as the standard method. The principle of giving systemic doses of penicillin large enough to penetrate into a body cavity has also been applied in the treatment of infections of the joints and the pleural cavity. Penicillin can be used in these large doses because of the work of the laboratory scientists in finding methods of producing it cheaply and in large quantity.[33]

In still another disease a change in the dosage made a profound change in the end result. Subacute bacterial endocarditis is an infection of the heart valves in which the course is more prolonged, lasting several weeks to months, than that of the acute forms of endocarditis caused by bacteria such as hemolytic streptococci and pneumococci. It is almost always caused by streptococci of non-hemolytic varieties, especially viridans, or green-producing, streptococci. Recovery without specific treatment is rare. Sulfonamide therapy sometimes prolonged the course of the illness, but only a few patients actually recovered. The early trials of penicillin in this disease had been disheartening. Although an occasional patient appeared to recover completely, more often the patients improved dramatically on penicillin therapy, losing their fever and gaining in weight and strength, only to relapse after a few days and follow the usual inexorable course to death. None of Ethel Florey's patients recovered, and Keefer's committee, knowing that they had to use their tiny stock of penicillin where it was most likely to produce results, directed investigators not to use penicillin in this disease. But by a quirk of fate in 1943 penicillin was given another chance.[34]

Leo Loewe of Brooklyn, on the basis of results obtained in experimental endocarditis in rabbits, believed that patients with endocarditis would recover if treated with the combination of an antibacterial drug and heparin, a substance derived from animal tissues that prevents coagulation of the blood. After using sulfonamides plus heparin without success, he wanted to try penicillin instead but could not obtain it through official channels because Keefer's committee was not authorizing its use in endocarditis. He then persuaded John L. Smith of the Charles Pfizer Company to give him enough of the antibiotic to treat a patient by his method. After this patient recovered, Loewe and his associates treated six others, and all recovered. When Dawson and Hunter heard of these results, they treated 16 patients with penicillin and heparin, 12 of whom recovered. It looked as if heparin was the key to endocarditis therapy, but heparin was producing its own problems. Dawson and Hunter believed that excessive bleeding from heparin had caused the death of one of their patients. Blake and his associates at Yale noted that Loewe's patients had been given larger doses of penicillin for longer periods of time than the Yale and Columbia groups had earlier employed. On the hypothesis that the larger doses of penicillin and not the addition of heparin had brought about the cures, they treated 12 patients for three to four weeks with 240,000 units of penicillin a day without heparin and cured 11 of them. Other clinicians, using the same or larger doses of penicillin without heparin, obtained similar results. Thus endocarditis was added to the growing list of diseases that could be successfully treated with penicillin.[35]

Because penicillin had such a powerful effect on bacteria as compared with the sulfonamides, doctors had high hopes that penicillin-resistant strains would not be a problem, but they were soon disappointed. Resistance to penicillin proved to be particularly significant in infections caused by staphylococci and gonococci.

The staphylococcus is a microorganism that lives continuously with man, usually doing him no harm but occasionally causing an infection which may be as mild as a pimple or as serious as an overwhelming bacteremia. In 1880 Pasteur cultured cocci from pus obtained from an abscess and produced abscesses in rabbits with them. In the same year Ogston cultivated staphylococci from abscesses, produced local abscesses in animals, and hypothesized correctly that bacteremia occurred when the host's defenses were unable to contain the bacteria at the local site. By 1885 staphylococci had been separated into three varieties according to the color of the pigment they produced: aureus (golden), albus (white), and citreus (lemon yellow). Most human infections are caused by *Staphylococcus aureus*, a few by *Staphylococcus albus*, while the citreus variety is only rarely responsible for an infection.[36]

Despite their ability to produce disease, staphylococci and people live together without any problem most of the time. These bacteria can be cultivated from the noses of about 50 percent of healthy adults, are frequent inhabitants of the skin, and are found in the feces. They can be cultured from the air of bedrooms and living rooms and from dust, clothing, and other inanimate objects. When they penetrate the outer defenses of the body, they commonly cause minor lesions, such as pustules (pimples) or abscesses (boils), and may produce more severe infections, such as osteomyelitis, pneumonia, meningitis, or a particularly severe form of bacteremia in which there are hundreds of small abscesses in various organs throughout the body.

Sulfonamides accelerated recovery from some staphylococcal infections, although they only occasionally cured the more severe ones. From the beginning, Florey's group concentrated on treating staphylococcal infections with penicillin with gratifying results. Others had the same experience. Among the first 500 cases reported by Keefer's committee, 228 patients had staphylococcal infections, 137 without and 91 with bacteremia. Although 80 percent of the first group improved on therapy, the effect of the penicillin was not always certain since many of these patients would have improved without therapy or with surgical drainage, which was often instituted at the same time. The results in staphylococcal bacteremia offered more solid proof. In a series of cases reported from the Boston City Hospital in 1941, 85 percent of patients who had received no specific treatment for this illness died and of those who had been given sulfonamides almost as many, or 79 percent, died. In contrast, Wesley Spink and Wendell Hall of the University of Minnesota reported a fatality rate of only 28 percent in patients treated with penicillin. Small wonder, then, that doctors expected staphylococcal infections to be soon brought under control. But they failed to take into account the remarkable adaptability of these bacteria.[37]

As early as 1942 occasional strains of *Staphylococcus aureus* recovered from human infections were found by laboratory tests to be more resistant to penicillin than the strains originally treated by the Oxford group, and as penicillin was used more widely, the proportion of resistant strains increased. At first investigators thought that bacteria originally susceptible to penicillin had become resistant during therapy, but although this may have occurred occasionally, it soon became clear that the vast majority of these strains had been resistant from the first. These naturally resistant strains were found by William Kirby of Stanford University to produce penicillinase, an enzyme which inactivated penicillin and which had been discovered by Abraham and Chain in 1940. In a patient treated with penicillin, the growth of a susceptible strain was suppressed while the penicillinase produced by a resistant strain allowed it to survive. Resistant strains then spread to other patients and to carriers. This was particularly true in hospitals. Doctors, nurses, and other attendants were likely to harbor penicillin-resistant staphylococci in their noses, whence they could infect others. Outside the hospital the situation was different. My associates and I demonstrated in 1952 that the carrier state of patients within and outside a hospital reflected that of the people they contacted (Fig. 1).[38]

The proportion of resistant strains of staphylococci cultured from infections in hospitalized patients rose steadily as the number of patients treated with penicillin increased and the doses grew larger. Mary Barber, a bacteriologist in Hammersmith Hospital in London, found that 12.5 percent of staphylococci cultured from patients were resistant in 1946, 38 percent in 1947, and 59 percent in 1948. In the Boston City Hospital, Finland reported that 73 percent of strains of *Staphylococcus aureus* isolated in 1953 were highly resistant. The increase in the number of resistant strains was reflected in the effect of penicillin therapy on staphylococcal bacteremia. Spink reported that in contrast to a 28 percent fatality rate observed at his hospital in 1945, the rate in the years 1951 to 1953 had risen to 54 percent, and in the period 1953 to 1955 it had reached 80 percent.[39]

When resistant strains spread within a hospital, patients with

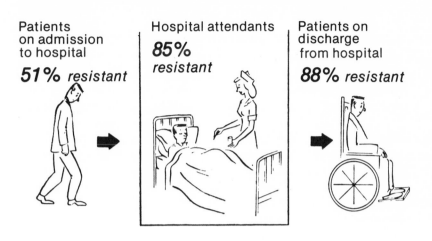

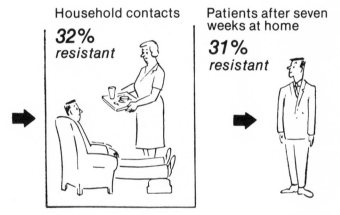

Figure 1 Percentage of penicillin-resistant staphylococci cultured from upper respiratory tracts of patients, hospital attendants, and household contacts.

Source: H. F. Dowling, M. H. Lepper, and G. G. Jackson, "Clinical Significance of Antibiotic-Resistant Bacteria," *Journal of the American Medical Association* 157 (Jan. 22, 1953): 329.

burns or operative wounds and newborn infants were particularly susceptible. Certain strains emerged in hospitals all around the world as particularly pervasive and virulent. Doctors and hospital officials were alarmed. Surveys were made and conferences were called to determine what could be done. Everyone agreed that rigid aseptic techniques should be mandatory in the management of patients with staphylococcal infections and that the overuse of penicillin should be stopped, although no one knew how to make sure that these principles would be followed. Indeed, the plight of those who were trying to prevent the spread of resistant staphylococci was reminiscent of the dilemma of health officers around 1900 who were trying to stop the spread of diphtheria. At that time a committee of public health officials chaired by Chapin had decided that it was useless to keep carriers isolated until they were rid of their diphtheria bacilli. The same conclusion was reached a half-century later when it proved impossible to bar carriers of staphylococci from hospitals or even from operating rooms and nurseries for any length of time. There were simply too many of them, and some carried resistant staphylococci in their noses for months. Other measures that were advocated included washing furniture and floors with antiseptic solutions, irradiating rooms with ultraviolet light, applying antibiotic ointment to the nostrils of carriers, and even inoculating newborn infants who were carriers of a resistant strain with a susceptible strain of staphylococcus of low virulence. The value of these procedures was highly conjectural, however, and they were too expensive and time-consuming for routine use. Eventually, by careful attention to aseptic techniques and with the aid of newer antistaphylococcal antibiotics, hospitals reached a state of uneasy equilibrium in which the spread of resistant staphylococci was kept at a minimum. Only when an epidemic threatened or occurred did they take more extreme measures such as closing wards, nurseries, or operating rooms.[40]

Hopes were high that penicillin was also the panacea that would finally bring gonorrhea under control. Keefer's committee reported 97 percent recoveries in 129 cases. Doses were small, totaling from 75,000 to 160,000 units per patient. The symptoms of gonorrheal urethritis in the male changed dramatically within a few hours after the first injection: local symptoms subsided, the secretion decreased in amount, and the number of gonococci diminished so that by the fourth to the sixth hour they could no longer be seen on a stained smear. Early gonorrhea in the female and vaginitis in children responded equally well to small doses. Deeper infections in the male and female required somewhat larger amounts and longer treatment for cure. Even gonococcal arthritis, which had sometimes required prolonged treatment with both sulfonamides and artificial fever, was cured when penicillin was given in doses large enough to penetrate into the joint tissues.[41]

The results of treatment fitted well with laboratory tests, which showed that gonococci were among the most susceptible of bacteria to the action of penicillin. Among 104 strains isolated from patients at the Boston City Hospital before 1947, all were inhibited by .02 micrograms of penicillin per milliliter of solution, a concentration well within the range achieved in the blood by the small doses given at that time. Yet as early as 1946 reports began to appear that less susceptible gonococci were being isolated from patients. At first, doctors were inclined to explain away these observations as erroneous. In 1948 it was suggested that when patients failed to recover on customary doses of penicillin, the infection was caused by some other microorganism, and that when gonococci had been reported as resistant, the laboratory studies had not been properly done. As late as 1958 Mahoney, who had been a pioneer in using penicillin in gonorrhea as well as syphilis, contended that "up to the present not any evidence of strain resistance has appeared."[42]

Meanwhile, as penicillin became more freely available, the doses used for gonococcal infections were progressively increased. By the late 1950s clinicians in Great Britain, Western Europe, and in the United States military forces in the Pacific were reporting failures in acute urethritis with doses of 0.3 to 1.5 million units of penicillin. At the same time laboratory tests were uncovering an increasing number of resistant strains. Some gonococci isolated in British cities from 1958 to 1960 required as much as one microgram of penicillin per milliliter for inhibition of growth, fifty times the largest quantity required to inhibit strains in Boston slightly more than a decade before. Strains from patients who had failed to respond to penicillin therapy tested at the Center for Disease Control in Atlanta showed a progressively increasing degree of resistance during the years 1955 to 1969, and at least three-fifths of these strains were more resistant than gonococci tested in the early days of penicillin.[43]

In response to these changes, clinicians shifted the doses of penicillin into higher ranges. Among 97 soldiers with acute gonorrheal urethritis at Fort Bliss, Texas, treated with 0.6 to 1 million units in 1962, only 86 percent were cured, while 97 percent of the 425 who received 2.5 million units recovered promptly. Thus, penicillin-resistance to gonococci developed gradually over two decades, and even at the end of this period most of the resistant gonococci were only partially resistant: patients infected with them still responded to large doses of penicillin. A more ominous note was sounded in 1976, however, when a few cases were reported to be caused by gonococci that were inhibited by concentrations of penicillin too high to be obtained in the blood and tissues. These gonococci, like highly resistant staphylococci, produce penicillinase, and if they spread widely, the control of gonococcal infections will have to depend upon other antibacterial drugs.[44]

Penicillin was also tried in gonococcal conjunctivitis and found to be highly effective both for prevention and for treatment. This raised the question of whether penicillin should be used routinely to prevent infection of the eyes of newborn children. Although prophylaxis with silver nitrate had unquestionably lowered the frequency of gonococcal conjunctivitis, the 1 percent solution required was highly irritating and often caused a chemical inflammation. Also, a number of tragedies had occurred when a stronger solution of silver nitrate was used by mistake, scarring the cornea, the transparent part of the eye in front of the pupil and iris. Doctors had long been dissatisfied with silver nitrate, but since milder, less irritating silver salts were not as consistently effective and since no other prophylactic agent was available, the hazards of silver nitrate had been accepted.

Studies showed that conjunctivitis occurred less frequently when penicillin was injected intramuscularly or placed on the conjunctiva as drops or an ointment than when silver nitrate was used, and penicillin caused practically no irritation to the conjunctiva. Soon doctors began advocating changes in public health regulations to allow them to substitute the newer drug. Some even recommended that preventive measures be eliminated entirely since gonorrhea in mothers was now infrequent and conjunctivitis could be treated so successfully if it did develop in an occasional infant. In 1956 the Board of Health of New York City repealed the provision in the Sanitary Code making prophylaxis mandatory. Several states changed their regulations to allow other substances to be used instead of silver nitrate. The relaxation of standards apparently went too far, because outbreaks of conjunctivitis were reported in institutions where routine prophylaxis was no longer required. In addition, penicillin had the disadvantage that local instillation could induce hypersensitivity to the antibiotic and thus preclude its use if needed later for a serious infection. A renewed focus on the problem produced a swing back to silver nitrate. In 1972 it was the sole drug used for prophylaxis in 37 states.[45]

Mahoney's announcement in 1943 that syphilis could be cured by penicillin came at a crucial time for the Allied military forces. Within a year and a half over 11,000 patients with early syphilis had been treated under 26 different treatment schedules by doctors in the armed forces, the Public Health Service, and selected clinics. The first studies with small doses of penicillin gave 90 to 97 percent cures in primary and secondary syphilis. In late syphilis, although the greater chronicity made for slower improvement, observations after a few weeks or months already showed favorable effects from penicillin therapy. Abnormalities in the spinal fluid improved in 74 percent of cases. In paresis 50 to 75 percent of the patients were improved. Similarly good results were observed in infections in other parts of the nervous system. Penicillin had two advantages over arsenicals, low toxicity and short duration of treatment. Both advantages helped persuade patients to undergo and complete the required course of treatment. Therapy was made still simpler when preparations of penicillin were made available that delayed absorption after injection. Optimal results were obtained in early syphilis with a total dose of 4.8 million units of procaine penicillin in three injections at three-day intervals. When benzathine penicillin, the preparation that liberated penicillin from the site of injection into the blood gradually during several weeks, was introduced in 1951, 2.4 million units given at one visit were thought to be all that was necessary for early syphilis. Ehrlich's dream of a single dose for the cure of syphilis, the *therapia sterilisans magna* or "great sterilizing dose," seemed to have come true. By 1953 penicillin had replaced other therapy for syphilis in the United States.[46]

Penicillin was first used in this country on a scale sufficient to have an impact on the public health in 1945 or 1946. Soon afterward the mortality rate from syphilis, which had been falling gradually for the past decade, began a steep decline. In 1940 it was 10.7 per 100,000 population; in 1950 it was 5 per 100,000, or less than half. By 1955 it had halved again, and by 1970 it was 0.2 per 100,000. The decline in infant deaths from congenital syphilis was even more spectacular, from 5.3 per 10,000 live births in 1940 to 0.57 in 1950 and 0.04 in 1968. The number of new cases reported reached a wartime high in 1943 at 447 per 100,000 population, excluding cases in the military services; in 1950 it was 154.2 and in 1970, 43.8 per 100,000.[47]

The decrease in the prevalence of syphilis forced changes in public health techniques. Mass blood testing programs uncovered so few cases that the time and expense that went into them were no longer justified. They were discontinued, except for certain groups where the incidence was high. More significantly, less syphilis meant less interest in syphilis. As early as 1953 R. H. Kampmeier of Vanderbilt University pointed out that the rapidity with which penicillin cleared up the manifestations of syphilis was making people neglect the post-treatment follow-up, and that the scarcity of relapses was making it seem economically unsound to utilize personnel for such purposes. As the rates for the discovery of new cases fell, so did federal appropriations, and it was assumed that the war against syphilis was almost over.[48]

When the number of cases of primary and secondary syphilis began to rise from a low of 3.8 per 100,000 population in 1957, few paid any attention. By 1960 this number reached 7.1 per 100,000, and by 1965 it nearly doubled again, reaching 12.3 per 100,000. The resurgence of syphilis was not confined to the United States. In the United Kingdom, Europe, and South America an increase in early cases took place at about the same time. Doctors, public health officials, and finally the public began to be concerned. Fortunately, penicillin-resistant treponemes were not the reason for the upsurge, for none was found after extensive search. In the absence of a technical explanation, there had to be a social one; since the behavior of the parasite had not changed, the cause had to be found in the behavior of people.[49]

The significant shift in the attitudes of society toward sex that had been taking place for several decades had accelerated in the past few years. The greater freedom granted to youth, the increased permissiveness allowed by parents, the toleration of unmarried persons living together in couples or in groups, the coming out into the open of homosexuality, and the wide use of oral contraceptives all opened the way for sexual promiscuity, which in turn made the spread of venereal diseases more likely. Added to these changes were the increased mobility of the population, the greater freedom of women, and the rapid urbanization of society. Although prostitution may have played a smaller role than in the past because of the easier availability of sexual partners, it was still present. The importance of promiscuity was documented by data collected from 34 states for the fiscal year 1970, which showed that among patients with infectious syphilis, 57 percent had sexual exposure to several members of the opposite sex, as compared with 19 percent who had exposure to only one person. This survey also confirmed the importance of homosexuality, since 14 percent of the patients reported exposure to members of the same sex.[50]

One evidence of changing mores was the increase in venereal disease in young people. The rate of acquisition of infectious syphilis for persons under the age of twenty-five more than doubled between 1956 and 1969, rising from 4.2 to 9.7 per 100,000 population; for gonorrhea the increase was even greater, from 174 to 372 per 100,000. Epidemics of syphilis were encountered in high schools. In the investigation of one such outbreak a positive serological test on a fourteen-year-old athlete led to the finding of 66 other students who had had intercourse with him or with his contacts. Fourteen of them had syphilis.[51]

In the face of the increasing threat of venereal disease and in response to pressure from doctors, public health officials, and lay groups, federal appropriations for combating these diseases were increased, old programs were rejuvenated, and discontinued programs were revived. Apparently as a result of these efforts, the rate of reported cases of primary and secondary syphilis declined slightly from its 1965 peak of 12.3 per 100,000 population to 10 in 1970. The increase in new cases of gonorrhea continued, however; in 1970 over a half-million new cases were reported, a rate of 285 per 100,000 population.[52]

The partial failure of control programs spurred the leaders in the fight to greater efforts. They now realized that temporizing with venereal disease control merely meant a resurgence of these diseases whenever public health measures were slackened. The watchword today is the eradication of syphilis in this country and throughout the world by using measures of proven value more widely and more effectively.[53]

Because Prontosil was first used in streptococcal infections, the sulfonamides were associated with these bacteria in the minds of doctors and laymen alike. Yet the sulfonamides were by no means a panacea in streptococcal infections. Some severe infections were still fatal, and the course of the more common streptococcal sore throat was not materially altered by sulfonamide therapy. In view of the scarcity of penicillin it was at first used in streptococcal infections only when the causative strain was resistant to sulfonamides. Among the 500 cases in the initial report of Keefer's committee, 23 were caused by hemolytic streptococci. Since most of these patients were seriously ill, the recovery of 13 could be considered a good result. As more patients were treated, the superiority of penicillin to sulfonamides in serious streptococcal infections became obvious.[54]

Streptococcal infections of the throat were easily diagnosed when accompanied by the rash of scarlet fever, and since these patients were segregated on special wards, they were readily available for trials of penicillin therapy. As early as 1945 Finland and his associates demonstrated the clear-cut superiority of penicillin over sulfonamides in this disease. Within forty-eight hours the infecting streptococcus had disappeared from the throats of seven of nine patients treated with penicillin, from only one of nine given sulfadiazine, and from none of the untreated controls. Symptoms disappeared more rapidly in the penicillin-treated patients, and that group alone remained free of complications. Although in this small study significant differences in the rate of the disappearance of fever could not be demonstrated, my associates and I later reported that in 86 patients with scarlet fever treated with penicillin the temperature was normal in fifty-two hours on the average, compared with ninety-nine hours in 123 patients who received no specific therapy. Also hemolytic streptococci disappeared from the throats of 92 percent of our patients who received penicillin, compared with 18 percent of those treated with drugs that merely relieved the symptoms. Complications occurred in 7 percent of the penicillin group and in 26 percent of the control group. Although in a parallel group of patients who received scarlet fever antitoxin the temperature usually fell more rapidly than in the penicillin-treated patients, their course and complications resembled those of the control group in every other respect. Similar results were obtained by doctors all over the world, both in scarlet fever and in streptococcal sore throat without rash. Penicillin rapidly became the accepted treatment, although in less affluent countries sulfonamides were used for a few years longer until the price of penicillin fell to comparable levels.[55]

When it became clear that penicillin eradicated hemolytic streptococci from the throats of patients and carriers, a natural step was to give it continuously to patients who had recovered from rheumatic fever to prevent recurrences. By 1954 penicillin was shown to be as effective for this purpose as the sulfonamides: rheumatic fever occurred in 0.6 percent of 740 persons who had taken penicillin orally, as compared with 8.7 percent of 932 controls who had received no penicillin. Since people who are not ill tend to forget to take medications, Gene Stollerman of New York University treated patients with a monthly injection of 1.2 million units of the long-acting benzathine penicillin. During a twenty-one month period no recurrences of rheumatic fever occurred in 145 patients who received these injections, compared with two recurrences among 111 patients given 0.2 million units of penicillin each day by mouth and five recurrences among 73 patients who received sulfadiazine. At present either penicillin or a sulfonamide is used for continuous prophylaxis against rheumatic fever.[56]

It would be impossible to give penicillin continuously to every child to prevent streptococcal infections, but it was found possible to prevent rheumatic fever by prompt treatment of streptococcal sore throat. One of the most comprehensive studies was conducted in 1949 by Rammelkamp and his associates at Warren Air Force Base in Wyoming. They gave penicillin to 1,178 patients with acute infections of the throat and observed 1,162 similar patients who received no specific therapy. During the succeeding forty-five days rheumatic fever developed in 28 of the controls and in only two of the treated group.[57]

Thus it appeared that the only requirement for wiping out rheumatic fever was to treat every patient having a streptococcal infection of the upper respiratory tract promptly with penicillin. But there was the rub. It was impossible to diagnose every streptococcal infection and to reach every patient. What would work in the armed forces, where men could be ordered to report for treatment and facilities could be mobilized for a predetermined purpose, would not necessarily be effective in the ordinary practice of medicine. Doctors were not consulted for every case of sore throat, and when they were, they could diagnose streptococcal infection in only one-half to three-fourths of the cases. Furthermore, many sore throats are caused by viruses, small particles ranging in size from that of a large protein molecule to that of the smallest bacterium, which multiply only within living cells by using some of the components of the cells.[58] Sore throats caused by viruses could not be distinguished with certainty from those caused by streptococci except by culturing the streptococci. Proper facilities for this procedure were not always available, nor did cultures invariably provide the answer, because a person who was a carrier of streptococci might become ill with a viral sore throat. Finally, the late complications of streptococcal infection were so rare in some communities that they could not justify penicillin therapy for all sore throats in view of the expense and the risk of adverse reactions it involved. Yet the effectiveness of penicillin was so clear and the need to prevent rheumatic fever so important that efforts were made to overcome the limitations.

Two different methods were tried in the private practice of medicine. One was to give penicillin to all patients with an acute sore throat. In a trial of this method five general practitioners in England gave either penicillin, a sulfonamide, or a placebo to all their patients with this diagnosis. Neither the patient nor the physician knew which preparation the patient had been given. On the third day of the illness 61 percent of patients who received the inert preparation were still ill, compared with about one-third of those who received either penicillin or the sulfonamide. More important, none of those receiving penicillin had a positive culture for hemolytic streptococci by the third day of therapy, compared with 46 percent of those given the sulfonamide and 57 percent of those who received no specific therapy.[59] Burtis Breese, a pediatrician in Rochester, New York, used the other method to try to overcome the limitations of diagnosis. He took cultures on all the 792 children seen in his practice between 1947 and 1952 with suspected streptococcal infections and thus was able to base his treatment on a bacteriological diagnosis, with the result that only one child developed rheumatic fever.[60]

Probably the most comprehensive project for controlling rheumatic fever in a civilian community was initiated in Casper, Wyoming, where every one of the 6,460 children in the primary schools was examined and a throat culture taken whenever a child had an upper respiratory infection. Any child from whom hemolytic streptococci were cultured was excluded from school until antibiotic therapy had been given or, if such therapy was refused, until hemolytic streptococci could no longer be cultured. Only five cases of rheumatic fever occurred in primary schoolchildren in two years, from 1955 to 1957. One of these was in a child who was out of town at the time, and another was in a boy who did not report his sore throat. During the second year new cases of rheumatic fever occurred at the rate of 0.3 per year per 1,000 schoolchildren in Casper, as compared with 2.4 and 4.1 per 1,000 in Cheyenne and Laramie, Wyoming, where no such program was in effect. Programs such as these not only verified the effectiveness of penicillin in preventing rheumatic fever but also showed the possibility of bringing the benefits of the antibiotic to the general population with proper planning.[61]

Prior to 1940 there was little in the way of community organization to help patients with rheumatic fever and its consequences. As funds became available from the federal government, states began to set up programs for the care of such patients. The American Heart Association, which was founded in 1922 and had become one of the most active voluntary health associations as a result of the increasing proportion of the population dying of heart disease, promulgated standards for the use of penicillin both in treating streptococcal infections and in preventing second attacks of rheumatic fever and persuaded physicians and health departments to adopt them.[62]

The results of the use of penicillin were difficult to measure. For several decades the frequency and severity of rheumatic fever had been diminishing. Before the sulfonamides and penicillin this decrease had been attributed to improvements in the standard of living and to the discovery of the role of the streptococcus with the consequent attempts to protect victims of rheumatic fever from further streptococcal infections. At the Good Samaritan Hospital in Boston between 1921 and 1951 the proportion of patients admitted with greatly enlarged hearts decreased from 30 to 14 percent and the proportion of patients who died within five years of the first observation decreased from 24 to 3 percent. Yet in 1970 it was found that the decrease in the frequency of first attacks of rheumatic fever since the thirties was only 30 percent, that approximately 100,000 new cases occurred in the United States each year, and that in 1969 the cost of physicians' visits to patients with rheumatic fever and rheumatic heart disease was $28 million. Obviously more needed to be done. It was recommended that facilities for obtaining cultures of streptococci be made available to every physician at minimal or no cost to the patient and that communities establish registers of all patients with rheumatic fever so as to follow them up and ensure that they remain on prophylactic therapy. Once more a technological change was demanding a social response. How complete that response will be still remains to be seen.[63]

When penicillin was used to treat streptococcal infections, it reduced the frequency of nephritis as well as rheumatic fever. In Casper, Wyoming, no schoolchildren who were treated with penicillin during one year developed nephritis, whereas the rates for Cheyenne and Laramie during the same year were 0.86 and 1.7 per 1,000 children, respectively. In an epidemic of nephritis caused by Type 12 streptococci at the Naval Training Station in Bainbridge, Maryland, 4.5 percent of those treated with penicillin developed nephritis, compared with 12 percent of those who received no specific therapy.[64]

Other streptococcal infections yielded to penicillin more rapidly and completely than to sulfonamides, including erysipelas, infections of the middle ear, and puerperal sepsis. In 1936 the maternal mortality rate in the United States was 568 per 100,000 live births. Thereafter it fell rapidly to 83.3 per 100,000 in 1950 and then more slowly, reaching 21.5 in 1970. Since many maternal deaths in the past have been caused by infections, an appreciable portion of this decline could be attributed to the effects first of the sulfonamides and then of penicillin. Yet occasional hospital epidemics of puerperal sepsis continued to occur, demonstrating that the hemolytic streptococcus could still be dangerous unless everyone remained vigilant. Once detected, these epidemics were successfully terminated by penicillin treatment of patients and carriers of Group A streptococci among hospital personnel with penicillin.[65]

Early laboratory studies had shown that penicillin inhibited the growth of certain other bacteria, but clinical confirmation was slow in coming for several reasons. In some of these infections other therapeutic agents were so effective that penicillin could not ethically be given except as an adjunct. Diphtheria was the best example. No additional benefit was observed when penicillin was added to the usual antitoxin treatment, probably because the important objective of therapy was neutralization of the toxin, upon which penicillin had no effect. Although in other countries a few patients were treated with penicillin alone, the results were not so definite nor the experience so extensive that doctors were willing to omit antitoxin. On the other hand, since penicillin speeded the elimination of diphtheria bacilli from the throats of patients and chronic carriers, it was adopted for this purpose.[66]

Infections caused by certain other penicillin-susceptible bacteria were encountered so seldom that proof of the antibiotic's effectiveness had to be built up laboriously case by case. An example was actinomycosis, a disease caused by a moldlike fungus and characterized by multiple abscesses, which often burrow their way to the surface. Penicillin greatly accelerated the healing of these abscesses, which in the past continued their course for months and years, sometimes ending in death. Finally, the value of penicillin in infections superimposed on traumatic or surgical wounds was hard to prove because the degree and kind of surgery performed varied from case to case. Surgeons in general, however, agreed that penicillin was superior in effectiveness to sulfonamides for infections caused by bacteria susceptible to the antibiotic. The recovery rate for British soldiers wounded in Europe in World War II and treated with penicillin was 95 percent, a result undreamed of in previous wars.[67]

Because the antibacterial action of penicillin was so much greater than that of the sulfonamides and its effect on human infections was so much more dramatic, it might be expected to have been adopted immediately and with no opposition. In general this was true; yet because the sulfonamides were effective in many cases and because the first cases reported by the Oxford group were not so convincing in print as they had been to the participants, some were hesitant "to cast the old aside" and take up penicillin. When Florey began more extensive clinical trials on his return to England in 1941, he was concentrating on surgical infections, partly because many of these were staphylococcal, in which sulfonamides seldom helped, and partly because of his conviction that wounded soldiers should have the first call on penicillin. But the surgeons, who were busy improvising various techniques for improving the results obtained with sulfonamides, were reluctant to try the new remedy on a patient until they were sure that the sulfonamides had failed. When that point had been reached, however, Florey complained that they were "asking me to resuscitate a corpse." Two years later, after R. J. V. Pulvertaft had reported success with penicillin in the treatment of wounded British soldiers in the Near East and Florey and Cairns had showed that even small quantities of penicillin could clear up infections that had been considered hopeless, surgeons in and out of the army swung over to penicillin.[68]

In the United States, although penicillin was used extensively in a number of civilian hospitals by 1943, Keefer's committee had no luck in prevailing upon the military to use it, according to A. Baird Hastings of the Committee on Medical Research of the Office of Scientific Research and Development. Because Keefer and Hastings believed that many of the wounded soldiers being flown back from South Pacific battlefields desperately needed penicillin, they persuaded Richards to call the surgeon general of the army and ask him to order its trial. Within hours of Richards' call, Lyons was directed to begin tests on war infections. His results, particularly in patients with chronic osteomyelitis, which in the past often continued to drain pus from the infected bone for years, were so convincing that from then on Keefer's committee had difficulty finding enough penicillin to supply the army's demands.[69]

But these delays were few and temporary. In general, penicillin was adopted enthusiastically, perhaps too enthusiastically. In large part this was because for a long time the supply could meet only a tiny fraction of the demand, a demand that was inflated by the lay and professional press alike. In 1944, when supplies of penicillin had increased somewhat, the Office of Scientific Research and Development threw the problem to the doctors by allocating a quota of penicillin to each of a thousand hospitals throughout the country, leaving it up to staff committees to decide which patients would receive it. When penicillin was finally released for sale through normal channels without limitation in 1945, the dam broke. So high were the hopes of doctors and patients for marvels to follow from the use of the "miracle drug" that it was given not only for illnesses where its effectiveness had been proved but also in diseases where laboratory tests showed that it would not work, as in typhoid fever. In fact, penicillin was tried in every illness that resembled an infection and for some that did not. I can remember seeing it given for viral infections such as hepatitis, where there was no evidence that it would have the slightest effect, and in conditions where its administration was the result of a forlorn hope plus wishful thinking, as in leukemia and other forms of cancer.[70]

The pharmaceutical industry was able to keep up with patients' needs. In 1945 they produced nearly 5 tons of penicillin, which by 1956 had increased to 530 tons. The price had continued to drop until in 1955 the retail price for a million units was only 60 cents.[71]

Adequate supplies and hence falling prices undoubtedly contributed to the overuse of penicillin, as did the widespread publicity, but another influence was also responsible, the illusion of complete safety. The Oxford group had predicted that "quite strong solutions should prove innocuous to tissue cells" and had found no toxic effects attributable to penicillin itself. In describing the clinical use of penicillin, Ethel Florey wrote, "Penicillin occupies a unique position [among drugs] in having, for practical purposes, no upper limit to the amount that can be given." In the enthusiasm over this amazing drug doctors tended to forget that every drug produces adverse effects in some persons under some circumstances. As larger and larger amounts were administered in an attempt to cure desperately ill patients, doctors finally reached doses that were toxic to the kidneys and the brain, but these doses were so high that they were seldom needed, especially when other antibiotics appeared. More important was another kind of hazard.[72]

Keefer's committee had noted that 14 of the first 500 patients treated in this country developed urticaria, or hives, which they postulated might have been caused by some impurity. When these and other types of allergic reactions actually became more frequent even after the manufacturers finally achieved a pure product, the most militant champions of penicillin had to admit that the antibiotic itself was responsible. The problem became more acute when reports of fatal allergic reactions began to appear in 1949. In a comprehensive survey by the Food and Drug Administration, which involved nearly one-third of the patients hospitalized in the United States during the years 1953 to 1957, 2,718 patients were found to have suffered from an allergic reaction to an antibiotic, of which 2,485, or 92 percent, were reactions to penicillin. Among 86 deaths from allergic reactions, 81 followed the administration of penicillin.[73]

When the medical profession and the public were faced with such figures, some of them wanted to abandon the drug altogether. For example, an article in the magazine *Coronet*, headed "Penicillin Turns Killer," related how "the wonder drug has become the blunder drug." The more significant figure, however, was the ratio of allergic reactions to the total number of patients treated. A rough estimate based on the population of the country amounted to four deaths per ten million inhabitants per year, though this did not include an unknown number that may have occurred in patients' homes or doctors' offices. In England and Wales deaths from hypersensitivity to penicillin occurred at the rate of one per ten million inhabitants during the years 1957 and 1958. Since figures from these countries were more accurate because of reporting through the National Health Service, their lower percentage probably reflected less use of penicillin. Studies in hospitals showed that the overall frequency of reactions to penicillin was not great. My associates and I found that only 24 of 1,303 patients who received penicillin in the Gallinger Municipal Hospital in Washington, D.C., had developed allergic reactions, none of which was life-threatening. Reactions were nearly ten times as frequent in patients who had received penicillin in the past as in those who were receiving it for the first time. The dosage was also important: 7.8 percent of patients who received more than six million units per day had allergic reactions, as compared with 0.6 percent of those given smaller doses. Such studies revealed that although allergic reactions were more of a hazard with penicillin therapy than with some other drugs, most patients escaped them, and doctors themselves could prevent many of these reactions by refraining from the use of penicillin when it was not needed, by avoiding excessive doses, and by identifying patients who had suffered from allergic reactions to penicillin in the past. These precautions and the appearance of other antibiotics that could be given to patients allergic to penicillin made it possible for doctors to give an antibiotic to practically any patient who needed one.[74]

Despite its limitations—of which every remedy has some—penicillin had proved to be the most effective cure for a number of common diseases and the only cure for several diseases. It was also effective in a larger number of infections than any other therapeutic agent. Yet it was not a cure-all. Some infections did not respond at all, and some bacteria escaped its action because resistant strains appeared. But penicillin had opened the door for the discovery of other antibiotics which could help to solve the problems that remained.

10. Streptomycin and the Chemotherapy of Tuberculosis

"Chance," according to Pasteur, "only favors the prepared mind." A chance observation that results in a discovery captures the public imagination while the "prepared" portion of Pasteur's phrase—the careful planning, the stubborn persistence, and the dull drudgery of a systematic search—tends to be ignored. Yet practically all of the specific remedies and preventives for infectious diseases have resulted from a carefully planned, systematic, relentless, long-continued exploration along an unmapped path. Specific serums and vaccines were perfected by this method, and so were the arsphenamines and the sulfonamides. Even in the case of penicillin, although a significant chance observation came first, the methodical search had to be made before its true value could be known. The next important antibiotic, streptomycin, was so obviously a product of planned, persistent explorations that when the chance discoveries came, they seem in retrospect to have been a matter of course.

The man behind the search was Selman A. Waksman, a Russian who emigrated to America in 1910. As an undergraduate student at Rutgers University, he began to examine soils, digging trenches at the college farm to collect samples of microorganisms growing in various places. Many of those he found were bacteria, as anticipated, but a few were not. When he showed some of the latter to his chief, he was told they were actinomycetes, microorganisms about which little was known but the name. Today, after much more study, they are still unique, occupying a position halfway between the true bacteria and the fungi, being slender, like the rodlike forms of bacteria, with a structure composed of several branches, like the fungi.[1]

Waksman never lost his interest in actinomycetes, returning to them after he had received his Ph.D. in biochemistry and had joined the Rutgers faculty. Emerging as one of the world's authorities in soil microbiology, he attracted large numbers of graduate students from the United States and other countries. One of these was Dubos, who came from France to Waksman's laboratory and later went on to the Rockefeller Institute. There his interest in soil microbiology merged with Avery's interest in the pneumococcus, with the result that together they found the soil bacillus that produces an enzyme which attacks the capsules of pneumococci. Later Dubos found another antibiotic, gramicidin, which was effective against streptococci, pneumococci, and staphylococci, but could only be used locally because it was too toxic when injected. Dubos' work encouraged Waksman to start a systematic study of soil microorganisms in the hope of finding antibiotics that could be used to combat human diseases. In 1940 he and his associates announced the isolation of actinomycin and in 1942, streptothricin. Unfortunately these antibiotics, like those found by Dubos, were too toxic to be used in human infections.[2]

In 1943 a farmer took a chicken to the Agricultural Experiment Station at Rutgers because others in the flock were dying. A strain of actinomycetes cultured from its throat was sent to Waksman, who identified it as *Streptomyces griseus*, a microorganism he had described several years previously. From this strain Albert Schatz, Elizabeth Bugie, and Waksman obtained a substance that inhibited the growth of a number of common gram-negative and gram-positive microorganisms, as had streptothricin. But unlike streptothricin, the new substance, which Waksman named streptomycin, showed minimal toxicity in laboratory animals. These promising observations were reported in January 1944, and news of even more significance followed shortly when the medical world was electrified by the announcement of William H. Feldman and H. Corwin Hinshaw of the Mayo Clinic that the new antibiotic was effective against tuberculosis in animals. Their discovery was also the result of a carefully planned and persistent search.[3]

Many attempts had been made to find a chemical substance that would cure tuberculosis, but without success. By the late thirties many people believed that the chemotherapy of tuberculosis was unattainable, but Feldman and Hinshaw refused to accept this hypothesis. In 1938 the ever-widening investigation of the sulfonamides at Johns Hopkins led to the observation by Arnold R. Rich and Richard H. Follis that sulfanilamide slowed the progress of tuberculosis in guinea pigs. Feldman and Hinshaw found that other sulfonamides were also effective but that all of the sulfonamides required such large doses to produce any effect that they could not be given to patients. Looking around for other possibilities, these investigators tried a related group of compounds, the sulfones, which had recently been advocated in England and France for use against streptococcal infections. Yet these compounds merely suppressed the growth of tubercle bacilli in animals rather than eradicating them, and although some patients improved more rapidly when treated with sulfones than would have been expected on bed rest alone, the drugs did not have a striking effect. The principal result of the use of these drugs was to convince Feldman and Hinshaw that tuberculosis could be attacked successfully with drugs. As Feldman later wrote, "We had in a sense 'got our foot in the door'; a door which previously had seemed an unpassable obstacle."[4]

In their original tests on streptomycin Waksman and his associates identified *Mycobacterium tuberculosis* as one of the bacteria inhibited by the antibiotic. Because Feldman had just visited Waksman and expressed an interest in trying out any promising new antibiotic that the group might find for tuberculosis, Waksman wrote him at once, offering a supply of streptomycin for antituberculosis tests in animals. Feldman accepted with alacrity. In their trials of sulfonamides and sulfones he and Hinshaw had developed a practical and efficient system for determining the ability of a drug to slow the course of tuberculosis in guinea pigs. They immediately began treating with streptomycin four guinea pigs that had been injected with human tubercle bacilli, holding another eight as untreated controls. Although the supply of streptomycin was exhausted after fifty-four days of treatment, striking differences were already observable. The eight untreated animals had severe and widely distributed tuberculosis, easily visible to the naked eye, while the amount of tuberculosis in the tissues of the treated guinea pigs was minimal and usually recognizable only under the microscope. More extensive experiments in animals confirmed these findings and showed no serious toxic effects from the streptomycin. With "a sense of impatient enthusiasm" Feldman and Hinshaw were now ready to try streptomycin in patients with tuberculosis.[5]

When they reported their observations to Waksman before publication, however, he confessed that the microorganism identified as *Mycobacterium tuberculosis* in the original tests on streptomycin had apparently been mislabeled; it was instead a nonpathogenic bacillus which had been used in his laboratory for some time for teaching students. He requested that the results of the trials in animals not be published until he had confirmed in the test tube the efficacy of streptomycin against pathogenic tubercle bacilli, which he and his associates were shortly able to do.[6]

At this point the ever-recurring question again arises of who should get the credit in scientific discovery. Dubos' strong belief that soil bacteria produced antibiotic substances that could be used in human infections and his partial success with gramicidin had helped to influence Waksman to begin a thorough exploration of soil micro-organisms for antibiotic producers. In this enterprise Waksman for years led a team of investigators, enlisting successive graduate students as they came to study under him. He persisted in the search for a better antibiotic even after actinomycin and streptothricin had proved to be too toxic for use in humans. When streptomycin showed its potential in animals, he arranged with Merck and Company for more extensive studies which could be done more rapidly and easily in their laboratories. In return, Waksman and Rutgers University agreed that Merck should have the right to apply for patents on techniques developed during these studies.[7]

The trustees of Rutgers created a foundation to receive the royalties and decided to give Waksman one-fifth of them; he voluntarily reduced his share to 10 percent in 1949. That Waksman should be the sole individual to share in the profits was questioned by Albert Schatz, who as a graduate student had been assigned to test the strain of *Streptomyces* that produced the first streptomycin. Schatz had been the senior author of the first report and the joint applicant with Waksman for the patent on streptomycin. Waksman at first opposed Schatz's suit for a share of the income; later he offered to divide the royalties with fifteen scientists, including Schatz, plus twelve laboratory assistants, clerks, and others who had participated in the studies on antibiotics. Athough the matter was eventually settled according to this formula, it left an unpleasantness that demonstrated once again the conflict between the pursuit of science and the pursuit of profits. Waksman acknowledged in court that Schatz was "entitled to credit legally and scientifically as co-discoverer of streptomycin." But the Nobel Prize committee thought otherwise: in 1952, two years after the settlement, they awarded Waksman the prize in physiology and medicine for the discovery of streptomycin.[8]

On the other hand, it seems that the credit for the significant revelation that streptomycin is effective in tuberculosis must be given to Feldman and Hinshaw. If they had not demonstrated its value by a carefully constructed method and persisted in following through on their observations, streptomycin might have lain on a laboratory shelf for years. At any rate, its effects on bacteria other than *Mycobacterium tuberculosis* would most likely not have been enough to justify its production during the exigencies of World War II. It is easy to second-guess Nobel Prize committees; yet one wonders whether Schatz, Feldman, and Hinshaw should not also have shared in the award.[9]

Once the value of streptomycin had been shown in animals, the path from that point, though by no means smooth, was not as difficult to traverse as Florey's path had been. He had had to break the way with only the help of his immediate team for many months. Now, especially because of penicillin's spectacular success, investigators in the laboratories of several pharmaceutical companies were working with antibiotics. After Feldman and Hinshaw had reported their findings, Merck and Company agreed to produce streptomycin in sufficient quantities for testing in patients with tuberculosis. Hinshaw enlisted the help of two physicians in a nearby sanitarium, Karl H. Pfeutze and Marjorie Pyle, who had been collaborating with him in the studies on the sulfones. On November 20, 1944, only ten months after the initial announcement of the isolation of streptomycin, it was given to the first patient, a twenty-one-year-old woman with progressive, far-advanced tuberculosis. Her case was so desperate that "everyone concerned, including the patient herself, was eager to try the new drug." The first doses were very small; yet even these, because of impurities in the early, crude preparations, caused muscular aching, chills, and fever. But the patient and the doctors persisted. The preparations became progressively purer and the doses were rapidly raised. Altogether the patient received five courses of only ten to eighteen days each. Yet even with this minimal and much-interrupted treatment, she improved and was eventually discharged from the hospital thirty-two months after her first dose. Seven years later she had married, was the mother of three children, and had no sign of active tuberculosis.[10]

Although streptomycin was apparently the determining factor in this woman's recovery, her long and tortuous course illustrated graphically the difficulty that would be faced in determining the effectiveness of the new drug in tuberculosis, as compared with acute illnesses such as diphtheria and pneumonia. Hinshaw and his associates made a preliminary report on 34 patients in 1945 and a fuller evaluation of 100 cases in 1946, showing that streptomycin favorably influenced the course of tuberculosis of the lungs, skin, bones and joints, meninges, and genito-urinary organs. Unfortunately, they found that in some patients streptomycin injured the organ of equilibrium in the ear. Deafness also followed streptomycin administration in some cases. Sometimes these adverse reactions were transitory, but often they caused permanent disability.

They were directly related to the dose of the antibiotic and the duration of therapy. Thus, although it appeared that streptomycin could reverse the course of tuberculosis, doctors would have to tread a narrow path between too much of the antibiotic, which could cause serious side effects, and too little, which would delay healing or fail to accomplish it altogether. In addition to determining the proper dose and duration of therapy, it was necessary to work out the proper relationship of therapy to surgical procedures. The question was whether surgery should be used as it was formerly, and if so, how doctors could decide whether improvement was the result of it or of streptomycin therapy.[11]

Meanwhile, the value of streptomycin in nontuberculous infections needed to be determined also. By the time streptomycin had been produced in appreciable quantities, the glow of infallibility surrounding penicillin had disappeared; many bacteria were not susceptible to its action, either initially or as a result of the appearance of resistant forms. Streptomycin might fill in the gaps. To find out, the producers of the new antibiotic offered nearly a million dollars to the National Research Council for clinical tests. Keefer was again drafted to conduct the investigation, in which approximately 3,000 patients were treated during 1946. The trials showed that streptomycin was effective in some infections where neither penicillin nor sulfonamides had been of value and that it was a useful adjunct to surgery in other infections.[12]

By 1948 eight companies were producing streptomycin in the United States, and the output for the year was over 80,000 pounds. Yet even if it had been used exclusively for patients with chronic tuberculosis, there would still have been enough to treat only about 1,000 patients with the doses recommended at the time, or only about 1/400 of the number suffering from the disease in this country alone. The result, as Hinshaw observed sadly, was that "desperate patients, their relatives, and physicians from many countries sought to obtain streptomycin when none was to be had." For the sake of all the victims of disease who would be attacked in the future, it was imperative that the scant supply be used to find out which patients would benefit from the antibiotic and how it could be used most efficiently. Four major clinical trials were undertaken in the United States.[13]

First, McDermott and his associates at Cornell Medical School put the new antibiotic to a stringent test by using it in two acute forms of tuberculosis. In the past practically every person affected either with tuberculous meningitis or with miliary tuberculosis, the generalized form of the disease in which thousands of tiny foci are scattered throughout the body, had died within a few weeks. All 17 patients suffering from these forms who were given streptomycin by the Cornell group improved dramatically, and although 12 relapsed several weeks or months later, five recovered completely—an unheard-of result, which by itself was enough to prove the superiority of streptomycin over any remedy previously advocated for tuberculosis.[14]

Impressive as these results were, the more important question was the effect of streptomycin in chronic pulmonary tuberculosis. Because streptomycin was expensive to produce, pharmaceutical companies had to be sure that it would be widely used in this, the most frequent form of the disease, before they were willing to lay out the large sums required for extensive production. The second major study was undertaken to answer this question. It was conducted by the American Trudeau Society, the medical section of the National Tuberculosis Association, beginning in 1946. Several investigators treated 332 patients with varying doses and for different periods of time. Patients given large doses of 1.5 to 3 grams a day fared best: 86 percent of them improved clinically and 82 percent improved according to x-ray films. Apparently streptomycin had a favorable effect on the disease, although the results were not completely convincing because no control patients had been observed at the same time. Also, the higher the doses were, the more frequent the toxic effects.[15] A safer dosage regimen had to be worked out if streptomycin was to be acceptable for routine use in tuberculosis. Fortunately these problems were being tackled in a third extensive collaborative study which became a landmark in clinical medicine.

Wars are ideal breeders of tuberculosis. By herding together thousands of men in a susceptible age group, they offer ideal conditions for the dissemination of tubercle bacilli. By subjecting men to stress, fatigue, and short rations, they lower resistance and thus allow the newly implanted bacilli to grow. In the period immediately following World War I, 12 percent of the 178,699 discharges for disability from the armed forces of the United States were for tuberculosis. During World War II more careful screening on entrance and earlier detection lowered the rate to 1.5 percent of discharges for disability. Still the number of tuberculosis patients in Veterans' Administration hospitals rose rapidly from about 5,000 in the early forties to approximately 9,000 in 1946, and it would increase even further to a peak of 16,000 in 1954.[16]

Working in outmoded, overcrowded, and poorly equipped hospitals, in an organization riddled with politics and bullied by veterans' associations, doctors in Veterans' Administration hospitals had struggled manfully with the flood of cases in the past. But now the tide had turned. In 1945 President Truman appointed General Omar Bradley administrator of veterans' affairs; Paul Hawley, who had been surgeon general of the army, became medical director. With them came sweeping changes: modern hospitals were built, usually near medical centers, consultants and attending physicians were recruited from nearby medical schools, research projects were initiated. Among the experts attracted to the Veterans' Administration was Arthur M. Walker, a pharmacologist who had written the chapter on penicillin in the book *Scientists against Time*, an account of the achievements during World War II of the Committee on Medical Research of the National Research Council. Thus the cooperative trials of penicillin in syphilis were connected by Walker to the program of cooperative studies on tuberculosis that the Veterans' Administration was about to undertake.[17]

When the program began in 1946, it involved only nine hospitals, but it was rapidly expanded until by 1948 it included one army, one navy, and 46 veterans' hospitals. From the first these trials had an advantage over those initiated by the American Trudeau Society, because the participating clinicians agreed to use certain regimens common to all. Thus they were able to gather rapidly enough cases to prove the value of the different dosage regimens of streptomycin in the various forms and stages of tuberculosis. In conferences held twice a year clinicians from all the hospitals presented the results of their studies, debated their significance, devised regimens to test new hypotheses, voted on which ones to use, and then went home to put the group's decisions into effect. The minutes of these conferences show how the doctors worked out their problems together, how Walker suggested and nudged but never ordered, and how in the process physicians whose expertise had been solely as clinicians became accomplished clinical investigators.[18]

The problem of untreated controls was raised in the first conference, and most of the participants were unwilling to apportion specified patients to a control group. Instead, they used each patient as his own control. The course of the disease during the 60 days before streptomycin treatment was compared with the course during 120 days of treatment. Bias was excluded by having x-ray films read by doctors unfamiliar with the cases. The results demonstrated that while only one-tenth of the patients were improving before therapy, three-fourths of them got better while receiving streptomycin. Also, the investigators were able to show that small doses of streptomycin were as effective as the larger doses used in the early cases and that vertigo and deafness were less frequent after the smaller doses. From the study of 2,780 patients it was concluded that streptomycin aided in the cure of tuberculosis and should be used in conjunction with other long-established measures, such as bed rest and appropriate surgical therapy.[19]

Meanwhile, investigators in Great Britain had successfully grasped the nettle of using controls in testing streptomycin. In the late nineteenth century British mathematicians had refined the statistical procedures developed by French mathematicians during the prior hundred years. One of the British mathematicians was Pearson, who had attempted unsuccessfully to persuade Almroth Wright to adhere to statistical principles in his experiments on typhoid vaccination. One of the disciples of this school, A. Bradford Hill, took the lead in planning the cooperative trials that were conducted by a committee of the Medical Research Council of Great Britain. He believed that, because Great Britain was able to purchase only a limited amount of the antibiotic at the time owing to currency restrictions, it was ethically acceptable to give it to some patients and not to others. Accordingly, the committee chose, on a random basis, some patients with pulmonary tuberculosis to receive streptomycin and others to be managed by the customary regimen of bed rest. X-ray films taken at intervals were interpreted by doctors who did not know which patients were receiving the antibiotic. In 1948 the council reported that 51 percent of 55 patients with pulmonary tuberculosis showed improvement on x-ray examination at the end of six months' treatment with streptomycin, in contrast to only 8 percent of 52 control patients, and that 7 percent of the patients given streptomycin died, compared with 27 percent of the controls.[20]

The fourth American group to initiate large-scale clinical trials of streptomycin in tuberculosis was the Public Health Service. Although their study was later in getting started because they had difficulty at first convincing clinicians that untreated controls should be used, their study was larger than the British trial, involving over 500 patients. They found also that improvement was faster and more consistent among the patients treated with streptomycin than in the controls.[21]

But streptomycin was by no means a panacea for tuberculosis. This became clear as early as 1946, when Guy Youmans and E. H. Williston of Northwestern University, together with the Mayo Clinic group, reported that the amount of streptomycin required to inhibit the growth of tubercle bacilli from some patients increased 1,000-fold during therapy. They were also able to induce resistance easily in tubercle bacilli by growing them in successively higher concentrations of streptomycin. Data from the clinical trials soon showed that the appearance of resistant strains was responsible for many of the failures of streptomycin therapy. Fortunately, at this time another therapeutic agent appeared which offered a way around the obstacle. The new drug was the product of applied research based on carefully reasoned theory.[22]

When Jörgen Lehmann of Gothenberg, Sweden, learned that salicylic acid stimulated the growth of tubercle bacilli, he reasoned that by modifying the salicylic acid, he could convert the stimulating into an inhibiting effect according to the principle of competitive inhibition, which had been shown to account for the action of sulfonamides on bacteria. In 1943 he demonstrated that para-aminosalicylic acid, later called PAS or PASA and officially named aminosalicylic acid, would inhibit the growth of tubercle bacilli. Clinical trials were begun in 1944, about six months before the first clinical trials of streptomycin. Although a controlled study showed that patients treated with aminosalicylic acid improved more rapidly than patients on bed rest alone, the new drug was not nearly so effective as streptomycin. Also it often caused nausea, vomiting, or diarrhea when given in doses large enough to produce maximum therapeutic effects.[23]

Doctors now had two specific remedies for tuberculosis where they had had none before, but each was only partially effective when given alone. The next logical step was to combine them. The Veterans' Administration began to investigate this possibility in 1949. Soon it became clear that patients given both drugs were more likely to improve than patients given streptomycin alone and that the addition of aminosalicylic acid delayed the emergence of resistant tubercle bacilli. The key to future antituberculosis therapy had been found in combination therapy.[24]

The success of aminosalicylic acid and the promising results obtained with the sulfones prior to the appearance of streptomycin had stimulated a search for other chemical agents to combat tuberculosis. Among the searchers was Domagk of Prontosil fame, who reported in 1946 that a synthetic chemical with a structure somewhat like that of sulfanilamide would inhibit the growth of tubercle bacilli in the test tube and counteract their effects in laboratory animals. Although this drug proved to be less effective than streptomycin and produced adverse reactions so often that it was never marketed in the United States, its antituberculosis properties encouraged others to explore along the same lines. The search led investigators in three pharmaceutical companies, two in the United States and one in West Germany, to find, independently and almost simultaneously, that the hydrazide salt of isonicotinic acid had striking antituberculosis activity in animals. This compound, which was officially named isoniazid and is often called INH, was first given to patients in June 1951 at Sea View Hospital on Staten Island. Later that year clinical trials were begun independently at Cornell Medical School and in Germany.[25]

When the three groups learned that they were all working with identical compounds, they arranged for simultaneous publication of their results in 1952. The reports were sanguine. Patients with advanced pulmonary tuberculosis treated with isoniazid at Sea View Hospital had gained an average of 20 pounds during the first four to fifteen weeks of therapy. Within this short time the x-ray films of half of the patients had already showed improvement. Even more persuasive were the results in patients with generalized miliary tuberculosis. One streptomycin investigator had admitted in 1948 that "early enthusiasm concerning the value of streptomycin in tuberculous meningitis was considerably dampened as one after another of the patients who were considered to be on their way to recovery . . . gradually lapsed into a comatose state or developed other complications and died." In contrast, all of the first ten patients treated with isoniazid by doctors at Cornell Medical School recovered promptly and remained well.[26]

Isoniazid not only appeared to be more effective than streptomycin but also could be given by mouth and rarely caused serious adverse reactions. But by this time doctors had learned that the precise value of a new drug in tuberculosis could only be determined by controlled trials. Within a few months the British Medical Research Council reported that isoniazid, while effective in pulmonary tuberculosis, was only slightly more so than the combination of streptomycin and aminosalicylic acid. They suggested that isoniazid would be more useful if given in conjunction with some other drug. Fortunately, this hypothesis was already being tested. Two months later investigators in the Public Health Service announced the preliminary results of the treatment of over 1,500 patients with pulmonary tuberculosis. Among patients who had streptomycin-sensitive bacilli at the beginning of therapy, 17.9 percent of those given streptomycin plus aminosalicylic acid had improved, compared with 22.6 percent of those receiving isoniazid alone and 30.3 percent of those who were given isoniazid plus streptomycin.[27]

The success of these three drugs stimulated a vigorous search for other antituberculosis antibiotics and chemicals. By 1973 an additional five antibiotics and three chemicals were regarded as useful in tuberculosis. One of the synthetics, ethambutol, was recommended for initial therapy. The others, because they were less effective or more toxic, were reserved for use if the first course of therapy failed, although one of the antibiotics in the group, rifampin, when combined with isoniazid, has since been shown in a controlled study by the Public Health Service to be as effective for initial therapy as a combination of isoniazid, streptomycin, and ethambutol. By the judicious use of combinations chosen from among four or five primary drugs as initial therapy, followed by combinations including the other drugs when necessary, along with proper surgical procedures in an occasional case, doctors can now achieve recovery in nearly every case of pulmonary tuberculosis, providing the patient cooperates fully.[28]

Soon after the introduction of isoniazid it occurred independently to investigators in various parts of the world that since it was given by mouth and caused so few adverse reactions, it might be used to prevent tuberculosis as well as to treat it. In the United States this conclusion was reached because of isoniazid's success in the treatment of primary tuberculosis. When a person, usually a child, is first infected with tubercle bacilli, the disease is usually either very mild or altogether without symptoms, being detected only by the change in the tuberculin test from negative to positive. In some cases, however, tubercle bacilli reach the bloodstream and are circulated to various organs where foci of infection develop. Very young children may become ill with miliary tuberculosis or tuberculous meningitis. Other individuals may have no symptoms during this period but at some later time may fail to contain the bacilli within these foci and may come down with active tuberculosis in the lungs or elsewhere.

In 1955 the Public Health Service initiated a study involving 2,750 children on the effect of isoniazid on primary tuberculosis. At the end of ten years, the local or general spread of the disease had occurred in 3 percent of the control children who had received an inert tablet as a placebo, compared with less than 0.4 percent of those who had been given isoniazid for one year. Also no serious complications, miliary or meningeal tuberculosis, had occurred in the isoniazid group.[29]

When this study showed that isoniazid could prevent a mild, usually asymptomatic infection from becoming a serious illness, the next logical step was to see whether it would prevent tuberculosis from developing altogether. For this purpose approximately 12,000 persons were selected who lived in the same household as a person with active pulmonary tuberculosis. Half of them were given isoniazid by random allocation, and the other half were given a placebo. At the end of ten years only 86 cases of active tuberculosis had appeared among the household contacts who received isoniazid, compared with 215 in the placebo group, a reduction of 60 percent. Studies carried out in mental institutions in the United States and in Kenya, the Philippines, and the Netherlands navy gave similar results. For the first time man had an effective preventive agent against tuberculosis that could be taken by the patient with no trouble and little possibility of adverse effects.[30]

The combination of fewer cases and shorter duration of hospital stay resulting from the use of the newer drugs brought a striking change in the availability of beds for tuberculosis patients. Waiting lists first shrank, then disappeared. The vacant beds, which became more numerous each year, were diverted to other uses. The number of beds allocated to tuberculosis patients decreased from 106,502 in 1951 to less than one-fourth of that number, or 23,401, in 1971. Some hospitals were closed entirely. In New York in the five-year period from 1957 to 1962 the number of public hospitals for tuberculosis decreased from 20 to 12. Even more private sanitariums were shut down, exemplified by the closing in 1954 of the very sanitarium Trudeau had founded seventy years before.[31]

Yet the picture was not as rosy as it seemed. When streptomycin and isoniazid first appeared, they had been welcomed enthusiastically by press and public. Patients with tuberculosis had rushed to be among the first to receive the new drugs. Doctors soon found that many of the burdens of antituberculosis treatment could be eased: the period of hospitalization was progressively shortened until it became only a few months for the average patient, the need for surgical procedures was reduced, and economic problems were lessened because patients could return to work sooner. Yet despite the much improved prognosis and the shortened, simplified therapy, some people remained indifferent to the dangers of tuberculosis. They did not seek help when confronted with suspicious symptoms, they refused treatment when the disease was diagnosed, they left the hospital without the approval of the doctors, or they failed to continue therapy or report for examinations after they returned to their homes. Ignorance, indifference, and lack of access to good medical care were more common in the lower socioeconomic classes. Many nonwhites were in this group, which partly explained their higher death rate from tuberculosis, amounting to 3.6 times the death rate of whites in 1950. Although hard work by antituberculosis forces reversed this trend so that by 1960 the ratio was only 2.8:1, there was still a long way to go.[32]

By the end of the fifties new cases of tuberculosis were becoming so scarce that it was no longer economical to conduct mass surveys. During that decade in New York State the average cost of finding one new case of tuberculosis through the survey method increased from $492 to $1,441. And when 17,873 persons were x-rayed in 1960 in one city in New York State which had a comparatively high mortality rate from tuberculosis, only one active case was found.[33] Programs had to be modified to meet the new conditions. Case finding was directed toward groups in which the disease was likely to be concentrated, such as the urban poor and migrant workers. Likewise, hospitals learned that as the number of beds for tuberculosis decreased, they had to improve their outpatient facilities. Finally, in view of the large number of persons alive with healed tuberculosis, programs for rehabilitation, both physical and social, had to be enlarged.

The natural tendency of the citizenry and of legislatures has been to cut appropriations when the major threat from a disease has passed. This attitude had been reflected in decreased appropriations for the control of venereal diseases after each world war and particularly after the advent of penicillin; it was now affecting funds for tuberculosis control. The average citizen was reluctant to spend tax monies to control a disease that did not seem to touch his life, since he no longer heard of relatives or acquaintances who were affected by it. The answer of the antituberculosis forces was to raise the banner for a new crusade. In the past the most that one could hope for was the control of tuberculosis; now the cry arose for complete eradication of the disease, not only in this country but throughout the world. This goal meant a vigorous campaign in which every person with active tuberculosis would be hunted down, as were the Typhoid Marys a generation before. It meant detection and treatment among the underprivileged minorities in this country and in the underdeveloped countries throughout the World. It meant intensification of measures for prevention by drugs or, if possible, the development of a vaccine more effective than BCG. The new crusade is an ambitious one. It has rallied and inspired the antituberculosis forces. Yet its success is by no means assured, in view of the all-too-prevalent complacency which holds that tuberculosis is conquered.[34]

While streptomycin was helping to bring tuberculosis under control, its value in other diseases was being widely tested. Early studies showed that a number of experimental infections in animals caused by gram-negative rods were prevented by streptomycin. The clinical trials carried out under the auspices of the National Research Council verified its efficacy in many of these. Gram-negative rods cause infections particularly in the abdomen, the pelvis, the urinary tract, and the meninges. While some of these infections had been suppressed or cured by the sulfonamides, streptomycin was in general more effective. In meningitis caused by *Hemophilus influenzae* only 20 percent of the patients who received streptomycin died, compared with over half of those who were given sulfonamides and practically all of those who received no specific therapy.[35]

Another disease that yielded to streptomycin was tularemia. This has been called a "purely American disease" because it was first observed in this country and because Americans discovered the causative agent and its transmission from one species of animal to another and to man. George W. McCoy of the Public Health Service described the disease in rodents in 1911 and a year later the microorganism that caused it, which he called *Bacterium tularense* after Tulare County, California, where it was first found. Humans acquire tularemia most often from contact with infected wild animals, particularly rabbits (hence its popular name rabbit fever), and occasionally from wild fowl. It may also be transmitted by the bites of flies and ticks which have fed on infected animals or by contaminated water. The usual features are an ulcer at the site of the bite or other initial contact, along with swelling of the lymph nodes draining the area. A more severe form is tularemic pneumonia, which follows inhalation of infectious material. Although death is infrequent, the illness and the disability that follows last for several weeks or months. Before the advent of streptomycin some doctors claimed that they had succeeded in shortening the disease by injecting a specific serum; others disputed this claim. The administration of streptomycin settled the argument, for it was followed by the prompt disappearance of fever, rapid healing, and recovery in nearly every case, even in the most extensive infections.[36]

In many infections of the urinary tract symptoms disappeared after streptomycin was given, and bacteria could no longer be cultured from the urine. Some of these patients remained well, while in others the infection returned and the rapid appearance of streptomycin-resistant bacteria often made further treatment useless. In other infections, such as typhoid fever, although streptomycin had been effective in animals, the results in patients were disappointing. Fortunately, the search for new antibiotics had intensified by this time, and within a few years of the introduction of streptomycin, other antibiotics became available which remedied most of its deficiencies.[37]

11. *Other Diseases and Other Antibiotics*

By the late 1940s it looked as if penicillin and streptomycin, with the help of the antitoxins and the sulfonamides, might soon bring the infectious diseases under control. Doctors now had the means to treat effectively the main killing and disabling infections— pneumonia, tuberculosis, diphtheria, scarlet fever, puerperal sepsis, the common varieties of meningitis, gonorrhea, and syphilis—the list reads like the bills of mortality of 1900. Yet there were other infections that were not susceptible to these agents, and these now stood out like evergreens in the forest when the hardwoods have shed their leaves. Several of these infections were caused by very small bacteria that differ in important respects from the familiar varieties of bacteria, such as pneumococci or tubercle bacilli.

The rickettsias form one of these groups. They are minute spherical or rod-shaped organisms which multiply only within living cells, unlike most bacteria which grow outside of cells. The first disease in the group to be recognized clinically was typhus fever, a severe infection that killed one of five of those attacked. The clinical features—high, continued fever, headache, delirium, general aching, and an extensive rash—can be attributed to a common characteristic of rickettsias, their tendency to multiply in the walls of the smaller blood vessels of the skin and internal organs throughout the body. Although typhus probably plagued man from the cave-dwelling days, not until the sixteenth century were descriptions sufficiently clear to be sure of its identity as the cause of epidemics. These were particularly likely to break out in jails, on ships, and during military campaigns—wherever unwashed people were herded together.[1]

Typhus accompanied the Europeans to the New World. It invaded Mexico and South and Central America alongside the Spaniards, and it landed in the United States with the colonists from northern Europe. Epidemics occurred in the colonies and in the new republic in the eighteenth century. The disease was endemic in American cities throughout most of the following century. The common form of typhus is transmitted from man to man by body lice, as shown in 1909 by Charles Nicolle, director of the Pasteur Institute in Tunis, and the disease is now called louseborne typhus in preference to its former designation of epidemic typhus. But long before the culpability of lice was known, the frequency of typhus was diminishing because of improvements in sanitary arrangements and personal cleanliness, such as the substitution of cotton for wool underclothing and the increased use of soap and water, aided by the appearance of bathtubs in this country after 1840.[2]

In 1898 Nathan E. Brill, a physician in New York City, described a mild form of typhus among recent immigrants. This outbreak was perplexing because apparently there had been no contacts with other cases. Hans Zinsser, a bacteriologist at Harvard, concluded correctly that these were recurrences of previous bouts of typhus sustained by the immigrants in their native countries. Their mildness was accounted for by the partial immunity left from the first attack. In the 1920s the form of typhus that occurred in the southern United States and Mexico, which was usually milder than louseborne typhus, was shown to be a natural infection of rats and mice which was transmitted to man by the rat flea. Accordingly it was named murine typhus.[3]

Beginning in 1873, doctors in Idaho and Montana recognized a disease which usually started suddenly with headache and chills, followed by fever lasting two to three weeks and a diffuse rash. This disease, also produced by rickettsias, was named Rocky Mountain spotted fever after the location of the early cases, although cases were later found elsewhere in the United States, in Canada, and in Mexico. Howard Taylor Ricketts of the University of Chicago, working in the Bitter Root Valley of Montana from 1906 to 1910, demonstrated that the disease is conveyed by the wood tick and transmitted the infection to animals by inoculating them with the blood of patients. Particularly as a result of the work of investigators in the Public Health Service, it was subsequently shown that the disease is carried from a number of wild and domestic animals to man by several species of ticks. Similar diseases also transmitted by ticks were traced to different species of rickettsias in other parts of the world: Africa, India, the Middle East, Russia, Mongolia, and Australia.[4]

Another rickettsial infection, which had been described in Japan over one hundred years ago, came into prominence during World War II when it attacked troops operating in the China-Burma-India theater. Its earliest name, tsutsugamuchi fever, so difficult for Western tongues to pronounce, was replaced by the designation scrub typhus after it was found to be transmitted to man by mites which had fed on small mammals living in the underbrush of fields and forests. Its clinical course is similar to that of typhus plus a local inflammation where the mite has bitten.[5]

The other important rickettsial infection, Q fever, was recognized in Australia in 1937 and almost simultaneously in the United States. During World War II the disease was identified among troops in the Mediterranean area and southern Europe. Cases were later found in all parts of the world. The disease differs from other rickettsial diseases in several ways: fever can last several months, pneumonia is a prominent feature, and rashes do not occur. The rickettsias are carried in livestock, usually without causing any signs of infection; humans are infected by direct contact with infected animals, by ingesting contaminated water or milk, or by inhaling air containing the rickettsias. Only rarely are they transmitted by ticks.[6]

The first of the rickettsias to be found was seen by Ricketts during his investigations on Rocky Mountain spotted fever and is called after him *Rickettsia rickettsi*. The causative agent of typhus was described by Ricketts, working in Mexico in 1910, and by a Bohemian, Stanislaus von Prowazek, in 1914, and it has been named *Rickettsia prowazeki* because they both died of the disease while investigating it. The causative agents of fleaborne and of scrub typhus are named *Rickettsia mooseri* and *tsutsugamuchi*, respectively. The agent that causes Q fever bears the name *Coxiella burnetii* in honor of Herald R. Cox of the Public Health Service and F. M. Burnet of Australia, who discovered it almost simultaneously in 1938.[7]

Knowledge of this important group of diseases was slow in coming both because rickettsias are so small that they are not easily seen under the microscope and because they are hard to culture. In 1938 Cox cultivated them in chick embryos by injecting them into fertile hens' eggs, thus facilitating both the study of the rickettsias and the production of a practical vaccine. Only one therapeutic agent had shown any promise, para-aminobenzoic acid. It had been tried by John C. Snyder of Harvard in 1942 because its chemical structure resembled that of the sulfonamides, which had been found to slow the growth of rickettsias in the test tube. Para-aminobenzoic acid suppressed infections in animals and in humans and caused the fever to fall slowly in some patients with typhus, scrub typhus, and Rocky Mountain spotted fever.[8]

Another group of small bacteria came into prominence after doctors observed that the sulfonamides and penicillin, while rapidly lowering the temperature of patients with pneumococcal pneumonia, did not affect certain other pneumonias. These pneumonias were characterized by a slower onset and by less likelihood of chest pain and bloody sputum than pneumococcal pneumonia and were more frequently accompanied by headache and general aching and more diffuse involvement of the lungs. This disease was first distinguished as an entity in the late 1930s and given the unwieldy name of primary atypical pneumonia. Like many other infections, it was a camp-follower: thousands of cases occurred among troops in World War II and probably in other wars, too, since the lungs of some soldiers who died in the Civil War showed similar pathological changes. In the early 1940s Monroe D. Eaton and his associates at the California State Department of Health isolated from patients with this disease a microorganism which was originally thought to be a virus because it was so small that it passed through filters that prevented the passage of the common bacteria. At first other investigators failed to detect the Eaton agent in cases of atypical pneumonia, mostly because it belonged to a group of bacteria about which little was known, so little, in fact, that some bacteriologists even doubted their existence. But in the 1960s Eaton was finally and completely vindicated, and the Eaton agent was established as a member of a class of microorganisms that are unique in that they lack cell walls. These are now called mycoplasmas, and the one responsible for primary atypical pneumonia is named *Mycoplasma pneumoniae.*[9]

Among the other small bacteria is a group named chlamydias, formerly called bedsonias, members of which are responsible for several widely different infections. Like the rickettsias, they grow only within cells and yet have the characteristics of bacteria; they contain both DNA and RNA and they reproduce by dividing into two organisms approximately equal in size. One species of chlamydias causes trachoma, an infection of the conjunctiva and deeper tissues of the eye, often leading to scarring and sometimes to blindness. Another species causes lymphogranuloma venereum, a sexually transmitted infection that progresses gradually and produces slowly advancing inflammation, destruction, and scarring of tissue on and near the genital organs of males and females. A third species causes an infection in parrots and certain other birds, hence the two names used for the disease in man, psittacosis and ornithosis. When transmitted to man, this chlamydia produces a pneumonia resembling mycoplasma pneumonia but usually more severe and more often fatal.[10]

Chlamydias responded to the sulfonamides and not to penicillin; mycoplasmas and rickettsias to neither. Other infections that were not affected by either of these antimicrobial agents were the dozens of diseases caused by viruses. Besides these atypical bacteria and the viruses, there were also some typical bacteria that did not respond to serums, sulfonamides, penicillin, or streptomycin. Among these were the brucellas. The disease they cause has been found in many countries and been given many names. It was originally called Malta fever because it was first described by a British army doctor who acquired it on that island. Later, as cases were identified in nearby countries, it was also called Mediterranean fever and the Neapolitan disease. In 1887 David Bruce, a British army surgeon, isolated a small, gram-negative microorganism which appeared to be spherical and thus a coccus. He named it *Micrococcus melitensis* after Melita, the Latin word for Malta. The means of prevention became apparent when Bruce's micrococcus was found in the milk of one-fifth of the goats on the island. Consumption of raw milk by the British garrison was stopped, and new cases of Malta fever in this group practically disappeared.[11]

A few years later, in 1895, Bernard L. F. Bang of Copenhagen, looking for the cause of abortions in cattle, cultured a short, gram-negative rod from aborting cows and named it *Bacillus abortus*. These disparate findings began to come together when Alice Evans of the Public Health Service showed in 1918 that Bang's coccus and Bruce's bacillus were closely related. A similar bacillus cultured in 1914 from aborting hogs by Jacob Traum of the United States Bureau of Animal Industry was also identified as a member of the group. Although all three microorganisms grew most often in the form of short rods, they sometimes took the shape of cocci; hence Bruce's original mistake. The names *Brucella melitensis*, *abortus*, and *suis* were given to those cultured from goats, cattle, and swine, respectively.[12]

The brucellas cause infections in wild and domestic animals throughout the world. In the United States *abortus* strains are most common and *suis* strains next. Prominent features of infection in humans are chills, fever often lasting for weeks or months, headache, backache, and pains in the joints. Brucellosis reaches humans via the milk of infected animals, by direct contact with them, or occasionally by air contaminated by them. In this country the elimination of raw milk and the prevention of the disease in animals by segregation of the infected ones and immunization of the others gradually diminished the incidence in humans. The number of reported cases in the United States fell progressively from about 3,500 in 1950 to 213 in 1970.[13]

Finally, the therapeutic agents available in the late 1940s failed in another area. Although the sulfonamides and the pioneer antibiotics affected a large number of bacteria, resistant strains were appearing in increasing numbers. Thus, despite the undoubted advances in the control of some infections following the introduction of these therapeutic agents, other infections were not well controlled or were not affected at all. There was every incentive to search vigorously for more antibiotics.

Penicillin had originated in the laboratories connected with two universities, streptomycin in the laboratories of a university and a pharmaceutical company. From this point on the commercial laboratories took over the search for antibiotics. The basic principle that substances to cure infections in humans could be obtained from microorganisms had been demonstrated by university scientists. Now the prospect of profits beckoned, and the pharmaceutical industry, with its superior resources and a commitment to practical research, responded. This industry had come of age in the United States when it was forced either to produce or to starve in World War I. Cut off from Germany by the Allied blockade, pharmaceutical companies had learned first to make for themselves the drugs originally devised in Europe and then to originate their own. They had mastered the production of serums, vaccines, arsenicals, and the sulfonamides. They were ready for the next turn of the wheel.[14]

By the late forties at least half a dozen companies had teams of investigators actively looking for antibiotics. One of these, Parke, Davis and Company, had established a research grant at Yale to enable a botanist, Paul Burkholder, to search through samples of soil for microorganisms with antibiotic potentialities. From soil collected in Venezuela he isolated a promising mold which was subsequently named *Streptomyces venezuelae*. Parke, Davis scientists extracted a substance from it which inhibited the growth of a number of pathogenic bacteria. The new antibiotic was named chloramphenicol and given the trade name of Chloromycetin.[15]

Parke, Davis scientists found that chloramphenicol was much more effective against rickettsias in the laboratory than para-aminobenzoic acid. They sent some to a leading rickettsiologist, Joseph E. Smadel of the Army Medical School in Washington, who was so enthusiastic about the results obtained in his laboratory that he advocated immediate trials in humans. Eugene H. Payne of Parke, Davis quickly took a team of scientists to Bolivia, where an epidemic of typhus was raging. In collaboration with local doctors, he treated 22 of the sickest patients, five of whom had been listed as certain to die. All recovered, including one for whom the death certificate had already been filled out and signed, awaiting only the insertion of the hour of death. Similar results were obtained in typhus by Smadel's group in Mexico. Smadel then arranged to try the new antibiotic on scrub typhus in collaboration with the British Institute for Medical Research in Malaya. Chloramphenicol was equally successful in this rickettsial infection; all 94 patients treated with it recovered. Temperatures became normal in thirty-one hours on the average, compared with an average of eighty-nine hours in 15 patients who received para-aminobenzoic acid and seventeen days in untreated patients.[16]

During the early trials in scrub typhus, Theodore Woodward, who was a member of Smadel's group, gave chloramphenicol to two young Malays just brought in from a rubber plantation where scrub typhus was prevalent. One was practically well twenty-four hours later, but the other was not noticeably improved. Suspecting that this patient might have typhoid fever instead of scrub typhus, Woodward looked for and was able to culture typhoid bacilli. Despite the change in diagnosis, he continued chloramphenicol therapy; on the third day the patient was definitely improved, on the fourth almost well. Of nine other patients with typhoid who were treated, all recovered, including one whose intestine had perforated and two who were bleeding from the intestines. Thus it happened that chloramphenicol was launched by American investigators working with three diseases that were uncommon in their own country.[17]

In contrast, another antibiotic discovered at the same time in the laboratories of a different pharmaceutical company was shown promptly to be effective in a number of infections indigenous to the United States. Lederle Laboratories began a search for antibiotics in 1939, in the same year as Parke, Davis. In 1944, Lederle appointed Benjamin Duggar, retiring professor of botany at the University of Wisconsin, to head the team conducting the search. Two years later this team discovered *Streptomyces aureofaciens*, from which they obtained an antibiotic that inhibited the growth of a wide variety of microorganisms in the test tube and suppressed these infections in animals. The antibiotic was given the trade name of Aureomycin and later the generic name of chlortetracycline. Lederle was able to bring its new antibiotic rapidly into large-scale production because it had already been using fermentation methods in the production of the vitamin riboflavin. The first patients were treated in 1948, one month after the initial clinical trial of chloramphenicol.[18]

Lederle had for many years specialized in remedies for infectious diseases, starting as a producer of diphtheria antitoxin, later producing more varieties of serums than any other company, and finally testing and marketing several sulfonamides, particularly those manufactured by another division of the parent American Cyanamid Company. In their previous work on infectious diseases its scientists had collaborated with several experts in university-affiliated hospitals. Thus the channels were already open for the rapid evaluation of chlortetracycline. Six short months after the first patient received the new antibiotic, clinical investigators reported success in checking several infections that had responded poorly or not at all to drugs and antibiotics previously available, such as Rocky Mountain spotted fever, Q fever, and many infections of the urinary tract. Good results had also been obtained in diseases that had yielded to penicillin, such as pneumococcal, staphylococcal, gonococcal, and meningococcal infections. These promising reports seemed to signal that a fourth important antibiotic was being launched. And so it proved, although Lederle had to share the honors and the profits.[19]

Not far behind Parke, Davis and Lederle investigators, a third team had begun a search for antibiotics. Chas. Pfizer & Company, because of its experience in producing organic substances by fermentation methods, was drawn into the penicillin and streptomycin programs at an early date. In 1949 investigators in its laboratories isolated *Streptomyces rimosus*, which produced an antibiotic to which was given the trade name Terramycin and which affected the same microorganisms as Lederle's Aureomycin. When its chemical structure was elucidated, the reason was plain: the chemical formulas are identical except for a chlorine atom at one point in the Lederle antibiotic and an oxygen-hydrogen atom group at another point in the Pfizer antibiotic. The basic structure, which contains neither of these radicals, was named tetracycline, from which the generic names for the other two antibiotics were derived, chlortetracycline for Aureomycin and oxytetracycline for Terramycin. When all three antibiotics were found to produce identical effects in human infections, tetracycline, which produced fewer adverse reactions, largely displaced the other two. These three, and other analogues subsequently developed, are known as the tetracyclines, and they, together with chloramphenicol, are called broad-spectrum antibiotics.[20]

With the entry of the pharmaceutical industry into the antibiotic field, conflicts over patent rights became a matter of course. While most investigators in universities believe that patents, if applied for at all, should be utilized for the public benefit, the captains of industry regard them as economic assets, to be used for increasing profits and funding new research. Consequently, commercial firms assert their claims and guard their rights to patents with all the vigor and legal skill they can muster. The fight for patent rights to tetracycline was one of the most hotly contested. Both Pfizer and the American Cyanamid Company claimed to have been the first to discover tetracycline, while the Bristol Laboratories sought a patent for developing a salt of tetracycline. After nearly three years of wrangling in the courts, the patent on tetracycline was eventually granted to Pfizer. They licensed American Cyanamid to produce it also, since it could be produced most cheaply from chlortetracycline, the patent for which was held by Cyanamid. Thus it turned out that the two new classes of antibiotics, the tetracyclines and chloramphenicol, were produced for a time by only three firms, Pfizer, Cyanamid, and Parke, Davis. In 1961 the Subcommittee on Antitrust and Monopoly of the United States Senate pointed to this concentration of production in so few hands as an example of monopoly power and claimed that it was a major cause of the high prices charged for these drugs.[21]

Regardless of the economic and legal problems they stirred up, the broad-spectrum antibiotics were a godsend to many patients, particularly those with diseases that up to that time had defied all therapy. Two important bacterial infections in this category were typhoid fever and brucellosis. Extensive trials of chloramphenicol proved its value in typhoid: it lowered the fatality rate to 1 or 2 percent, compared with 10 to 15 percent in patients treated by other methods.[22]

Both the sulfonamides and streptomycin had suppressed the growth of brucellas in the test tube and in animals, but they had little or no effect upon human infections even when given together. In contrast, the effect of the tetracyclines and chloramphenicol was striking. Within forty-eight to seventy-two hours after the beginning of treatment, the temperature was usually normal and the pains and aches had disappeared. Although relapses sometimes followed, they were arrested by another course of the same antibiotic.[23]

Because the tetracyclines were effective in other pneumonias, they were tried in mycoplasma pneumonia also. Although the first reports were favorable, controlled studies gave conflicting results. But once the causative agent was known and this pneumonia could be differentiated from those caused by viruses, which produced a similar disease, the value of the tetracyclines could be tested definitively. In 1961 doctors from the navy and the National Institutes of Health reported on a carefully controlled clinical trial conducted during an epidemic at a marine base in North Carolina. Either demecycline, an analogue of tetracycline, or a placebo was given by random allocation to marines with atypical pneumonias. Tests of antibodies showed that 59 of those who received the antibiotic and 50 who received placebos had had mycoplasma pneumonia. The temperature dropped more promptly, symptoms disappeared more rapidly, and the infiltration in the lungs cleared more speedily as determined by physical examination and x-ray in the patients who were given demecycline. The results of this model study were so conclusive that the tetracyclines became the routine therapy for mycoplasma pneumonias.[24]

The tetracyclines were also effective in the chlamydial infections, lymphogranuloma venereum, trachoma, and ornithosis. Treatment of ornithosis in birds does not always eliminate the carrier state so that prolonged therapy is needed. The number of human cases in the United States steadily declined after 1956, when 568 cases were reported, to 35 cases in 1973 as a result of careful surveillance of birds kept as pets and fowls raised for food. In 1974, however, the number of cases rose somewhat as a result of epidemics in turkey-processing plants.[25]

In some bacterial infections the broad-spectrum antibiotics served as reserve troops. Although generally less effective than penicillin in infections caused by streptococci, staphylococci, pneumococci, gonococci, and meningococci, they could be used in its place when the infection did not respond to penicillin or when the patient was allergic to it. Similarly, although streptomycin produced the most rapid improvement in tularemia, the broad-spectrum antibiotics were also effective and could be given instead. Broad-spectrum antibiotics, however, became the drugs of choice in several other bacterial infections. For instance, the tetracyclines were clearly superior in dysentery caused by shigellas, and they were preferable in infections of the urinary tract because they were often effective against a greater variety of bacteria than streptomycin and resistant forms did not appear so rapidly.

By the time doctors were becoming familiar with the relative values of penicillin, streptomycin, chloramphenicol, and the tetracyclines, the situation was complicated again by the introduction of still more antibiotics. The pattern for discovery had been laid down: obtain microorganisms from every possible source, find one that inhibited pathogenic bacteria, extract the active substance, and test it for effectiveness and lack of toxicity in lower animals and finally in humans. The process was like passing a mixture through a series of sieves having progressively smaller holes. Many potential antibiotics failed to pass through each successive sieve, so that thousands of bacteria and fungi had to be screened to obtain one antibiotic that was either more effective or less toxic than those already available, was safe to use, and could be produced cheaply and efficiently. Burkholder, for instance, tested 7,000 samples of soil before finding the mold that produced chloramphenicol. The equipment and man-hours needed to produce a new antibiotic required an investment of millions of dollars. Yet the rewards of success were high. Parke, Davis, for instance, sold $9 million worth of chloramphenicol in 1949, $28 million worth in 1950, and over $52 million worth in 1951. Lederle's sales of chlortetracycline reached a peak of $61 million in 1952. Such a potential for profit attracted pharmaceutical companies, while university groups and research institutes abandoned attempts to find antibiotics because they could not afford to make the necessary investment. As many firms took up the challenge, the results began to come in. Polymyxin, which suppresses the growth of a number of gram-negative rods, was developed in the Wellcome Research Laboratories in England. Eli Lilly and Company of Indianapolis produced erythromycin, which acts on cocci, especially pneumococci and streptococci, and on *Mycoplasma pneumoniae*. Antibiotics in the aminoglycoside group, which includes neomycin, kanamycin, and gentamicin, inhibit the growth of many gram-negative rods not affected by streptomycin or the broad-spectrum antibiotics. Other antibiotics were found that are effective in infections caused by fungi.[26]

By this time the law of diminishing returns was beginning to operate; the results obtained from massive screening programs did not always justify the outlay. The pharmaceutical industry turned to other methods of developing antimicrobial drugs. Several new chemical agents for treating infections of the urinary tract were synthesized. But the greatest progress came from combining the techniques of antibiotic production and chemical synthesis. As was so well illustrated by the sulfonamides, changing the side chains attached to the basic nucleus of a drug could improve its effectiveness or diminish its toxicity. Chain, who had become "fed up with screening" for new antibiotics as early as 1946, suggested to the Beecham Research Laboratories around 1956 that it try to modify the penicillin molecule. Within two years its investigators had isolated the nucleus of that molecule, 6-aminopenicillanic acid. They then began to add various chemical configurations to this nucleus. The 1,241st compound that they "semisynthesized" in this way looked particularly promising. It was highly active against gram-positive cocci, was not readily broken down by the enzyme penicillinase, and was of low toxicity for animals. Clinical trials demonstrated its effectiveness in infections caused by penicillin-resistant staphylococci, and in 1960 this compound was marketed as methicillin. Subsequently a number of other penicillins that were not vulnerable to penicillinase were produced.[27]

A different side chain added to the basic penicillin molecule was shown to increase the effectiveness of the antibiotic against gram-negative bacteria without materially reducing its action against gram-positive bacteria, making it into a broad-spectrum antibiotic. The first of this group of semisynthetic antibiotics, ampicillin, was rapidly followed by others with selective action against particular gram-negative rods. Ampicillin proved to have an advantage over chloramphenicol in typhoid: in addition to being effective in the treatment of the disease, it often eliminates the chronic carrier state —something that chloramphenicol had been unable to do.[28]

The intensive efforts to modify the penicillin molecule brought dividends in another area. In 1948 Florey received a letter from a friend in Italy telling him of an antibiotic produced by a fungus isolated from sewage by Guiseppe Brotzu of the University of Cagliari, Sardinia. Investigators in Florey's group, under the leadership of Abraham and G. G. F. Newton, found that this fungus actually produced several antibiotic substances. One of these, which Brotzu had not detected because it was present in the growth medium in only small amounts, was named cephalosporin C. It turned out to be the most promising of all because it was not broken down by penicillinase. Meanwhile, the British government, not wanting to be burnt a second time after its unfortunate experience with patents on the production of penicillin, had established the National Research and Development Corporation to obtain and administer patents in the public interest. Eli Lilly obtained a license from it to work on the cephalosporins. Scientists in this firm's laboratories were able, after many difficulties, to obtain the nucleus of the cephalosporin C molecule, 7-aminocephalosporanic acid, and from this to semisynthesize the first of a new series of antibiotics, which was named cephalothin. It proved to be highly resistant to the action of penicillinase, the reason for which is evident from an examination of its chemical configuration. The basic nucleus of penicillin is composed of two rings, a four-sided and a five-sided ring; the nucleus of cephalothin is closely similar, being composed of a four-sided and a six-sided ring. The six-sided ring of cephalothin is not attacked by penicillinase, which breaks down only the five-sided ring of penicillin, except in those semisynthetic penicillins, like methicillin, where a side chain is added on to protect the five-sided ring. Clinical trials with cephalothin, begun in 1962, showed that it is effective not only against penicillin-resistant staphylococci but also against several gram-negative rods. Thus it resembles both groups of semisynthetic penicillins.[29]

But all of the advances in the treatment of resistant infections were not made by discovering new antibiotics or devising new compounds. Soon after streptomycin became available, several bacteriologists tried adding it to penicillin in the test tube to see whether the combination would be more effective than penicillin alone. The results were variable; although adding streptomycin usually did not increase the potency of the penicillin, in some cases the effectiveness of the mixture was greater than that of penicillin alone. In humans, combinations of antibiotics were first tried in endocarditis, a disease that had been almost invariably fatal before the advent of penicillin. Although adequate treatment with this antibiotic would cure about 70 percent of patients with the most common variety, the one caused by viridans streptococci, the remainder died no matter how high the dose nor how long the treatment. In a less common variety, caused by Group D streptococci, which are often called enterococci because they are found in large numbers in normal intestines, the results were even poorer, less than half of the patients responding, even to massive doses of penicillin. Thomas Hunter found that a combination of penicillin and streptomycin killed enterococci in the test tube when neither antibiotic would do so alone, and in 1950 he reported that this combination was successful in the treatment of enterococcal endocarditis. Later he tried moderate-sized doses of penicillin and streptomycin in patients with the viridans variety of endocarditis and was able to cure 94 percent of them with a course of therapy lasting only two weeks. Both results were amply confirmed by others. This phenomenon of synergism, in which two or more antimicrobial agents are more effective than one alone, was later found to operate in several other infections when the proper combinations of antimicrobial agents were used. Combination therapy could be employed not only to achieve a synergistic effect, but also to delay or prevent the appearance of resistant bacteria.

Such regimens were found to be particularly successful in tuberculosis, where they are now the accepted method of therapy.[30]

The contrary phenomenon to synergism, antagonism, in which poorer results are obtained when two or more antibiotics are given than when the most effective of them is given alone, was first encountered clinically in the treatment of pneumococcal meningitis. In the past this infection too had been almost invariably fatal. Although many patients recovered when penicillin was given in large doses, about 30 percent still died. Lepper and I, seeking a way to reduce the deaths, gave chlortetracycline and penicillin to some patients and penicillin alone to alternate patients. Much to our surprise, the fatality rate among the patients receiving the combination was more than twice as high as in patients given the same dose of penicillin by itself. Later, antagonism was observed in some cases when chloramphenicol was added to penicillin for the treatment of pneumococcal meningitis, and occasionally when certain combinations of antibiotics were given for other infections. Extensive laboratory studies by Ernest Jawetz of the University of California established two groups of antibiotics. When antibiotics from Group I, to which penicillin and streptomycin belonged, were combined, synergism would sometimes occur, whereas when an antibiotic from Group I was combined with one from Group II, which included the tetracyclines and chloramphenicol, synergism would not occur but antagonism might. Jawetz's system became a guide for clinicians in the treatment of infections that could not be controlled by safe doses of a single antibiotic. Combination therapy to achieve a synergistic action was needed less often as new antibiotics were found that affected bacteria formerly not amenable to therapy. Perhaps its greatest significance today is that unnecessary therapy with combinations is one of the factors contributing to the overuse of antibiotics.[31]

When the public first heard of the astonishing cures obtained with penicillin, the general reaction was accurately depicted by a cartoon showing a little girl and her mother standing outside a movie theater which was advertising a film on the miracles at Lourdes. The child was saying, "But mother, that must have been before the days of penicillin!" The general public was likewise swayed by the gusts of hyperbole that followed the discovery of each new antibiotic. Chloramphenicol was labeled "one of the world's greatest medicines," and chlortetracycline was hailed as "God's Gift to the Doctors." As the newspapers and magazines echoed the cry of "miracle drug" whenever a new antibiotic appeared, people began to expect antibiotics to cure every ill of man and took for granted that a visit to the doctor would end up with a prescription for one of the "wonder drugs." Nor were doctors always as conscientious as they should have been in giving antibiotics only when they had good reason to do so. They used progressively more antibiotics in their practices. For instance, community pharmacies in the United States filled on the average 89 prescriptions for antibiotics for every 100 persons in 1971, compared with 50 prescriptions per 100 persons in 1964. The quantity of antibiotics produced in this country for medicinal purposes increased correspondingly, from 3 million pounds in 1960 to 9.6 million pounds in 1970.[32] A study of seven community hospitals in 1968 and 1969 showed that 30.6 percent of the patients received antibiotics. In university hospitals, where illnesses were generally more complicated, the usage was even greater. At the Johns Hopkins Hospital in the winter of 1964, 38 percent of the patients received at least one antibiotic and 6 percent received three or more.[33]

Unfortunately, many prescriptions for antibiotics were unnecessary. For example, in one survey involving physicians in private practice half of the patients diagnosed as having a common cold were given a prescription for an antibiotic or a sulfonamide, despite the fact that none of them has any effect on the duration or intensity of that illness and experts in infectious disease condemn their use. Evidence of overusage was also seen at the Boston City Hospital. The amount of ampicillin used during the period 1966 to 1968, when a specialist in infectious diseases had to be consulted before that antibiotic was prescribed, averaged 7,127 grams per year. During the next four years, when the use of ampicillin was unrestricted, an average of 62,613 grams was used each year. In the period when a consultation was not required before chloramphenicol could be given, from 19,593 to 23,544 grams were used each year; after the consent of a specialist was required, the average quantity used per year fell to 2,900 grams.[34]

The overuse of antibiotics added to the hazards of therapy. The adverse effects of penicillin and streptomycin were only two examples. As each new antibiotic was used extensively, its peculiar adverse reactions became evident. Untoward effects from the tetracyclines are usually not serious, although these antibiotics occasionally cause severe diarrhea, which can be fatal in a patient weakened from other causes, such as an extensive operation. Some of the more serious reactions, such as toxic effects on the kidney, which are produced by several of the less frequently used antibiotics, require that these antibiotics be used with caution and only in infections that would otherwise be fatal.

Unfortunately, doctors did not always balance the risk of adverse reactions from a particular antibiotic against the risk of the disease for which it was given. This was shown best in the experience with chloramphenicol. In 1950 reports began to appear in which patients who had been treated with this antibiotic developed aplastic anemia, a condition in which the cells in the bone marrow and blood are drastically reduced in numbers, resulting in weakness, bleeding under the skin or into internal organs, and diminished resistance to infection, culminating in death in about half the cases. By 1952, 23 cases of aplastic anemia associated with chloramphenicol had been reported in the United States and three in Great Britain; 24 of the 26 were fatal. In that year a survey by the Food and Drug Administration identified 198 patients in this country who had developed a serious deficiency of the blood cells after receiving chloramphenicol. Doctors were warned of the dangers of chloramphenicol, and the Food and Drug Administration required that a warning appear on labeling and advertisements for chloramphenicol that it should not be used indiscriminately or for minor infections. As a result, the amount of chloramphenicol produced fell from over 33 million grams in 1952 to under 6 million grams in 1953. Yet it increased again over the next few years, so that by 1960 it reached a record of nearly 55 million grams. In that year the Council on Drugs of the American Medical Association called attention to the dangers of the indiscriminate use of chloramphenicol and the Food and Drug Administration required a stronger warning on the label. Again in 1961 the production of chloramphenicol dropped to 27 million grams, and again it rose in subsequent years, reaching over 47 million grams in 1966. The increase was explained in 1968 when a Subcommittee of the United States Senate chaired by Gaylord Nelson heard testimony to the effect that the producer, Parke, Davis, had tried to convince doctors that chloramphenicol was a safe drug in spite of the warnings required on the labels.[35]

Following the Nelson Committee's hearings and further curbs on the promotion of the antibiotic imposed by the Food and Drug Administration, the quantity produced dropped in 1968 to less than half of the 1967 volume. It seemed likely that the volume would remain low, because by this time public opinion had been aroused by several developments. Hearings in Senate committees over the past two decades had made many people look critically at how doctors were prescribing drugs. Newspapers and magazines, largely as a result of these hearings, had given widespread publicity to chloramphenicol and its dangers. And finally, relatives of persons who had died from aplastic anemia had started lawsuits against the producers of chloramphenicol and against the doctors who had prescribed it. Their actions were representative of the new era in which patients were no longer content to take a doctor's word on faith and obey his instructions unquestioningly. The new science of medicine had made the leaders of the medical profession more critical of therapy with drugs; it was now making the public more critical also.[36]

While congressional committees and the public made forays into medical fields at times, doctors had the job of continuously evaluating their success, or lack of it, in the control of infections. After the initial excitement of watching gravely ill patients improve within a few hours after their first dose of penicillin, doctors were forced back to reality by having to reckon with the cases where the course did not proceed smoothly to recovery. So it was with each new antibiotic. Besides adverse reactions, which inevitably occurred in some persons following the administration of any antibiotic, the problems of resistant microorganisms and superinfection were the most persistent and the most baffling. Resistant strains of a particular microorganism tended to displace susceptible strains because they could grow in the presence of the antibiotic while the susceptible strains could not. Superinfection acted similarly: a new species replaced the one that had caused the original infection because it could grow in the presence of the antibiotic being administered. As an example, some patients given penicillin for pneumococcal pneumonia responded with a fall in temperature and subsidence of the pneumonia, only to develop fever a day or so later and a spread of the pneumonia. This time the causative agent would most likely be a staphylococcus or a gram-negative rod resistant to penicillin. At first it appeared that the broad-spectrum antibiotics would be the answer, but soon strains and species resistant to them as well were causing infections or were complicating existing infections, creating a special menace in hospitals.[37]

The problem of resistance took on a grimmer aspect when Japanese investigators found in 1959 that resistance to several antibiotics at the same time could be transferred from one gram-negative rod to another, even to ones of a different species, during conjugation, when two bacteria fused temporarily to exchange nuclear material. The transfer factor was shown to be a nonchromosomal element which carried with it a genetic code that programmed the bacterium receiving it to become resistant. Aside from the scientific interest this discovery created, it helped explain why antibiotic-resistant gram-negative rods were becoming so widespread.[38]

The replacement of antibiotic-susceptible microorganisms by nonsusceptible ones produced a drastic shift in the infections most frequently causing death. David E. Rogers of Cornell University reported that in the years just before the advent of penicillin the infections contributing to death were caused mainly by pneumococci, streptococci, tubercle bacilli, and staphylococci. In the years 1957 and 1958, with the exception of staphylococci, these bacteria were rarely a factor in fatal illnesses, whereas the fungi and gram-negative rods, which had played a minor role in the past, were now, alongside the staphylococci, the chief causes of death from infection. Similar changes in the kinds of microorganisms causing infections were revealed by the extensive studies of Finland and his associates at the Boston City Hospital. Also, a number of species which had seldom or never been reported to cause disease in humans were found to be responsible for infections. One reason for the emerging infectiousness of certain species was the wide range of microorganisms no longer competing with them because they were suppressed by the many antibiotics in use. Another was that, as antibiotics cured more people, the patients in hospitals were more likely to be persons weakened by old age or a chronic disease and therefore more susceptible to infection with microorganisms that were not usually pathogenic. For instance, Rogers found underlying disease in 60 percent of patients dying from infections before the antimicrobial era and in 93 percent of such dying patients in the years 1957 and 1958.[39]

To counteract the spread of antibiotic-resistant microorganisms, specialists in infectious diseases recommended both that doctors use a single antibiotic unless additional ones would definitely improve the results and that they stop using antibiotics in the vain attempt to prevent infections except in the few diseases where such prophylaxis had been proved effective, such as giving penicillin to prevent recurrences of rheumatic fever. Finally, they insisted upon scrupulous asepsis in situations where bacteria could spread easily, as in operating rooms, newborn nurseries, and places where wounds and burns were dressed—in other words, a return to the principles of Semmelweis and Lister. Although the shift in the nature of infections presents a challenge to develop antibiotics to control the newer infections, it is unlikely that man will ever devise drugs to combat all pathogenic bacteria. Even if he did, experience has shown that other previously nonpathogenic microorganisms would take their place.

As a partial answer to the increasing numbers of antibiotic-resistant bacteria, many physicians and others advocated controls over the antibiotics that doctors prescribed. A conference in 1958 on nosocomial infections, those that occur in patients who are already ill, led to the requirement by the Joint Commission on Accreditation of Hospitals that hospitals establish an infection control committee. These committees sometimes succeeded in stopping epidemics in their hospitals. They also educated doctors to use only those antibiotics that were necessary and to prescribe the least harmful ones when a choice was available. But in many hospitals these committees existed in name only; they failed to act, either from inertia, lack of expertise, or the belief that no one should tell doctors how to treat their patients. This belief, championed by many doctors, also prevented actions on the state or national level to limit the use of antibiotics that caused problems. Some doctors and laymen advocated that state licensing boards require doctors to take postgraduate courses so that they would know how to use antibiotics properly or to be reexamined every few years. Progress in science had increased the expectations of the public; the public in turn was demanding more from the doctors.[40]

Despite the problems attendant upon antibiotic therapy, it is important not to lose perspective. Many doctors used antibiotics wisely, and even those who did not always do so accomplished more good with them than harm. The improvement in the control of infectious diseases from the use of antibiotics is obvious and striking. This is clearly seen in the lowered case fatality rates for many diseases, such as the drop in deaths from pneumococcal pneumonia from one in three without specific therapy to one in twenty when effective antibiotics were given, or the change from nearly 100 percent deaths in the commonest form of endocarditis to nearly zero with proper antibiotic therapy. The improvement is also evidenced by the fall in mortality rates from infectious diseases as a whole in the United States. In 1958 Carl Dauer of the National Office of Vital Statistics estimated that 1.5 million lives had been saved during the preceding fifteen years as a result of the accelerated decline in mortality from these diseases, most of which was the result of the use of sulfonamides and antibiotics—an achievement that should be a source of pride to all who took part in their development and trial.[41]

12. The Newer Vaccines

Successes in the treatment of infections with serums, sulfonamides, and antibiotics gave rise to much satisfaction, even a little smugness, along with the expectation that before long a cure might be found for every infectious disease. But some infectious agents resisted control by specific therapy. Not until chloramphenicol and the tetracyclines were introduced was a satisfactory treatment for the rickettsial infections available, and none has yet been found for any of the viral diseases. Also, even when specific cures are available, they are never 100 percent effective; some patients still die. Furthermore, many patients do not report to the doctor in time for proper treatment or are not seen by a doctor at all. For all these reasons prevention of an infection is preferable to treatment, and one of the most effective preventive measures is vaccination.

Although early methods of vaccination against smallpox, rabies, typhoid fever, diphtheria, and tetanus were either successful or held out much promise of success, similar results in other infections were slow in coming. More basic knowledge and improved techniques were required before real progress could be made. The rapid strides in these areas beginning in the 1930s led to the development of newer types of vaccines for protection against certain rickettsial, viral, and bacterial infections.

In 1918, a few years after rickettsias were identified as the cause of typhus fever, animals were successfully immunized with vaccines made from infected lice. Yet this was hardly a feasible procedure since infection of lice was accomplished by inoculating each one through the rectum! Vaccines made from ticks infected with the rickettsias of Rocky Mountain spotted fever, introduced in 1924 by scientists of the Public Health Service, were more widely used but were still difficult to produce. Nor did ground-up organs of infected animals make satisfactory vaccines because they contained proteins from the animals' organs which often produced unwanted reactions. Not until 1938 when Cox cultivated rickettsias in the yolk sac of the chick embryo was a feasible method for the large-scale production of rickettsial vaccines available. Vaccines of this type were given during World War II to several million troops who were sent to areas where there was danger of contracting typhus. Among American troops immunized during the war, only 64 developed typhus, all of whom had mild infections and recovered. Vaccines made by the Cox method with the rickettsias of murine typhus or Rocky Mountain spotted fever were also used to protect individuals whose occupations exposed them to the possibility of infection. Thus, once a satisfactory method of preparing these vaccines was found, immunization against most of the rickettsial diseases became possible. In contrast, control of the viral diseases by immunization was a more gradual process, and suitable vaccines for some viral infections are still not at hand.[1]

Because viruses are generally so much smaller than bacteria and because they require living cells for their growth, knowledge about them was slow in coming. They had been used in vaccines for the prevention of smallpox and rabies before such an entity as a virus was known to exist. The process of identifying them was made simpler by Charles Chamberland, one of Pasteur's pupils. In 1884 he reported having devised a filter which had pores small enough to hold back the common bacteria while allowing smaller particles such as viruses to pass through. Eight years later a Russian, Dmitrii Ivanowski, reported that the infectious agent of tobacco mosaic, a disease of plants, passed through such a filter, and soon afterward filter-passing agents were shown to be the cause of other infections. In 1898 two Germans, Loeffler and Paul Frosch, attributed an animal infection, foot-and-mouth disease, to an agent that passed through small-pore filters, and in 1901 Walter Reed of the United States Army reported that a filter-passing agent caused yellow fever in humans.[2] Once viruses were recognized as the causative agents of a number of diseases, much effort was expended in trying to devise vaccines. Among these attempts none was more extensive or more challenging than the work on influenza.

In 1657 Thomas Willis described an illness that was sweeping across England: "The particular symptom of this disease . . . was a troublesome cough, with great spitting, also a Catarrh [excessive secretion] falling down on the palat, throat and nostrils, also it was accompanied with a feaverish distemper, joyned with heat and thirst, and want of appetite, a spontaneous weariness, and a grievous pain in the back and limbs." Little can be added to this description today except that the bronchial tubes are also affected, opening the way for pneumonia, which is the main complication and is often fatal. Willis was also impressed by the rapidity with which influenza spread. He wrote: "About the middle of April, suddenly a distemper arose as if sent by some blast of the stars, which laid hold of very many together; that in some towns in the space of a week about a thousand fell sick together."[3]

A disease like this, which suddenly strikes a number of persons, is labeled "epidemic," a word meaning literally "upon the people." Such diseases seem to come from nowhere and to descend suddenly upon their helpless victims. They have been a source of fear and terror since ancient times. By the early twentieth century two epidemic diseases, cholera and yellow fever, had been shut out of the United States by quarantine and sanitary measures, and smallpox had been largely brought under control by vaccination. During the next few decades two others were gradually limited to occasional small outbreaks, typhoid fever by the application of sanitary measures and diphtheria by immunization. Yet some infections continued to "descend upon the people," and occasionally one of them would spread so rapidly and so widely that it aroused fear akin to the terror that accompanied the bubonic plague in the Middle Ages. Such a one was influenza. Its name signifies "influence" in Italian; originally the word probably meant the influence of the stars. With the advent of the bacteriological era doctors began to look for a different kind of "influence," but not until well into the twentieth century was the causative agent finally identified as a virus.[4]

Influenza appears not only as an epidemic, but also as a pandemic, affecting many persons in countries throughout the world within the space of a few months. The first pandemic of influenza to be clearly recognized as such occurred in 1173. At least thirty were recorded between 1510 and 1930. In the spring of 1918 a few small outbreaks of a relatively mild influenza appeared in this country and elsewhere, but they were not severe enough to divert Americans from the war effort. During the same period epidemics were also occurring in Japan, China, and Africa. In contrast, a more virulent form of the disease spread like a prairie fire through the Allied armies in France and invaded England in June. In August this form of the disease appeared in America, first at ports where soldiers from Europe were disembarking, then in army and navy camps, then all across the country. Some idea of the virulence of the wave can be obtained from Cole's description of Welch's behavior when he was called to investigate this "mysterious epidemic": "The only time I ever saw Doctor Welch really worried and disturbed was in the autumn of 1918, at Camp Devens near Boston . . . It was cold and drizzling rain, and there was a continuous line of men coming in from the various barracks [to the hospital], carrying their blankets, many of the men looking extremely ill, most of them cyanosed [blue from inadequate oxygen intake via the pneumonic lungs] and coughing. There were not enough nurses, and the poor boys were putting themselves to bed on cots, which overflowed out of the wards on the porches. [Later in the autopsy room] when the chest was opened and the blue, swollen lungs were removed and opened, [Doctor Welch] turned and said, 'This must be some new kind of infection or plague,' and he was quite excited and obviously very nervous . . . [I]t shocked me to find that the situation, momentarily, was too much even for Doctor Welch."[5]

The experience in other army camps was similar. During four weeks in Camp Grant, Illinois, 9,032 soldiers were admitted to the hospital, or about one-fourth of the camp. Twenty-six percent of the hospitalized patients developed pneumonia, and of these 43 percent died. In the country as a whole the death rate for influenza and pneumonia rose from 165 per 100,000 in 1917 to 589 per 100,000 in 1918. House-to-house surveys in several localities revealed that 28 percent of the population were attacked during this wave of the epidemic. Another smaller wave followed in early 1919. Altogether about half a million persons died in the United States from influenza or its complications during the epidemic, and 21 million throughout the world, almost three times as many as were killed in World War I. Microbes had been deadlier than bullets.[6]

The suddenness of the onslaught, the severity of the symptoms, the death of many persons in the prime of life, and the days and weeks of prostration which often followed the acute illness all had a devastating effect on morale. In many homes everyone was laid low at once so that the less sick members of the family had to nurse the seriously ill as best they could. In army camps the overflow from the hospitals had to be housed in tents. On October 21 an observer in Washington wrote: "Conditions today might be copied from a history of the plague in the Middle Ages, an average of 1500 new cases and 95 deaths a day. Nurses and doctors are broken down and dying of it, hospitals overflowing so that many buildings have been commandeered for hospital use. Soldiers by the hundreds are detailed to dig graves, and the demand for coffins is so much greater than the supply that cemetery chapels are filled with bodies waiting for burial."[7]

While dismay and fear gripped the people, frustration prevailed among public health officials, for all attempts to control the spread of the disease failed. In Newark, New Jersey, for example, when influenza first became a threat, the mayor appointed a committee to help mobilize health resources against the disease. The committee started off sensibly, warning people to avoid unnecessary exposure, to dress warmly, to refrain from spreading the disease by coughing and sneezing, and to go to bed and call a physician at the first sign of a respiratory infection. As the epidemic increased in size and seriousness, public gatherings were banned, and schools, churches, places of recreation, and businesses were closed. A warehouse was turned into a hospital, and a pool of doctors was organized to respond to calls from patients who could not obtain doctors. Everyone with training in nursing, no matter how slight, was pressed into service. When these measures failed to stem the epidemic, however, self-interest began to operate. Under the pretext that liquor was a part of treatment, the mayor permitted saloons to reopen, and further pressure was being put on him to reopen churches and businesses when the epidemic subsided two months later.[8]

In retrospect, it is doubtful that the closing of the institutions had any effect on the spread of the disease. Chapin, who was wrestling with the same problems in Providence, quickly saw the futility of such measures and grasped its reason. He wrote: "When a community is pretty well sown with cases of the disease, as is true of our New England cities today, the disease will run its course. When the susceptible material [the pool of nonimmune individuals] is used, the disease will stop." In other words, influenza traveled too fast and spread too widely to be stopped by measures of isolation and quarantine. The epidemic had to run its course. Nor could the doctors and nurses who had been mobilized so expertly change materially the course of the patient's illness. Their capabilities were limited mainly to making him more comfortable. Doctors had no medicines to alter the outcome of pneumonia, which was the principal cause of death.[9]

Doctors did manage, however, to tackle the question of what caused the pneumonias. The consensus at the end of the 1889 pandemic had been that *Hemophilus influenzae,* then called *Bacillus influenzae,* caused both influenza and the pneumonia which accompanied it. Although subsequently these bacteria were frequently found in persons who did not have influenza, the idea that they were responsible for that disease still prevailed in 1918. This assumption was supported by the experiments of Blake and Cecil while in the

United States Army, in which they produced an influenzalike disease in monkeys by swabbing influenza bacilli in their noses. Their superior officer told them: "there can be no doubt that you have proved that the influenza bacillus is the cause of influenza." Yet during the 1918 epidemic doctors were unable to culture influenza bacilli consistently from the air passages of patients with influenza even though bacteriological techniques had greatly improved. In patients with pneumonia other bacteria, such as pneumococci, streptococci, and staphylococci, were cultured from the lungs more often than *Hemophilus influenzae.* By the end of the epidemic, although some still clung to the idea that the influenza bacillus was responsible, the majority believed that influenza was caused by some as yet unidentified agent and that the pneumonia that followed it was produced by whatever pathogenic bacterium accompanied the causative agent from the respiratory passages into the lungs.[10]

Attempts to show that influenza was caused by a filter-passing virus were at first ineffectual. For instance, investigators in the Public Health Service, in collaboration with the navy, failed to transmit the disease by means of secretions from patients with typical symptoms. The causative agent was finally found through the instrumentality of animals. In 1918 a "new disease" had appeared in swine which closely resembled human influenza and which occurred in close association with outbreaks of the human disease. In 1931 Richard E. Shope of the Rockefeller Institute obtained from swine a filterable virus that produced a mild influenzalike illness in other swine. This stimulated investigators to try harder to recover a virus from humans. During the epidemic of 1932 Wilson Smith, Christopher Andrewes, and Patrick Laidlaw of the National Institute for Medical Research in England tried without success to induce the disease in rabbits, guinea pigs, mice, hamsters, monkeys, and even hedgehogs. Then, because a mysterious illness had developed in a ferret colony in another laboratory while the staff were coming down with influenza, Smith inoculated ferrets with throat-washings obtained from Andrewes, who was ill with influenza. When respiratory symptoms developed in these animals, the investigators were able to transmit the disease from ferret to ferret and subsequently to mice. The modern knowledge of influenza can be dated from this point.[11]

Later, during an epidemic that prevailed in 1940, Thomas Francis of the Rockefeller Institute and his former associate, Thomas P. Magill, independently isolated an influenza virus that was different from those found earlier in that the antibodies it produced would not protect against infection with the previous viruses, and vice versa. In other words, the new virus possessed different antigens, complex chemicals that are capable of inducing immunity. The virus isolated by Smith, Andrewes, and Laidlaw was labeled Type A and the new one Type B. Later a third type was discovered, designated Type C.[12]

The identification of specific influenza viruses made it possible to trace the spread of the disease. In addition to the pandemics and larger epidemics, it became clear that influenza occurred in smaller outbreaks and as isolated cases. From 1918 through 1973, 37 epidemics were recognized, those caused by the A virus occurring every two or three years, and Type B epidemics, which tended to be milder, at intervals of three to six years. Type C virus has caused only a few small outbreaks.[13]

Just when virologists believed that they had the knowledge to control influenza, their hopes were dashed during an epidemic which began in 1947. The virus isolated at that time, while definitely belonging to Type A, differed significantly from previous Type A viruses, so much so that vaccines prepared from the original strains gave no protection against the new one. A major shift in the chemical composition of one of the antigens had occurred, the first of several such changes that were destined to take place over the next few decades. A second shift in the antigens of the A virus occurred in 1957, generating a pandemic that began in China and spread around the world. Another minor shift occurred, and another pandemic followed in 1968. The viruses isolated between 1935 and 1947 were designated A0, the virus of the 1947 epidemic A1, and those isolated from 1957 on were called A2. Antigenic shifts away from the original Type B strain have also occurred.[14]

Changes in antigenic composition explained many of the features of influenza epidemics. A major change from previous strains of A virus must have occurred in 1918. The new strain infected people everywhere because so few had any immunity to it. That strain must also have been unusually virulent in view of the high mortality rate. For the next three decades only localized epidemics occurred because most persons were immune to the prevalent strains. When another antigenic shift occurred in 1947 and produced the A1 virus, it too caused a pandemic. Yet, while large numbers of people came down with influenza, fewer died than in the previous pandemic, presumably because this strain was less virulent. The pandemics of 1957 and 1968 were also less lethal. Since the major cause of death in the past was pneumonia and since most of the pneumonias are caused by bacteria and only a few by the viruses themselves, sulfonamides and antibiotics also must have played a part in the lower mortality of the later pandemics. Beginning in 1957 a valuable method was introduced for studying the effect of influenza upon the death rate: calculation of the excess mortality, the degree by which the mortality during an influenza epidemic exceeds the expected mortality for that period. By the use of this method it was found that, whereas 90 percent of the excess deaths during the 1918 epidemic were directly attributable to infections of the respiratory tract, this percentage gradually decreased until in the years 1957–1960 only about 30 percent were attributable to these infections. At the same time the proportion of excess deaths attributable to diseases of the heart and blood vessels increased to 50 percent. Just how influenza viruses cause these deaths has not been determined.[15]

Soon after the influenza viruses were discovered, it was found that antibodies to them developed following an attack of the disease and that vaccination with the viruses could produce similar immunity. Utilizing one of the A1 strains, Francis and Magill reported in 1936 that injections of live influenza virus produced antibodies in humans in quantities comparable to those found in patients convalescent from influenza. British investigators, however, were unable to produce immunity with a vaccine when the virus had been inactivated with formalin. In 1942 the Commission on Influenza of the Army, later the Armed Forces, under the guidance of Francis began an intensive program to develop effective vaccines. The clinical trials of these vaccines were carefully controlled, in contrast to the early experiments on typhoid vaccines. In each trial observations were made on an equal number of unvaccinated persons, who were chosen at random from the same group as the vaccinated. Vaccines composed of formalin-inactivated A0 strains reduced cases of influenza to less than one-third the number occurring in control groups. After the prevailing strain changed to A1 in 1947, the original vaccines were of no value. Until the next antigenic shift occurred in 1957, however, vaccines composed of A1 strains decreased the frequency of the disease to from one-third to one-eighth of that occurring in unvaccinated controls. Thus, potent vaccines could be prepared that would protect persons during the smaller epidemics that followed a pandemic, but when another antigenic shift occurred, the previously effective vaccines were useless.[16]

Investigators tried many procedures to overcome this formidable obstacle. They made more potent vaccines, but these still gave little protection against virus strains of different composition. Antibodies to the A2 strain isolated in the 1957 epidemic were found at the beginning of the epidemic in the serums of many persons in their seventies and eighties, in other words, those who had been alive during the pandemic of 1889. This seemed to indicate that the number of antigenic variations was limited and that the 1889 variant had appeared again in 1957. On this assumption some investigators, particularly Francis, now at the University of Michigan, and his associate, Fred Davenport, championed the idea that repeated immunizations with the known strains of the virus, beginning at a young age and continuing throughout life, might build up an immunity broad enough to protect against any new strain that might appear. Another school of thought, led by Andrewes, contended both that the number of antigenic variants may be endless and that the evidence for the production of a broad immunity through repeated vaccination is flimsy. They placed their hopes in a network of laboratories that had been established in various countries to identify new variants as soon as they appeared and to send these variants promptly to central laboratories so that a vaccine could be made in time to immunize the population of a country before the full force of the epidemic reached it.[17]

Unfortunately this system failed in the 1957 epidemic, when it proved impossible to make enough vaccine in time to check the epidemic. Although the United States government "launched the fastest medical mobilization ever attempted against an epidemic disease," the incidence rate rose sharply while the production of vaccine increased only slightly. As a result, almost half the nation's population was confined to bed with an influenzalike illness between July 1 and November 9 of that year.[18]

When another pandemic began in 1968, the rate of vaccine production was accelerated. The first cases were identified in Hong Kong on July 13, and four days later a strain of virus was sent to laboratories in England. On August 16 the strain was reported to be antigenically related to but different from previous A2 strains, and on November 15 the first lot of vaccine made from the new strain was cleared for release by the National Institutes of Health. Although 20 million doses of the vaccine had been cleared by January 1, 1969, there were 1,151 excess deaths in 122 cities in the week ending January 11. By mid-January the epidemic began to recede, although in March the deaths from influenza and pneumonia were still above the usual number. Clearly the accelerated program was still not fast enough.[19]

In February 1976 at Fort Dix, New Jersey, a new strain appeared.

Viruses obtained from 13 soldiers with influenza, of whom one died, differed from all the strains found in recent epidemics but resembled instead the influenza virus obtained from swine by Shope in 1931. Although this virus had been found frequently in swine since then and occasionally in humans in direct contact with swine, these infections had not spread to other persons. Tests in the serums of soldiers at Fort Dix indicated that approximately 500 had been infected with the new strain. It was postulated that a recombination of genetic material from the virus that had been present in swine since 1918 and the A/Victoria strain which was prevalent in the human population in 1976 had produced a swinelike virus which was able to spread from man to man and that a pandemic might be in the making.[20]

In March 1976 President Ford, on the advice of doctors in the Department of Health, Education, and Welfare and outside consultants, announced the National Immunization Program, whereby the federal government would provide vaccine made from the new strain for the population of the United States. Four pharmaceutical companies prepared the vaccine, and field trials showed that the vaccines produced antibodies and were followed by few adverse reactions, none of them serious. The program was launched in September, but on December 16 it was suspended after a number of persons who had received the vaccine developed the Guillain-Barré syndrome, a disease of the nervous system characterized by weakness or paralysis involving the lower limbs and usually ascending to the upper limbs and sometimes to the facial muscles. Between September 26 and December 31, 571 cases were reported in the United States, of which 291 occurred in persons who had received influenza vaccines. Although it was estimated that only 1.55 cases occurred per million vaccinated persons, the reported rate in unvaccinated persons was only one-ninth as great. For this reason and because no further epidemics of swine influenza had appeared, it seemed doubtful by the end of 1976 that the immunization program would be resumed unless further epidemics occurred.[21]

Fortunately, much of the knowledge and many of the techniques learned during attempts to immunize against influenza have been used successfully in the prevention of other infections. One of the most striking examples is poliomyelitis. One afternoon in 1921, after a few days of inexplicable lassitude, Franklin D. Roosevelt had a chill and went to bed early. When he awoke the next morning, his left leg felt heavy and lagged behind when he tried to walk. During the next several days he had a high fever and pains in his lower limbs and became paralyzed from the waist down. The pains and fever left, and he regained control of his bladder, but for the rest of his life he could walk only a few steps and these only with the aid of braces and a strong supporting arm. His was a typical case of severe poliomyelitis, a viral disease characterized by fever, headache, and sometimes stiffness of the neck and back. In paralytic cases the virus attacks the motor nerve cells in the spinal cord, which govern the muscles. If the muscles needed for breathing or swallowing are affected, the patient may die. Unless weakness or paralysis of the muscles in the limbs is minor or improves after the acute attack, the victim will be crippled for life, a living memento of man's powerlessness to check the course of the disease once it has begun. And until recently nothing could be done to prevent it either.[22]

Deformed limbs, which could well have been caused by poliomyelitis, have been found in skeletons dating back as far as 3700 B.C., and a convincing representation of a young man with an atrophied leg, typical of the end stage of the disease, appears on an Egyptian stele dating from the Eighteenth Dynasty (1580–1350 B.C.). Cases resembling poliomyelitis have been described from the time of ancient Greece and Rome. In the late eighteenth and early nineteenth century descriptions of cases became more definitive, culminating in 1840 with a classic monograph by Jacob Heine, a German orthopedic surgeon, who depicted the disease so accurately that a modern medical student could learn to diagnose it from his description.[23]

In the early nineteenth century, however, doctors were reporting poliomyelitis entirely in the form of single cases. A new phase of the disease apparently began around 1835, when John Badham described four cases that had appeared within a few days of each other in a town near Sheffield, England. Other small outbreaks were observed at about the same time on the island of St. Helena and in rural Louisiana. Yet no large epidemics were reported until Karl Oscar Medin described 44 cases in Stockholm in 1887. This was the beginning of a series of epidemic waves that swept periodically across many parts of the globe, especially northern Europe, North America, and comparable areas in the Southern Hemisphere.[24]

Heine's complete description of individual cases and Medin's report of an epidemic earned poliomyelitis the name Heine-Medin disease, by which it was known for several years. Medin observed that the fever of the early phase, when symptoms were generalized, sometimes fell before rising again with the onset of paralysis, which showed that the first rise represented a systemic phase of the infection and the second the invasion of the central nervous system. In 1905 Sweden suffered the largest epidemic seen to that time, and Medin's pupil, Ivar Wickman, traced the spread of over a thousand cases. He found that for every case in which the nervous system was involved there were one or more cases where the disease stopped at the stage of a minor, systemic illness and correctly concluded that the disease was spread largely through these minor cases because the disease was not diagnosed and the patients were not isolated.[25]

By this time large epidemics were occurring in other countries. In the United States the first one of any size appeared in 1894, involving 132 cases in Rutland County, Vermont. Increasingly larger outbreaks followed, culminating in 1916 in an epidemic composed of at least 29,000 cases in which there were 6,000 deaths. During the same period another change was occurring. Heretofore the victims had been mainly infants and small children; hence a common name for the disease was infantile paralysis. Now it began attacking older children and adults in growing numbers. For instance, in Sweden in 1895 only 3 percent of Medin's patients were over six years of age and only 1.5 percent were over fifteen years, whereas Wickman reported that in the 1905 epidemic 61 percent were over the age of six and 21 percent were over fifteen. This turn of events was especially alarming because the disease tends to be more severe and more extensive in older patients.[26] Thus a disease that half a century before had been relatively unknown was now attacking hundreds and thousands each year, especially in the economically advanced countries, and was obviously getting worse. An observant layman could not go many days without seeing at least one child hobbling about in braces or dangling a useless arm or an adult stumping around with one lower limb shorter than the other, reminders that the disease was not under control.

Not that doctors were not trying. French pathologists had shown in the 1870s that the chief abnormality lay in the anterior or front portion of the gray matter of the spinal cord, where the main part of the motor nerve cells lie. From this location came the name anterior poliomyelitis, meaning inflammation of the gray marrow, later shortened to poliomyelitis. But the greatest step forward was the demonstration that poliomyelitis was caused by a filterable virus. This was reported in 1908 by a Viennese immunologist, Karl Landsteiner, and his associate, Erwin Popper. They ground up the spinal cord of a boy who had died during the acute stage of poliomyelitis, passed it through a small-pore filter, and injected it into the peritoneal cavities of two monkeys. Both died, one after developing paralysis of both legs, and their spinal cords showed changes that closely resembled those of human poliomyelitis. Although more precise identification of the causative agents was not possible for three decades, it was generally accepted that the disease was caused by the poliomyelitis virus, or poliovirus for short.[27]

As epidemics became more frequent and more extensive, doctors and public health officials tried many procedures to stop them. First they used the time-honored measures of isolation and quarantine. Such attempts reached their peak in the United States during the 1916 epidemic when New York City went so far as to require persons under sixteen leaving the city to produce a certificate showing that their homes were free of polio. These measures failed to check the epidemic because, as was later shown, most of the spread is from persons with mild or completely symptomless infections. In 1937 an entirely different method of control was tried, spraying children's noses with chemicals. The only result was to abolish their sense of smell, in some cases permanently. This procedure was based on the faulty assumption that the virus usually reached the central nervous system by traveling along the nerves from the nose. Simon Flexner had suggested as early as 1910 that this was the route of infection, basing his theory on experiments in which he transmitted the disease by rubbing the virus into the membranes of monkeys' noses. As the years passed and no other route of infection was proved, Flexner's concepts became firmly entrenched in the minds of most investigators.[28]

Since poliomyelitis could not be prevented, the concentration was on treatment. As Sir Walter Scott wrote of his own childhood paralysis: "When the efforts of regular physicians had been exhausted, without the slightest success, my anxious parents, during the course of many years, eagerly grasped at every prospect of cure which was held out by the promise of empirics, or of ancient ladies or gentlemen who conceived themselves entitled to recommend various remedies, some of which were of a nature sufficiently singular." This was the experience of all too many victims. In their zeal to make a child whole again, parents chased any will-o'-the-wisp, and doctors were not far behind.[29]

Samuel D. Gross told his medical students in 1872 that although the common remedies of bleeding or blistering were of value, he preferred the application of a red-hot iron over the affected area until an ulcer was formed. Milder remedies consisted of applying ice to the spine or rubbing in mercury ointment to produce blisters during the acute stage, and then at the end of the first week of paralysis administering a course of electrical treatments in which faradic current was applied. A more rational therapy was tried during the 1916 epidemic in New York, after the disease had been shown to be infectious, when serum from convalescent patients was given to patients in the acute stage of the disease. But because doctors were neither familiar with the techniques of a controlled clinical trial nor convinced of the necessity for one, no proof of the efficacy or inefficacy of serum was obtained. Further trials during the twenties also produced no valid evidence one way or the other. Finally in 1931, W. Lloyd Aycock and his associates at Harvard compared patients who had received convalescent serum with those who had not and found that the serum failed to shorten the acute phase or prevent paralysis.[30]

In the early decades of this century the wisest doctors advised rest during the acute stage of the disease along with warm baths and gentle massage of the affected muscles. In the later stages the emphasis was placed upon rebuilding the strength of the muscles. Sometimes braces had to be resorted to so that the patient could walk, and occasionally the motion of a limb could be improved by a surgical procedure. Because these techniques were in the province of the orthopedist, this specialist was looked to in the twenties and thirties for advice on the treatment of poliomyelitis even in the acute stages. Toward the end of this period orthopedists were increasingly recommending that every affected limb be immobilized in a cast as soon as the diagnosis was made, with the object of attaining maximum rest. As a result, the muscles of the limb atrophied, and rehabilitative procedures, such as massage and passive exercise, were often delayed so long that the damage was irreversible.[31]

The impetus that moved doctors from this extreme position came from an unlikely source. Sister Elizabeth Kenny, a nurse working on her own in the Australian bush country, unaware that splinting was the accepted practice, applied warm, moist packs to the affected limbs and found that many of her patients improved. She evolved a system of treatment in which this procedure was combined with passive exercises, followed after the acute stage by intensive efforts to reeducate the patient to use to the fullest the muscular power that remained. In 1940 she came to the United States, where she was warmly welcomed by some doctors, strongly opposed by others. The American public, hungry for any crumb of evidence that poliomyelitis could be cured, lapped up the stories about Sister Kenny's "victories" over poliomyelitis and her "persecution" at the hands of conservative doctors. Unfortunately, her innovations were not confined to new techniques. She claimed, for instance, that "training of muscles prevents 'mental alienation,' distortion of mental pathways so that a muscle does not respond to voluntary function." Here she was on shaky ground.[32] Unprejudiced doctors, after trying her techniques, concluded that she was correct in her view that prolonged immobilization was detrimental because it delayed the starting of measures to improve muscle function; they also agreed that in the retraining stage patients needed to be continually encouraged to use the remaining muscles to the utmost. On the contrary, the hot packs, highly uncomfortable for the patient and time-consuming for the nursing staff, were found to be needed in only a few cases and then only intermittently to relieve muscular spasm and pain. The treatment of poliomyelitis returned to a regimen of rest and support for affected limbs, followed as soon as possible by exercise and reeducation. Sister Kenny's contribution was in pointing out that orthodox medicine had put too much emphasis upon rest at the expense of other necessary therapy.[33]

The paralyses in the upper or lower limbs, though serious enough, were not lethal, as were some other complications. As more older children and adults were attacked, extensive involvement of the nervous system was seen more often. When the muscles used in breathing were affected, the patient had to struggle to take in enough air, and frequently he lost the battle. To substitute for the ineffective muscles, respirators were devised, ranging from a large tank that enclosed all the patient's body except his head, known as the "iron lung," to a vestlike apparatus that was placed over his chest. Rocking beds were also used, which aided the action of the diaphragm as the patient's body was alternately tilted up and down. If the upper part of the brain stem was affected in bulbar poliomyelitis, the nerves governing respiration and swallowing could fail to function. The swallowing difficulty was treated by frequent suction of the excessive secretions either from the throat or through a tube placed in an opening in the trachea; respiratory failure required the use of a respirator. By the 1950s the average poliomyelitis ward was a crowded sea of patients struggling to keep afloat in respirators or on rocking beds with nurses hastening from one to the other, encouraging, adjusting, and ministering. Care of poliomyelitis patients was difficult, expensive, heart-rending, sometimes rewarding, all too often hopeless. Fortunately, medical science was on the way to finding a means to end all this.[34]

As is so often the case in research, the solution of the poliomyelitis problem had to await the development of better techniques. In 1939

Charles Armstrong of the National Institute of Health reported that he had transmitted a strain of poliovirus—the Lansing strain, obtained from a patient in Lansing, Michigan—to cotton rats and from them to white mice, which were common and cheap laboratory animals. Armstrong's discovery opened up many new possibilities. Virologists tried passing poliovirus to other rodents, which led to the realization that not all strains of poliovirus could be transmitted to mice and cotton rats. Max Theiler of the Rockefeller Institute, by passing the Lansing strain through a series of mice, was able to diminish its pathogenicity for man. But the more paths that were opened for exploration and the more techniques devised, the greater the sums of money required. Fortunately, a new organization appeared at this time which soon satisfied a large part of the need.[35]

It is probable that Franklin Roosevelt's attack of poliomyelitis had more influence on the control of a disease than any other illness in history. People were willing to give largely of money and effort to support research on poliomyelitis because their President symbolized this tragic disease and reminded them of the desperate need to conquer it. An effective connection between the public and Roosevelt's illness was made by the National Foundation for Infantile Paralysis. Its birth was royal, and the promise made at its nativity was princely. For its founding, under the guidance of Basil O'Connor, Roosevelt's former law partner, was announced in 1937 over a nationwide radio hookup from the White House. President Roosevelt pledged that the foundation would make "every effort to ensure that *every responsible research agency in this country is adequately financed* to carry on investigations into the cause of infantile paralysis and the methods by which it may be prevented." Specific questions that remained to be answered were whether there was one poliovirus or many, how the viruses entered into, traveled through, and left the human body, whether vaccines could be made that would produce antibodies, and if so, whether they would prevent the disease. The foundation promptly began to support investigations on these problems.[36]

Important information came from studies on antibodies. During an epidemic of poliomyelitis in 1910, J. F. Anderson and Wade H. Frost of the Public Health Service had collected samples of blood from persons who had recovered from attacks of the generalized infection without ever having had symptoms of the nervous system phase and found that their serums contained antibodies against poliovirus. In 1912 investigators at the Rockefeller Institute reported that antibodies to poliovirus were present in the serums of healthy adults in quantities comparable to those in patients convalescent from the disease. In the 1930s the Poliomyelitis Unit at Yale, headed by Trask and John R. Paul, began measuring antibodies in the serums of healthy persons of various ages and those convalescent from the disease. Data gradually accumulated by them and others showed that the serums of most adults contained antibodies against poliovirus and that in areas where environmental sanitation and personal hygiene were poor, these antibodies usually appeared in the first few years of life, whereas in the more advanced countries the percentage of persons with antibodies rose more slowly so that many older children and young adults remained susceptible.[37] These findings provided an explanation for the paradoxical rarity of paralytic poliomyelitis in countries where crowding and primitive sanitation prevailed and the increasingly frequent epidemics of the disease in populations with the highest standards of living. In the former, polioviruses were widely disseminated, the disease remained endemic, and virtually all children had acquired infection and immunity by the age of five years. The price for this immunity was a few cases of true infantile paralysis, but by far the majority of infections were inapparent. In such a population no epidemics occurred; indeed, poliomyelitis was hardly recognized. In contrast, in Sweden, the United States, and similar economically advanced countries where children were protected from exposure and infection in early life, epidemics of mounting size and severity began to appear among the highly susceptible children and young adults.

An important serologic survey was carried out in 1948–1949 by Paul in two isolated Eskimo villages in northern Alaska. Poliomyelitis had been absent from the area for two decades, and all persons under the age of twenty were found to lack antibodies, while most of those over twenty were antibody-positive. The significance of these findings lay in the demonstration that once immunity was acquired through a single minor or symptomless infection, it persisted for years, as antibodies continued to be produced. If natural infections could produce such immunity, it was possible a vaccine could too.[38]

Another important question was whether antibodies would prevent the disease from developing. This was unlikely if the virus traveled directly from the nose along nerve pathways to the brain, as Flexner had contended and many agreed, because there would not have been time for the antibodies to act. As late as 1941 Rivers wrote that some investigators believed antibodies were "nothing more than by-products of infection and play no significant role in resistance to poliomyelitis." But the hypothesis so strongly defended by Flexner had been coming under increasing attack. Observations made during epidemics did not square with the concept of immediate direct invasion of the central nervous system. In 1905 Wickman had identified cases in which there was systemic disease with little or no involvement of the nervous system, and in 1912 a team of Swedish virologists headed by Carl Kling reported finding poliovirus in the contents and walls of the small intestines of patients. This work was generally ignored, and for some reason the observations were not followed up vigorously for another quarter-century. Other facts that did not square with Flexner's hypothesis forced a reexamination. For one thing, Medin's observation that a generalized phase of the disease preceded involvement of the nervous system was verified by Wickman in 1905. Another Swedish doctor reported in 1912 that poliomyelitis was not likely to return to a community for several years after it had been visited by an epidemic of the disease, from which he concluded that enough persons had been immunized by inapparent infections to protect the community for some time afterward. Yet these isolated findings were not enough to convince many people, and the techniques used were too crude and too costly, involving as they did a large number of expensive monkeys, to be tried often.[39]

By 1932, however, enough evidence had accumulated so that some students of poliomyelitis were ready to assume that the virus gained access by way of the digestive tract as well as the nose. Others went further. In 1941 Albert B. Sabin of Cincinnati culminated a long series of observations by himself and others on the pathology of poliomyelitis by reporting, with Robert Ward, that in a thorough search of various tissues obtained from eleven patients at autopsy, they had failed to find poliovirus in the olfactory tissues or the membranes of the nose. The virus was distributed more widely in the alimentary tract than anywhere else except the central nervous system. This pattern of virus distribution, the authors concluded, "pointed to the alimentary tract as the primary site of attack by the virus." Although not immediately accepted by everyone, this concept eventually became the basis for the effective control of poliomyelitis. Almost at the same time Howard Howe and David Bodian of Johns Hopkins showed that when chimpanzees were fed poliovirus by mouth, they became infected and occasionally developed a disease which closely resembled human poliomyelitis clinically and pathologically.[40]

The substitution of the alimentary tract as the route of entry for the virus instead of direct spread from the nose to the brain explained several formerly puzzling observations. The early febrile phase of the disease represented the period when the virus was proliferating in the throat, intestinal wall, and nearby tissues; and the antibodies found in the blood of some patients soon after they had developed symptoms of involvement of the nervous system were the product of this early phase. Thus a localized infection in the throat or intestines enabled a person to develop immunity in the course of a minor illness or in one so mild as to produce no symptoms at all. The picture was completed in 1954 when Dorothy Horstmann at Yale and Bodian and his associates independently demonstrated poliovirus in the blood of patients before the onset of symptoms attributable to the nervous system and proposed that invasion of nervous tissue occurred via the bloodstream.[41] Now there was every reason to drive ahead to develop a vaccine because it was likely that the antibodies produced by a vaccine could stop the infection before it reached the central nervous system.

To make a satisfactory vaccine, however, it was necessary to know how many types of poliovirus there were. Bodian and his associates had reported in 1949 that they could differentiate three distinct types, but no one knew how the multitude of strains isolated in various parts of the world were related to these types. The job of typing them would be time-consuming, expensive, and unappealing to investigators. Yet it had to be done. The National Foundation not only provided the funds in this case but, more significantly, directed the program. Portions of the typing program were parceled out to virologists in five universities—Bodian and Howe at Johns Hopkins, Jonas Salk at Pittsburgh, Herbert Wenner at Kansas, L. P. Gebhardt at Utah, and J. F. Kessel at Southern California—to be coordinated by the foundation. When the work was completed in 1951, it showed that the development of a vaccine would be less complex than some had supposed, for all the strains fell into the three types: 82 percent were classified as Type I, 10 percent as Type II, and 8 percent as Type III.[42]

Yet several other obstacles remained in the pathway to a vaccine. One of them was psychological, the memory of the tragic summer of 1935, when two different poliomyelitis vaccines had been tried in this country. Maurice Brodie, working in Park's laboratory, used formalin for the purpose of killing the virus. After successfully immunizing 20 monkeys with the resulting vaccine, he gave it to 3,000 children. Working independently, John A. Kolmer of Temple attempted to produce a vaccine composed of live viruses. He believed that he had sufficiently diminished the pathogenicity of a strain of poliovirus by passing it through a succession of monkeys, but to make certain, he treated the virus with a chemical, sodium ricinoleate. After testing the vaccine made with this attenuated strain in 42 monkeys and 26 persons without mishap, he gave it to a number of doctors to try. During the first half of 1935 about 20,000 children received this vaccine. It soon became apparent that all was not well: reports began to come in of poliomyelitis among vaccinated persons. James P. Leake of the Public Health Service traced down a dozen of these cases, half of them fatal, which occurred in localities where no other poliomyelitis had appeared that year and which were therefore almost certainly caused by the vaccine. When Leake revealed these facts at a meeting in November of that year, Kolmer is said to have replied that he wished the floor would open up and swallow him. As for Brodie's vaccine, it appeared to be ineffective in preventing poliomyelitis and may have caused some cases in addition. Cox, then at the Rockefeller Institute, explained these results when he tested Brodie's method of vaccinating monkeys with poliovirus. He found that if too little formalin was used, the vaccine was dangerous because live virus remained; if too much, the vaccine would not generate enough immunity to protect the animals against the disease. After this double debacle, practicing doctors and polio experts shied away from the idea of vaccinating against polio as from a rattlesnake.[43]

The odds were strongly against the success of any vaccine in the mid-thirties because techniques for cultivating poliovirus were too crude to allow for accurate measurement of dosage, much less for precise titration of the amount of a chemical needed to kill all the virus present without destroying the ability to immunize. Even if vaccines could have been successfully freed of live virus, they were grown at that time only in the nervous tissues of animals, and injections of these tissues sometimes produced serious reactions. Finally, since it was not known that there were several types of poliovirus, the use, for instance, of a Type I vaccine, followed by the appearance of cases caused by Type II among the vaccinated persons, would have led to the assumption that the vaccine had failed.

One by one the barriers to a vaccine were removed. A major breakthrough occurred in 1949, when John F. Enders, Thomas H. Weller, and Frederick C. Robbins of Harvard reported that they had successfully cultivated one strain of poliovirus in cells of nonneural tissues. Later they grew other strains also. For this achievement, which cleared the way for a poliomyelitis vaccine, they were awarded the Nobel Prize.[44]

One by-product of the typing program was that Salk, in the course of this "chore-work," became fired with the idea that a safe and effective vaccine could be produced. He had been introduced to the problems of developing a vaccine when he worked with influenza vaccines, first under Francis at Michigan and then on his own at Pittsburgh. The National Foundation responded to his enthusiasm and ability by backing his research financially and later by planning the work of other investigators so that it fed into his vaccine program. From 1950 on the foundation put more and more effort behind the development of a vaccine until eventually this became its major goal.[45]

In 1951 Hilary Koprowski and his colleagues at Lederle Laboratories immunized monkeys with a strain of poliovirus attenuated by passage in cotton rats. It was given orally to 20 volunteers without ill effects; all the vaccinees became infected, and antibodies developed in those who had had none before. Work on inactivated vaccines was also being carried on. Isabel Morgan of Johns Hopkins had shown in 1948 that antibodies could be produced in monkeys by intramuscular injections of poliovirus grown in the spinal cords of other monkeys and inactivated with formalin at low temperatures, and in 1952 Howe in the same laboratory reported that similar vaccines containing all three types of poliovirus given to ten chimpanzees and six humans were followed by good antibody response to one type and, after a booster injection, to all three types.[46]

Salk, capitalizing on the work of Enders' group and others on tissue cultures, selected monkey kidney cells as the medium in which to grow viruses for his vaccines. He prepared the vaccines by Morgan's method and gave them first to patients who had recovered from poliomyelitis and then to other persons. He was able to induce antibodies against all three types of poliovirus when they were emulsified in mineral oil, a technique he had first developed in working on influenza vaccines, but only against Type II when he eliminated the oil. In the course of his experiments, which involved 161 persons, he found no evidence that the vaccines were infectious.[47]

When Salk reported his results to the National Foundation in January 1953, wider trials in humans were requested, so Salk extended his experiments to include more than 5,000 persons. By this time pressure had built up within the National Foundation to develop a vaccine as rapidly as possible. Its sense of urgency was related to the source of its funds. To obtain money from a large segment of the public, the foundation had been able to capture and hold their interest for a time with a continuous barrage of publicity through radio, television, and the press, but after a while it became necessary to show results. The very success of the fund-raising campaign, which brought in an average of $25 million a year from 1938 to 1962, made it difficult to repeat indefinitely.[48]

Thus, the foundation's decisions tended to be made on the basis of the speed with which a vaccine could be developed. By this time the president, O'Connor, had taken over the decision making on major scientific questions as well as on fund raising and administration. When experts advocated that the vaccine be given to a still larger number of persons under careful observation before a wide-scale clinical trial was undertaken, the foundation disregarded their advice and instead appointed a new group, the Vaccine Advisory Committee, to monitor the development of an inactivated vaccine made according to Salk's methods. By the summer of 1953 the foundation and Salk were ready to make a wide-scale clinical trial, thus bypassing further preliminary trials. Joseph Bell, who was brought from the Public Health Service to conduct the trial, objected to the use of mineral oil in the vaccine because some authorities believed that it might cause cancer. He also demanded a stricter allocation of controls than the foundation wanted. As a result of these disagreements, Bell resigned. In the words of his biographer, "Propaganda prevailed over science."[49]

Fortunately the foundation was able to persuade Salk's former mentor, Francis, to supervise the trial instead, but only on his conditions: that it be carried out entirely under his direction and that double-blind controls be employed. Arrangements were made for the necessary quantities of inactivated vaccine to be prepared by two pharmaceutical companies according to Salk's specifications and for each batch to be tested for safety not only by the firm that prepared it but also by Salk's laboratory and by the National Institutes of Health. In the spring of 1954 a trial of this vaccine was conducted in 84 areas in 11 states. In the states where the foundation was already committed to injecting second grade pupils and observing first and third grade children as controls, this procedure was followed for 222,000 children who were vaccinated and for 321,000 children who were observed as controls. The incidence of poliomyelitis was 25 and 54 per 100,000 population in these two groups, respectively. In other states, over 200,000 children selected in a randomized fashion were vaccinated and a like number injected with a similar solution that contained no viruses. Poliomyelitis occurred in these groups at rates of 28 and 71 per 100,000 population, respectively. Francis and his associates concluded that the vaccination was 60 to 80 percent, or perhaps as much as 90 percent, effective against paralytic poliomyelitis. Reactions were minimal and were identical in the placebo and vaccinated groups. The vaccine appeared to be both effective and safe.[50]

These results were announced on April 12, 1955, which was also the tenth anniversary of President Roosevelt's death, in an atmosphere of scurrying reporters and popping flash bulbs. The Public Health Service immediately licensed several manufacturers to market the poliomyelitis vaccine prepared by them according to Salk's method, some of which had already been produced, and a national immunization campaign was started in which these vaccines were used. Thirteen days after the announcement, however, reports began to come in of cases of poliomyelitis in persons who had received the vaccine. Prompt and efficient investigation by epidemiologists under Alexander Langmuir at the Communicable Disease Center of the Public Health Service established that cases attributable to the vaccine were confined to persons who had received lots produced by one pharmaceutical company, the Cutter Laboratories. On April 27 all of this firm's vaccine was recalled, and on May 7 the surgeon general of the Public Health Service recommended that all injections of poliomyelitis vaccine be suspended. In less than a week it was clear that vaccines of other manufacturers were not implicated, and immunization with their vaccines was resumed. Altogether 79 cases of poliomyelitis, 61 of them paralytic, occurred in persons who had received the Cutter vaccine; an additional 105 cases appeared in family contacts of children who had been given the vaccine, of which 80 were paralytic. It was estimated that at least 10 percent of the vaccinees without adequate antibodies became infected. For a while the public and the medical profession remained wary of the vaccine. But once the methods of manufacture had been improved and the direct testing of vaccines had been instituted by a new division within the National Institutes of Health, confidence was restored.[51]

A scapegoat for the Cutter incident was sought. Some thought Salk was to blame because live viruses were found in some lots of vaccine prepared by the Cutter Laboratories, which had followed Salk's instructions exactly, and because Salk had insisted on using a potent strain of Type I poliovirus, rather than a less virulent strain, against the advice of three members of the Immunization Committee. Others blamed the Cutter Laboratories for failing to detect the live viruses, since the other commercial producers had detected the presence of live viruses after following Salk's technique and had thus been able to discard those batches of vaccine. The scientific community in general tended to put most of the blame on the National Foundation for rushing the testing program too fast and for allegedly pressuring the Public Health Service to release the vaccine prematurely.[52]

Fortunately, the vaccination program went well thereafter. No further cases of poliomyelitis occurred attributable to Salk-type vaccines. The episode of the Cutter vaccine was forgotten, and Salk became a popular hero. He had accomplished something the public could understand. Formerly there had been no vaccine, now there was one; polio seemed to be conquered. The politicians, quick to sense the public mood, responded by heaping him with honors. Salk's own protests that he had built on the work of others were lost in the hubbub.[53]

The incidence of poliomyelitis in the United States fell dramatically, from an average of 14.6 cases per 100,000 population for the period 1950 to 1954, to 1.8 cases per 100,000 for the period 1957 to 1961, representing a reduction of 87.4 percent. Yet the poliomyelitis experts and public health officials knew that the facile assumption that polio was conquered was false. Once the more alert and knowledgeable parents had brought their children to be vaccinated, the numbers of new vaccinees dwindled. Paul wrote in 1961: "In spite of all-out efforts, in many areas half the population less than forty years of age remains unvaccinated. An estimated 40,000,000 persons need vaccination." If it was difficult to persuade everyone to be vaccinated the first time, it was even harder to persuade them to return after a year or more for a booster injection. These problems were multiplied a thousandfold in developing countries.[54]

Some investigators had contended all along that a vaccine composed of inactivated viruses would not be as satisfactory as one composed of attenuated, live viruses because people would be more willing to take a vaccine by mouth and because an attenuated vaccine would produce longer lasting immunity. Live viruses multiplying in the intestines would stimulate the production of such high levels of antibodies that protection would remain for many years, as shown by Paul's observations on natural infections in Alaskan Eskimo villages. But the burning question with regard to a live vaccine was whether its safety could be assured or whether the attenuated strains might still cause paralysis in some persons or might regain their virulence for nervous tissues by spreading to non-immunized persons. This could only be determined by giving the oral vaccine to humans.

Although Koprowski had reported success in feeding an attenuated strain of poliovirus to 20 volunteers in 1951, real progress had to await the discoveries of Enders and his associates. In addition to cultivating all three types of poliovirus in nonneural tissues, they showed how to recognize their presence and to measure it quantitatively by detecting certain changes in the cells, and they were able to attenuate poliovirus by passing it through a series of cell cultures. These observations raised the hopes of those interested in an oral vaccine, and before long a spirited contest to be the first to develop a safe, effective vaccine was under way among Koprowski, now at the Wistar Institute in Philadelphia, Sabin, and Cox, who had now taken personal charge of Lederle's poliomyelitis team. By 1959 millions had received oral vaccines prepared from strains developed by one or another of these groups. Since so many persons had already been immunized in the United States and northern Europe, the large trials were conducted mostly in Russia, Czechoslovakia, Poland, the Belgian Congo, and Latin America. The attenuated viruses were found to multiply in the alimentary tracts of those who received the vaccines and to be capable of causing a symptomless infection in other persons in the same household and occasionally in outside contacts. A slight increase in the virulence of some strains occurred, although there was not a return to the virulence of naturally occurring strains.[55]

Sabin's vaccine was the most thoroughly tested. When approximately 11 million children had received oral vaccines prepared from Sabin's strains without any evidence that paralysis had been induced in a single person, these strains were accepted for general use. In 1961 the Public Health Service licensed the first of the Sabin vaccines. Because some cases of poliomyelitis occurred in Florida and West Berlin after oral vaccine prepared from the Lederle strains was given, Cox and his group discontinued work on them, and the Lederle Laboratories decided to use Sabin's strains, thus recouping little or nothing of the $10 million they had expended on poliomyelitis research.[56]

Once the Sabin-type oral vaccine was licensed, it was used widely in the United States. By 1964, 100 million doses of vaccine of each of the three types of poliovirus had been distributed. But this vaccine also had its misfortunes. In 1962 several cases of paralytic poliomyelitis occurred that were considered to be "compatible with the possibility of having been induced by the vaccine." In the next two years 57 other cases were observed. Advisers to the surgeon general stressed that even if every one of these cases had been caused by the vaccine, the risk was so low that its use should not be discontinued. They recommended that live vaccines be restricted to persons under the age of eighteen years in whom the risk was less and the need greater than in older persons. During the years 1965 through 1967 over 70 million doses were given, while only eight cases of poliomyelitis occurred that might have been caused by the vaccine, or 0.11 cases per million doses administered. Epidemiologists from the Communicable Disease Center concluded that oral poliomyelitis vaccine presented only a "minor public health problem." At present no one is entirely satisfied with the situation. The majority of experts apparently believe that in order to obtain universal or nearly universal immunization in the United States, one must accept a minimal risk.[57]

The widespread use of the vaccines has practically eliminated paralytic poliomyelitis from the United States. In 1974 only seven cases were reported in the entire country. Results have been similar in other countries where vaccines have been used extensively. The Sabin-type vaccine has other advantages over the Salk-type besides being given by mouth. It apparently produces a greater and more lasting immunity because the multiplication of attenuated viruses in the intestines results in local immunity, reducing the chance of reinfection with naturally acquired viruses. It can be used to control an epidemic already in progress since it can be given to many persons within a short period of time. For these reasons it is used more widely in this country and abroad than Salk-type vaccines, although Sweden, using the Salk-type alone, has achieved results similar to those obtained in countries where the oral vaccine is used.[58]

The conquest of poliomyelitis had all the elements of an American success story: the focus on an objective so simple in concept and so obviously beneficial to mankind as to resemble the quest for the holy grail; the enlistment of all groups and classes in the crusade; the use of the mass media for publicity; the appeal to sentiment; the thorough organization of the campaign; the unwillingness to leave the final decisions on research support in the hands of the experts; and the popularization of a hero as David, the killer of the giant. Yet sweet as was the victory, it left a slightly bitter taste in many mouths. The obvious scrambling of several participants to be the first to win the race and the intemperate attacks by some who took up the cudgels on behalf of one or another of the chief actors smacked of the arena rather than the scientific laboratory. Perhaps this is the American way to success; if so, one can only hope that we will eventually learn to concentrate more on conquering nature than on outshining each other.

Scientific discovery has its own momentum, as does a series of wins in football: one victory predisposes to another. The concepts and techniques learned in devising vaccines for influenza and poliomyelitis were quickly applied in other viral infections. One of these was measles, a highly contagious disease, seen in the United States mainly in children, but capable of infecting adults or sweeping through entire populations when there is no immunity. Although its most prominent feature is a generalized rash, inflammation of the upper respiratory tract and the conjunctivae is always present. Sometimes the infection goes deeper into the respiratory tract, and serious results follow. Pneumonia may be caused by measles virus, but more often by bacteria that take advantage of the break in the body's defenses made by the virus. A rarer complication is encephalomyelitis, inflammation of the brain and spinal cord.

Although measles was described two thousand years ago, it was not clearly separated from smallpox and scarlet fever until the seventeenth century. The confusion probably arose because measles tended to be more severe in the past and was sometimes characterized by a hemorrhagic rash which was difficult to distinguish from similar rashes that accompanied other infections. Early attempts to transmit measles with filtered secretions were not always successful, but eventually in 1921 Blake and Trask conclusively demonstrated that measles could be transmitted by bacteria-free secretions, and in 1938 the virus was first cultivated in tissue culture.[59]

Impressed by the immunity induced by smallpox inoculations, an Edinburgh physician, Francis Home, in 1758 injected blood taken from measles patients at the height of the fever into children with the object of producing the disease and rendering them immune. Of 12 children inoculated ten developed mild measles, and at least one of the other two had already had the disease. Because some doctors were unable to reproduce Home's results, however, the procedure was never widely used. A few children were immunized in the 1940s with a vaccine made from measles virus grown in chick embryos, but this method was unsatisfactory because the quantity of virus present could not be easily measured. The difficulty was overcome by Enders and Thomas C. Peebles, who reported in 1954 that measles virus produced visible changes in the cells when grown in tissues from human or monkey kidneys. These cytopathogenic effects, as they are called, made it possible to measure the quantity of virus present in a culture and thus to standardize the dose of a vaccine. In 1963 a vaccine composed of live attenuated virus was licensed in the United States. Widespread use of this vaccine reduced the number of reported cases in the United States from nearly 482,000 in 1962 to a low of 22,000 in 1968, but then the number rose again, reaching 75,000 in 1971. The reason was not hard to find, for the quantity of vaccine distributed in the United States had increased progressively through 1966 and then had fallen each year through 1970. Surveys showed that the percentage of children actually receiving the vaccine began to drop after mid-1969. The immunization program tended to falter particularly in the low income groups. Some blamed the resurgence of measles on the curtailing in 1969 of federal funds for immunization programs. But whatever the reason, more efforts were obviously needed to immunize those who were not protected. Health departments renewed their immunization campaigns, with considerable success. Once more the price of health had proved to be eternal vigilance.[60]

Soon after measles vaccine appeared, a vaccine for another common childhood infection followed. Rubella resembles measles except that the rash is pinker, more variable, and more fleeting, and the general symptoms are milder. For a long time it was dismissed as a trivial infection. Its true significance was revealed by the brilliant detective work of an Australian ophthalmologist, Norman Gregg, who suspected a connection between cataracts in newborn infants and rubella in their mothers during pregnancy. In 1941 he reported 68 cases in which cataracts developed in infants whose mothers had contracted rubella when pregnant, and he pointed out that some of these infants also had a low birth weight or defects in the structures of the heart. Observations by others indicated an ever wider scope for this "trivial" disease, but not until investigators at Harvard and Walter Reed isolated rubella virus in tissue cultures could all the complications of rubella be determined.[61]

Like measles, rubella in the United States most often affects young schoolchildren, but unlike measles it also attacks many persons between the ages of fifteen and thirty-five. While it causes little discomfort in children, the fever and general symptoms tend to be more severe in adults. In women frequently, and in men occasionally, rubella is complicated by arthritis, usually transient but sometimes lasting for weeks or months. Other complications are encephalitis and diminution in the number of blood platelets, the small discs that function especially in bringing about coagulation. Serious complications are found in fetuses, where malformation or malfunction of almost every organ can occur if the maternal infection occurs early in pregnancy when various fetal organs are undergoing development. The most important disabilities are malformations of the heart and large blood vessels, deafness, and defects of the eyes and the brain. Once rubella virus had been cultured, it was isolated from the tissues of infected fetuses and from the throats and particularly the eyes of infected children for months or even years after birth.[62]

As the tragic tale unfolded, the pressure for a method of prevention grew. In 1966 virologists at the National Institutes of Health reported that they had prepared a live, attenuated vaccine, and three years later the first vaccine was licensed for distribution. By 1971 nearly 20 million children had been vaccinated. Whereas in 1966, the first year that rubella was reportable in the United States, nearly 47,000 cases were recorded, in 1974 only about 12,000 cases were reported.[63]

Another viral disease for which a successful vaccine was recently developed is mumps, which had been recognized by Hippocrates as a mild epidemic illness characterized by swellings in front of the ears and sometimes enlargement of one or both testes. Doctors later identified the facial swellings as inflammation of the parotid and other salivary glands and learned that internal organs, such as the ovaries, pancreas, and nervous system, could be affected also. Although mumps is generally a benign disease, protection against it is desirable because sterility may rarely result from involvement of the testes. Using techniques similar to those found successful in other viral diseases, investigators first developed a vaccine composed of inactivated virus. But because the immunity produced was short-lived, other virologists concentrated on a vaccine made from live, attenuated viruses. Immunity produced by this vaccine has been shown to persist for four years and gives promise of lasting many years, perhaps a lifetime. Though it has not been assigned as high a priority as measles and rubella vaccines because the disease it protects against is not so life-threatening, mumps vaccine is currently recommended as part of the regular immunization schedule of children in this country.[64]

Influenza, poliomyelitis, measles, and rubella may have tragic sequels, but for sheer annoyance, nothing excels the familiar infections of the upper respiratory tract that are generally labeled the common cold. The symptoms include nasal discharge and obstruction, sneezing, dryness or soreness of the throat, cough, chilliness, headache, and general aching, or combinations of these. The uncomplicated cold lasts less than a week, although the illness may be prolonged by inflammation of the sinuses, middle ears, bronchi, or lungs. Preschool children have as many as six to twelve colds per year, older children somewhat fewer, and adults an average of two or three.[65]

Colds were first transmitted by filtered secretions in 1914 by Walther Kruse of Leipzig. Attempts to isolate the viruses that cause infections of the upper respiratory tract met with many failures and finally with some successes. In 1953 Wallace P. Rowe, Robert J. Huebner, and their associates at the National Institutes of Health isolated a virus from human adenoids which became the first of a group eventually named the adenoviruses. Various members of this group cause illnesses such as sore throat, a mild influenzalike infection, conjunctivitis, and pneumonia. Although they sometimes produce illnesses that resemble the common cold, it soon became clear that they are not a major cause of colds.[66]

In 1957 another group of virologists, headed by Robert M. Chanock at the National Institutes of Health and Robert H. Parrott at the Children's Hospital of the District of Columbia, isolated two other viruses from children with respiratory infections. Eventually named parainfluenza and respiratory syncytial viruses, they were found to cause a few of the upper respiratory infections of children and adults. Their main role, however, was as incitants of frequently serious infections of the larynx, bronchial tubes, and lungs in children. Later other agents, including the influenza virus and *Mycoplasma pneumoniae*, were identified as causing a few upper respiratory infections in both children and adults. Yet all of these together, while responsible for illnesses in other parts of the respiratory tract and elsewhere in the body, account for only a minority of common colds in adults.[67]

In 1954 Winston H. Price at Johns Hopkins isolated a virus that causes typical colds in adults. This proved to be the first of a group now named the rhinoviruses, from the Greek word for nose. In 1960 D. A. J. Tyrrell and his associates in the Common Cold Research Unit established by Andrewes in Salisbury, England, discovered that rhinoviruses grow best in an acid medium and at a lower temperature than is used to cultivate most microorganisms and thus opened up the field for intensive research. It now appears that there are at least 60 different antigenic types of rhinoviruses, and when these are added to the other viruses that can cause the common cold, there are over 100 altogether.[68]

Before the various viruses were identified, the reason that persons suffered from one cold after another was a puzzle. Some evidence that immunity played a part was gleaned from observations of isolated communities where colds tended to die out and then explode in epidemics when contact was made with persons from the outside. In 1959 George Jackson and I, working at the University of Illinois, reported that among volunteers who had been inoculated with secretions obtained from patients with the common cold, only 9 percent developed a cold if they had been given the same secretion before, whereas among those who had been given a different secretion the first time, 43 percent developed colds. When it became possible to cultivate the viruses that cause colds, these results were in general verified. Antibodies appeared in the blood following natural or experimental colds and remained for several months or more. In most instances these antibodies appeared to protect against subsequent infections by the same virus. This finding raised hopes that vaccines could be devised to protect against colds, but as more viruses were isolated from persons with colds and many different antigenic types were identified, these hopes dimmed. Experience has shown that when too many different bacteria are included in a vaccine, the immunity generated is poor. This is apparently also true for viruses. Thus it appears impossible with present techniques to immunize against all or even most of the viruses that cause colds. This is the problem with which investigators are now wrestling.[69]

By contrast, viral hepatitis, or inflammation of the liver, appears to be caused by only a few different viruses and thus lends itself more readily to control by vaccines. Epidemics of jaundice, which were observed since Greek and Roman times, were the first clues to the existence of such a disease, but an understanding of its course and character did not come until bacteriological techniques were developed. In 1896 certain cases of jaundice with involvement of the kidneys as well as the liver were found to be caused by a spiral-shaped bacterium, *Leptospira icterohaemorrhagiae*. When this entity, called Weil's disease, was separated off, there remained a distinct group of illnesses in which jaundice was associated with nausea, abdominal pains, swelling of the liver, and general symptoms of malaise, fever, and headache. Careful studies of epidemics and experiments in volunteers in the 1940s and 1950s demonstrated that hepatitis often occurs without jaundice; especially in children the disease is frequently mild and sometimes without symptoms altogether.[70]

Hepatitis is a significant illness in the United States. In the decade from 1964 through 1973 an average of over 50,000 cases were reported each year, and on the basis of known underreporting it is estimated that about 300,000 persons may actually develop the disease each year. More than 8,000 deaths were recorded during these ten years, and this may be only about one-fourth of the actual number.[71]

Although the occurrence of epidemics made it likely that the disease was infectious, proof was a long time coming. Intensive research over three decades, mainly by British and American investigators, differentiated two varieties: hepatitis A, previously called infectious hepatitis, and hepatitis B, formerly called serum hepatitis. The first is almost always transmitted by the fecal-oral route and the latter by the injection of infected blood, serum, or other blood derivatives, although Type A can follow the injection of infected blood and Type B can be transmitted by the oral route. Another distinguishing characteristic is the length of the incubation period, which is usually 15 to 40 days in hepatitis A and 60 to 160 days in hepatitis B.[72]

In 1967 Baruch Blumberg and his associates at the Philadelphia Institute for Cancer Research detected by serological methods an antigen in the serums of patients with Type B hepatitis. Later this antigen was seen by electron microscopy to consist of small particles which are believed to be viruses or portions of viruses. In the same year Friedrich Deinhardt and his associates at the University of Illinois reported that hepatitis A could be transmitted to marmosets by inoculation of serum from patients with that disease. Later they infected marmosets orally as well as by injection and with feces from infected patients as well as with infected serum. Virus particles were seen in the infected serum, feces, and liver tissue. Thus, it appears that the viruses that cause both types have been identified.[73]

Prevention of hepatitis A up to the present has stressed the avoidance of contaminated water and food, while to prevent hepatitis B, doctors have recommended careful sterilization of syringes, needles, and surgical instruments and the rejection of prospective blood donors who have recently had hepatitis. Persons exposed to hepatitis A have been injected with gamma globulin, pooled from several healthy donors to ensure the presence of antibodies. If this is given early enough, the ensuing disease will be mild, although it is questionable whether it can be prevented altogether. Recent investigations indicate that gamma globulin will give similar results in persons exposed to infection with hepatitis B. But gamma globulin is expensive and is not always effective; vaccination would obviously be a much better method of protection. The recent rapid progress in characterizing and identifying viruses causing hepatitis makes it likely that within a few years' time both hepatitis A and B will be added to the growing list of viral diseases that can be prevented by vaccination.[74]

Early attempts to immunize against bacterial infections with vaccines composed of whole bacteria had been fairly successful in preventing typhoid fever, whereas vaccines composed of whole pneumococci, meningococci, or other bacteria had given equivocal results at best. Although interest in bacterial vaccines declined after effective therapeutic agents were introduced, a fundamental discovery made at the Rockefeller Institute opened the way for a new concept in vaccination. In the early 1920s Avery began trying to separate out some of the chemical fractions of the pneumococcus, but without success. One day, after producing a gummy mess, he walked into the laboratory of Michael Heidelberger, a chemist, and shook the tube containing the incomprehensible mixture under his nose, saying, "See this, Michael. If you can find out what this is, we'll learn a great deal about pneumonia." Heidelberger told him to go away because he was busy with another problem, but later, when Avery repeated the performance, he capitulated. An important result of their collaboration was the extraction of a complex carbohydrate from the capsules of each type of pneumococcus studied. These carbohydrates were slightly different in chemical structure for each type; in other words, they determined the specificity of the type of pneumococcus.[75]

Hoping to put this discovery to practical use, investigators at the Rockefeller Institute tried repeatedly to immunize rabbits with these capsular carbohydrates but were unable to do so, although German bacteriologists had some success in using them to immunize mice. In 1930 Tillett and Francis, then at the Rockefeller Institute, found that patients convalescing from pneumonia, when injected with these substances, developed antibodies against the corresponding type of pneumococcus. They were loath to attribute these antibodies to the injection of the carbohydrates, but after Finland and Whelan Sutliff of the Boston City Hospital showed that antibodies were also produced in persons who had not had pneumonia, it became clear that these substances did have immunizing powers. The carbohydrate extracted from each individual type of pneumococcus was found to produce immunity against its specific type. Since extraneous substances had been removed in the process of extraction, adverse reactions were minimal. More important, only a fraction of a milligram was needed to produce immunity, which meant that carbohydrates derived from several types of pneumococci could be combined in a single injection.[76]

Among the several trials of immunization with capsular carbohydrates in which large groups of people were injected, two tests in the 1940s stand out because they were rigidly controlled and were therefore definitive in their results. The first involved the immunization of 5,750 older persons with the capsular carbohydrates from Types I, II, and III pneumococci, and the simultaneous observation of 5,153 controls of the same ages. During the six years of the study pneumonia occurred in 44 percent of the controls and in only 17 percent of those vaccinated. Among the controls 19 percent died from pneumonia, as compared with 6 percent of those vaccinated. In the second study capsular carbohydrates from Types I, II, V, and VII pneumococci were injected into 8,586 soldiers attending a technical school where high rates of pneumonia had been prevalent. During the ensuing six months four cases of pneumonia caused by these types of pneumococci occurred in this group, as compared with 26 cases among the 8,449 soldiers who served as controls.[77]

As a result of these and other investigations, there has been a reawakening of interest in vaccination of susceptible persons against pneumonia, particularly the aged and those with chronic diseases. Robert Austrian and Jerome Gold compiled additional data which supported the need for immunization. They found that in Kings County Hospital in Brooklyn over one-third of the patients with pneumonia who had pneumococci in their blood died within twenty-four hours of admission to the hospital, which is often too soon for even the best of medicines to effect a cure. The only way to have saved these persons' lives would have been to immunize them so that the pneumonias would have been milder or better still prevented altogether. Convinced that immunization would lower the mortality rates from pneumonia, the National Institutes of Health has in recent years been financing the production of vaccines made from the capsular carbohydrates.[78]

Like pneumococcal pneumonia, meningococcal meningitis appears to be well controlled by sulfonamides and antibiotics. Yet some deaths still occur. While most patients recover following proper treatment, the disease sometimes progresses so rapidly that the patient is dead before he can receive a sufficient amount of a curative drug or even before the diagnosis can be made. Prevention is the only defense against such catastrophes. Since epidemics among military recruits have been a continuing problem, Malcolm Artenstein and his associates at Walter Reed developed from the carbohydrate antigen derived from the capsule of Group C meningococci a vaccine which produced high levels of antibody in volunteers. Among 28,000 recruits injected with this vaccine, only two developed Group C meningococcal infections, while 72 cases occurred among 115,000 controls who were not vaccinated. Since 1971 this vaccine has been given to recruits in all three military services, and concomitantly there has been a fall in the number of cases of meningitis caused by Group C meningococci. After promising results were also obtained with a Group A vaccine, both vaccines were recommended for the control of epidemics.[79]

Success in immunizing with antigenic fractions of pneumococci and meningococci spurred interest in vaccines composed of similar fractions derived from other bacteria, such as the gonococcus, *Hemophilus influenzae,* and the dysentery bacilli. It is likely that several vaccines of this kind will be perfected in the next decade or two.

Vaccination today is a necessary appendage to civilized life. Instead of allowing diseases to smolder in or sweep through a population so that the survivors may be immune, communities seek to clean up the environment, thereby diminishing the chance of natural infection and at the same time the number of persons who are immune. Accordingly, they must strive for universal immunization by inoculation or ingestion of vaccines. Like other systems in our complex world, such as telecommunication, air transportation, or the sanitation of metropolises, protection through wholesale immunization requires sophisticated technology, careful surveillance, and proper regulation, but when these are present, the method works well. The future will undoubtedly see more vaccines added and prevention by this method extended to more people both in the United States and abroad.

13. *The Continuing War*

In the first three-quarters of the twentieth century more specific remedies and preventives were devised for the control of infectious diseases than during the entire history of mankind before that time. The health professions acquired the knowledge and skills and developed the organization to use these remedies and preventive measures. What were the results, and what more remains to be done?

The effectiveness of a chemical or biological agent in preventing a disease or, once illness has begun, in shortening the course, decreasing complications, and saving lives can be determined accurately if enough patients are observed, particularly if they are compared with properly chosen controls. Yet "there is many a slip" between adequate knowledge and its proper use. As an example, a patient with pneumonia may sicken and die without seeing a doctor, or he may report to a doctor after it is too late for an antibiotic to help him. The doctor, in turn, may fail to diagnose the pneumonia, may make another diagnosis through carelessness or ignorance, or, having diagnosed the pneumonia correctly, may prescribe the wrong medication. Even if the right remedy is prescribed in the proper dosage, some member of the health team may make a mistake in dispensing or administering it. Finally, the patient may not receive the remedy because he is unable to pay for it or because he refuses to be treated, or he may fail to take the medication in the proper dose or for the prescribed period of time. Similar obstacles may interfere with the effective use of a vaccine or drug for preventing an infection. In addition, other factors may affect the number or the severity of infections. An epidemic of influenza or a rise in the proportion of elderly persons in a community may increase the number of cases of pneumonia. Mobilization of troops may skyrocket the incidence of gonorrhea.

Thus the results of a particular therapeutic or prophylactic measure observed in a test situation cannot be applied to the nation as a whole. Despite the demonstration that the case fatality rate for pneumococcal pneumonia in particular hospitals fell from approximately 30 percent before any specific measures were available to approximately 5 percent in patients treated with penicillin, it would be unrealistic to expect the mortality rate from pneumonia in the United States to fall to one-sixth of its previous level following penicillin therapy.

Finally, no matter how carefully data on the causes of death are collected and tabulated, their main weakness is at their source. The doctor who fills out a death certificate may be mistaken as to the cause of death or may be unfamiliar with the system of terminology; in either case the cause of death will be incorrectly tabulated. Furthermore, as medical knowledge increases, it becomes necessary to change the classification of certain diseases, so that diseases listed in a certain category at one time may later be placed under another. Yet despite these limitations, and especially if allowance is made for them, mortality rates are a good measure of progress in the control of infectious diseases, and if properly analyzed, they can sometimes reveal the effect of a particular therapeutic or prophylactic measure.

A comparison of the ten infectious diseases that caused the most deaths in 1900 with the same diseases in 1970 shows a dramatic decline in the death rate in every case (Table 1). In fact, six of them caused less than one death per million population in 1970. Mortality rates for several other important infections dropped correspondingly. Syphilis accounted for 2.7 deaths per 100,000 population in 1900, compared with 0.2 per 100,000 in 1970; appendicitis and its complications caused 8.8 deaths per 100,000 in 1900 and only 0.7 per 100,000 in 1970; while the mortality rate for tetanus, which was 2.4 per 100,000 in 1900, was only .04 in 1970.[1]

Among the ten leading causes of death in 1900, four were infectious, and three of these led all the rest—pneumonia, tuberculosis, and diarrhea—while diphtheria was tenth on the list. In 1970 only one infectious disease, pneumonia, was among the ten leading causes of death, and it was fifth, causing 30.9 deaths per 100,000. Two other infections were seventeenth and twentieth among the top twenty causes of deaths in 1970: infections of the kidney, which were responsible for 4.0 deaths per 100,000, and tuberculosis, which accounted for 2.6 per 100,000.[2]

While it is clear that infectious diseases caused far fewer deaths in 1970 than in 1900, it is difficult to demonstrate the effects of specific remedies or prophylactic measures on this decline. The reason for the uncertainty is that death rates for a number of infections were falling for many years before 1900. Some diseases seem to have

Table 1 Death Rates for Common Infectious Diseases in the United States in 1900, 1935, and 1970

	Mortality rate per 100,000 population		
	1900	1935	1970
Influenza and pneumonia	202.2	103.9	30.9
Tuberculosis	194.4	55.1	2.6
Gastroenteritis	142.7	14.1	1.3
Diphtheria	40.3	3.1	0.0[a]
Typhoid fever	31.3	2.7	0.0
Measles	13.3	3.1	0.0
Dysentery	12.0	1.9	0.0
Whooping cough	12.0	3.7	0.0
Scarlet fever (including streptococcal sore throat)	9.6	2.1	0.0
Meningococcal infections	6.8	2.1	0.3

Sources: Carl C. Dauer, Robert F. Korns, and Leonard M. Schuman, *Infectious Diseases*, American Public Health Association, Vital and Health Statistics Monographs (Cambridge: Harvard University Press, 1968), p. 1; Robert D. Grove and Alice M. Hetzel, *Vital Statistics Rates in the United States, 1940–1960* (Washington, D.C.: U.S. Department of Health, Education, and Welfare, Public Health Service, National Center for Health Statistics, 1968), pp. 578–587; National Center for Health Statistics, *Facts of Life and Death*, DHEW Publication no. (HRA) 74–1222 (Rockville, Md.: U.S. Department of Health, Education, and Welfare, Public Health Service, Health Resources Administration, 1974), pp. 15, 39.

[a] 0.0 = quantity more than 0 but less than 0.05.

been cyclic in prevalence or virulence. For instance, mortality records of an isolated town in Romania, available from 1607 to the present, indicate that measles was severe during three periods: up to 1668, from about 1800 to 1873, and again around 1916. In the intervals the disease was mild. Likewise, serious epidemics of scarlet fever occurred in the early nineteenth century and again in the early twentieth century. Figures from England and Wales and from Massachusetts show decreasing death rates from tuberculosis, typhoid, scarlet fever, and diphtheria in the latter half of the nineteenth century. Some observers attribute the declining mortality to improvements in the standard of living, which would act to diminish crowded living and working conditions and to improve nutrition and personal hygiene. Yet these cannot account for cyclic periods of severity, as in measles. Thus some, and perhaps even a large part, of the variation in mortality rates must be attributed to changes in host-parasite relationships, such as an increase or decrease in the virulence of the microorganism or in the immunity of a population, or both.[3] The death rates for certain infections in the United States after 1900 should be examined in the light of these long-term changes.

Mortality rates from typhoid declined almost continuously from 1900 on (Fig. 2), and data from Massachusetts show that the decline began in the 1870s.[4] Since the half-century following 1870 corresponded with the major thrust to provide pure water and milk and proper disposal of sewage, these factors must have contributed to the fall in the death rate. In addition, better personal hygiene and improvements in the preparation and refrigeration of food un-

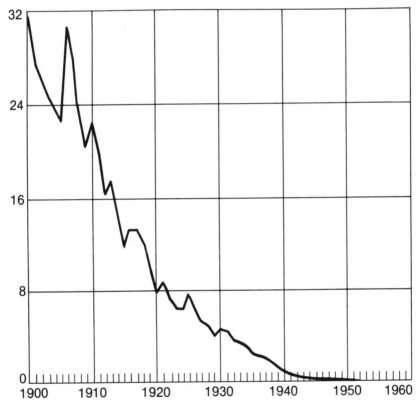

Figure 2 Death rates for typhoid fever in death registration states, 1900–1932, and United States, 1933–1960 (per 100,000 population).

Source: Robert D. Grove and Alice M. Hetzel, *Vital Statistics Rates in the United States, 1940–1960* (Washington, D.C.: U.S. Department of Health, Education, and Welfare, Public Health Service, National Center for Health Statistics, 1968), p. 82.

doubtedly played a part. Whether changes in host-parasite relationships also took place cannot be determined.

Diphtheria mortality presents more of a problem. Data from Massachusetts, Michigan, and New York show a gradual decline in mortality rates from about 1880 on, followed by a steep drop in both morbidity and mortality beginning around 1925. Since the case fatality rate decreased progressively in all three states from about 1890 to about 1930, part of the early decline in mortality has been attributed to antitoxin and the subsequent rapid fall to immunization (Fig. 3). This concept seems to be supported by data which show that the decline in both incidence and mortality was greater in the north than in the south and that these declines paralleled the percentage of children immunized against diphtheria in each region. Furthermore, an abrupt drop in mortality occurred in England and Wales after 1940, when widespread immunization was introduced.[5]

This is not the only interpretation that can be made, however. When the mortality rate in the United States as a whole is considered (Fig. 4), a rapid fall is seen to have occurred between 1922 and 1925. Some have claimed that this decline was the result of widespread immunization. Yet it may have been merely the expected fall in a cycle of higher mortality that had reached its peak in 1921, for if this and previous peaks are disregarded, the decline in mortality appears to have been continuous from 1900 to 1940 in the United States and from 1878 on in Massachusetts. More recent studies have also shown that the parallels between the proportion of children immunized and the decline in mortality do not hold for every region in the United States. At the same time, additional support for the positive effect of immunization upon mortality has come from observations during epidemics that immunized persons usually develop no illness or only a mild one and that, in general, in those who do become ill, the higher the quantity of antitoxin in the blood, the milder the illness.[6]

Since controlled studies were not done when diphtheria antitoxin and immunization were introduced, it is impossible to be sure of the precise effect of these procedures on mortality rates. Apparently some undetermined factors, probably those resulting from improvement in the standard of living plus those caused by changes in the host-parasite relationship, operated to lower the morbidity and mortality rates from diphtheria over the past century. In addition, immunization and, to a lesser extent, therapy with antitoxin must have

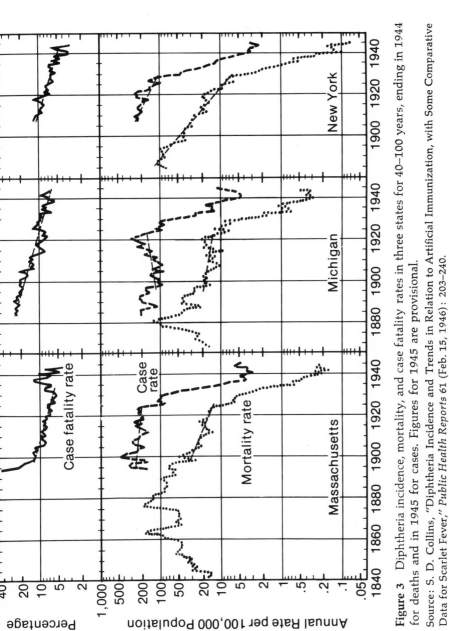

Figure 3 Diphtheria incidence, mortality, and case fatality rates in three states for 40–100 years, ending in 1944 for deaths and in 1945 for cases. Figures for 1945 are provisional.

Source: S. D. Collins, "Diphtheria Incidence and Trends in Relation to Artificial Immunization, with Some Comparative Data for Scarlet Fever," *Public Health Reports* 61 (Feb. 15, 1946): 203–240.

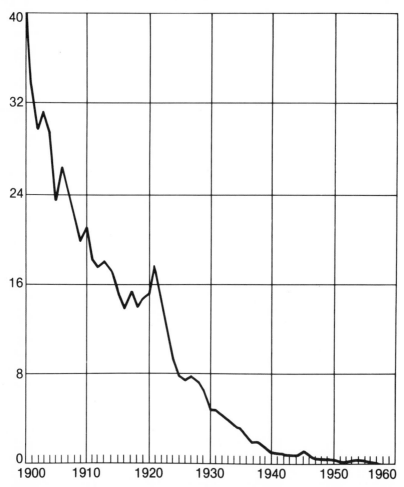

Figure 4 Death rates for diphtheria in death registration states, 1900–1932, and United States, 1933–1960 (per 100,000 population).

Source: Robert D. Grove and Alice M. Hetzel, *Vital Statistics Rates in the United States, 1940–1960* (Washington, D.C.: U.S. Department of Health, Education, and Welfare, Public Health Service, National Center for Health Statistics, 1968), p. 84.

had some effect on death rates, although how much cannot be determined.

The picture for tuberculosis is much the same. The mortality rate in the United States declined progressively from 1900 to the mid-fifties (Fig. 5), and figures for Massachusetts, available from 1861 on, show that this fall was the continuation of a long-term trend.

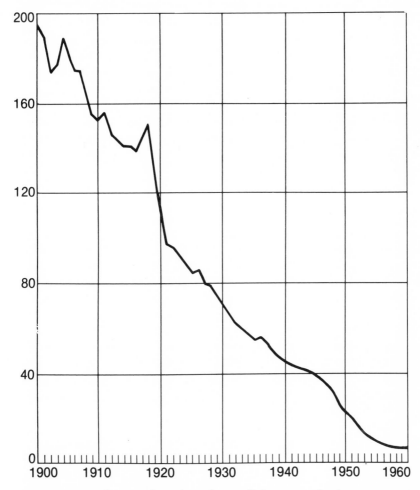

Figure 5 Death rates for tuberculosis, all forms, in death registration states, 1900–1932, and United States, 1933–1960 (per 100,000 population). Source: Robert D. Grove and Alice M. Hetzel, *Vital Statistics Rates in the United States, 1940–1960* (Washington, D.C.: U.S. Department of Health, Education, and Welfare, Public Health Service, National Center for Health Statistics, 1968), p. 80.

In England and Wales the decline had begun by 1838 and probably before that.[7] This persistent fall in the death rate can be attributed to improvements in the standard of living and in personal hygiene, to a favorable change in the relationship of the human host to the tubercle bacillus, and to a lesser extent to specific procedures aimed at the prevention of the disease. The sharper drop after 1945 may represent merely the accelerated decline following the adverse effect of World War II on the mortality rate, or it may indicate in part the effect of the introduction of streptomycin, followed by isoniazid and other specific curative agents. As in diphtheria, the effect of the use of specific agents cannot be proved by data on the mortality of the total population.

The adoption of rest, sanitarium treatment, and collapse therapy as standard procedures for the therapy of tuberculosis took place gradually between the early years of this century and the 1940s, and they were not taken up with the same alacrity nor utilized to the same extent in the various parts of the country. These procedures could not therefore be expected to affect the national mortality rate all at once; instead, their effect would have been perceived gradually, over a period of decades. Another reason that the effect of a therapeutic measure on tuberculosis mortality could be expected to be gradual is that the disease tends to be chronic, sometimes extremely so. For instance, a person may die of the disease forty or fifty years after his first symptoms. Among persons with the more chronic forms of the disease, therefore, the influence of a specific therapeutic measure would not show up in the death rate until years or decades after it was first employed. Thus the influence of therapeutic agents upon tuberculosis mortality would tend to be gradual, and such gradual effects might well merge into a previous downward trend so as to be indiscernible, particularly when large numbers of persons are considered.

Data from smaller areas are more helpful. In New York City, for instance, the mortality rate fell irregularly from 49 per 100,000 population in 1941 to 40 per 100,000 in 1947 and 1948, then fell sharply, reaching 27 per 100,000 in 1951. Another rapid drop began in 1952, which resulted in a mortality rate of 13 per 100,000 by 1955 (Fig. 6). The first rapid fall in mortality coincided with the introduction of streptomycin and the second with isoniazid. Evidence of the effectiveness of therapeutic and prophylactic agents must come from studies in localized areas such as this and from controlled studies made when the various agents are introduced. In any event, as a result of the operation of various factors, known and unknown, tuberculosis mortality in the registration area of the United States, which was 194.4 per 100,000 population in 1900 and 39.9 in 1945, reached 2.6 per 100,000 in 1970.[8]

Mortality rates from certain other diseases lend themselves more readily to an interpretation of the effect of individual measures.

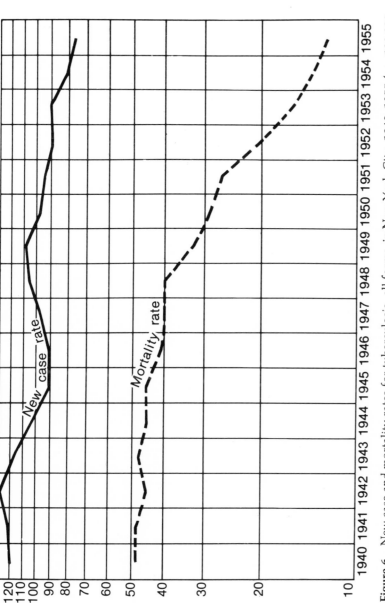

Figure 6 New case and mortality rates for tuberculosis, all forms, in New York City, 1940–1955 (per 100,000 population).

Source: Anthony M. Lowell, *Tuberculosis in New York City, 1955: Prevalence of Disease in an Urban Environment* (New York: New York Tuberculosis and Health Association, 1956), following p. 5.

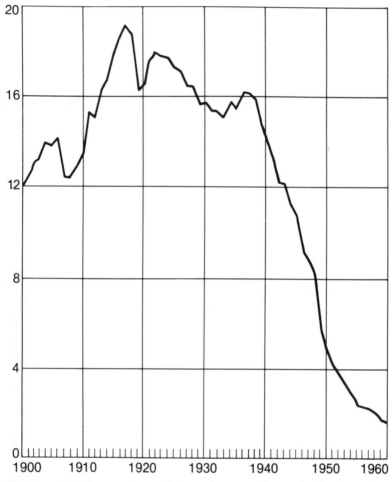

Figure 7 Death rates for syphilis and its sequelae in death registration states, 1900–1932, and United States, 1933–1960 (per 100,000 population). Source: Robert D. Grove and Alice M. Hetzel, *Vital Statistics Rates in the United States, 1940–1960* (Washington, D.C.: U.S. Department of Health, Education, and Welfare, Public Health Service, National Center for Health Statistics, 1968), p. 81.

Syphilis in the twentieth century is much less malignant than it was in the fifteenth, and its severity may still have been decreasing in the decades immediately before the introduction of arsphenamine in 1910, since the incidence of syphilis in eleven of the major armies of the world showed a decrease between 1870 and 1913. Yet in the United States the death rate from syphilis actually rose from 1900

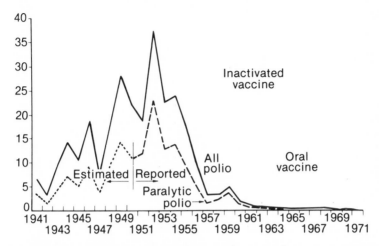

Figure 8 Case rates for poliomyelitis in the United States, 1941–1971 (per 100,000 population).

Source: U.S. Department of Health, Education, and Welfare, Public Health Service, Health Services and Mental Health Administration, *Immunization against Disease—1972* (Atlanta: Center for Disease Control, 1972), p. 34.

to 1918 and decreased only slightly for the two decades after 1918 (Fig. 7). The abrupt drop after that coincided with the public health campaigns to find contacts and to treat patients and contacts adequately with the arsenicals and later with penicillin.[9]

The incidence of paralytic poliomyelitis apparently began to increase in the United States and Europe toward the end of the nineteenth century, and widespread outbreaks continued until the epidemic that peaked in 1952 (Fig. 8). After inactivated vaccine was introduced in 1955 the number of cases decreased dramatically and has remained low since that time. There seems to be no question that this decrease is due to the use of vaccine.

Whatever the relative effects of the various causes, the mortality rate for Americans fell rapidly and almost continuously from 1900 to 1954 after which it tended to level off (Fig. 9). When the ratios are adjusted for changes in age composition of the population, the decline is more obvious. The decrease in the death rate is reflected in the increase in life expectancy. It is estimated that the average remaining lifetime for children born in Massachusetts and New Hampshire in the years prior to 1789 was 35.5 years. For children born between 1900 and 1902 in the Death Registration Area of the United States the life expectancy was 49 years. In contrast to a gain of 13.5 years over the course of these eleven decades, there was

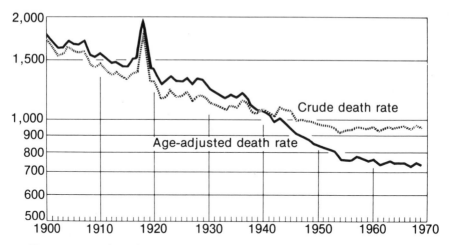

Figure 9 Crude and age-adjusted death rates in death registration states, 1900–1932, and United States, 1933–1969 (per 100,000 population). The age-adjusted death rate is a convenient summary index that "corrects" for differences in age composition.

Source: A. Joan Klebba, Jeffrey D. Maurer, and Evelyn J. Glass, *Mortality Trends: Age, Color, and Sex, United States, 1950–69*, DHEW Publication no. (HRA) 74–1852 (Rockville, Md.: National Center for Health Statistics, 1973), p. 1.

a 22-year gain in the seven decades after 1900, for in 1971 the life expectancy at birth was 71 years. The control of infectious diseases played a large part in these gains. The increase in longevity benefited the childhood years particularly, although it was not confined to that period. As an example, the average expectation of further life at age 65 was 12 years in the period 1900–1902 and 15 years in 1971, a 25 percent increase, as compared with a 45 percent increase for the newborn child.[10]

The decreased mortality from infections caused a pronounced shift in the major causes of death. Pneumonia and tuberculosis no longer vie for the rank of "Captain of the Men of Death." Their place has been taken by diseases of the heart, cancer, cerebrovascular diseases, and accidents.[11] The increase in the average duration of life also greatly changed the age distribution of the American people. In 1900 persons over the age of sixty-four comprised about 5 percent of the population in the Death Registration Area; in 1970 the proportion of persons in this age group in the United States was double that figure and is still increasing.[12]

More important still has been the change in attitudes engendered by the anticipation of a longer life. The stones in an old graveyard are mute evidence that in the past many of those who died were infants or young children. Parents had to be prepared for this. Throughout the nineteenth century there was little reason for them to feel differently from Voltaire, who wrote: "I lost two or three children as nurselings not without regret but without great grief." Children hardly counted until they reached a sufficient age so that they were likely to survive.[13] As a corollary, a mother was expected to bear a number of children, provided she lived long enough, and if she did not, the husband married again, for there had to be enough children so that some would live to take over the family farm or business and support the parents. Today the family is smaller and parents save for their children's college education with every expectation that the children will grow up to use the savings. Middle-aged couples plan for several years of retirement and usually live to enjoy them. The lives of Americans are fulfilling the ancient prophecy:

> "Thou shalt come to thy grave in a full age,
> like as a shock of corn cometh in his season."[14]

Americans also feel more secure because the fear of epidemics is almost gone. People no longer have to bar their doors to friends or flee their cities in droves at the approach of a dread disease. This change in attitudes has an interesting by-product. It has been said that the fear of epidemics was the motive force for public health measures. The question now is whether the public will be disposed to support these measures adequately when it is no longer afraid. It is too early to tell, but many believe that the diminishing status of state and city public health departments is evidence that interest will not long survive fear.

The high expectations engendered by the spectacular results obtained by sulfonamides and antibiotics affected people's attitude toward doctors. From a low point in the early and mid nineteenth century the doctor's prestige rose gradually, then shot up to an all-time high after penicillin appeared.[15] Dimly aware of the many diagnostic tools available to doctors and hearing that "miracle drugs" had now been placed in their hands, people expected doctors to cure just about everything. When this proved to be impossible, the patient and the family tended to blame the doctor, attributing his failure to stupidity or lack of interest. So much has been discovered; people could not conceive that so much was still unknown.

At first the health professions too were swept along in the wave of optimism that inundated the public. They enthusiastically tried each new antibiotic in fields where it had not been used, and whenever the available drugs were ineffective, they looked expectantly for new ones that would fill the void. Such an attitude prevailed particularly in the United States, where the idea of progress, the concept that the condition of mankind will improve progressively and continuously if only we put our minds and hands to it, had been the guiding philosophy of the nation from its beginning.[16]

But doctors and other health scientists soon became disillusioned. Microorganisms resistant to antibiotics appeared, first in a few species, then in a great many. When antimicrobial drugs suppressed the dominant microorganisms, other bacteria, viruses, and fungi that were rarely pathogenic in the past were able to multiply profusely and cause disease. It became clear that drugs alone were not going to control infections. Then came a renewed interest in vaccines, several of which were successful in combating viral infections that had previously been uncontrolled. Yet antibodies, whether developed in response to vaccination or natural infection, do not invariably protect against infections. For instance, viruses that live harmlessly in a person's body for years, such as the *Herpes simplex* virus which causes the common fever blister, can break loose, even though antibodies are present in the host, and produce serious or fatal illnesses when the general defenses of the host are weakened. Furthermore, certain previously unexplained diseases have been shown to be infectious, as demonstrated by the recent discovery by D. Carleton Gajdusek and his associates at the National Institutes of Health that kuru, a disease of certain tribes in New Guinea which causes a gradual deterioration of the function of parts of the central nervous system, is a viral disease. This discovery has stimulated a search for other "slow virus" diseases affecting humans. A gradually progressive form of encephalitis, which may last for years before killing its victim, was recently shown to be caused by the measles virus. Another example of a slow virus infection is the multiplication of rubella virus over the course of several years in some congenitally infected children.[17] Such observations have convinced medical scientists that the infectious diseases are by no means conquered and that new frontiers of research need to be opened.

Nature yields ground to man reluctantly and is always on the alert to recoup her losses. Anyone who has tried to keep a garden weeded or has seen how quickly a deserted farm becomes overgrown with vegetation knows this. Similarly, when the growth of a strain or species of microorganism is diminshed by some means, other species or strains tend to take its place. Hospitals, harboring as they do a high concentration of patients and microorganisms within a limited area, are especially favorable environments for the spread of infections, as Semmelweis and Lister emphasized. Roux wrote in 1894 that he was appalled at the way pneumonia attacked patients on the crowded diphtheria wards, and doctors on contagious disease wards at the beginning of this century frequently saw patients recover from the infection for which they had come to the hospital only to be attacked by another. What is new in recent years is the wholesale suppression of a microorganism by an antibiotic or chemical agent, which upsets the ecological balance in an individual. Thus other pathogenic microorganisms are free to proliferate rapidly, and species that seldom or never produced disease in the past are able to attack vital areas and cause disease. What is also new is the invasion by surgeons of ever more vulnerable areas of the body, such as the heart and the blood vessels, and the use of foreign organs or other substances within the body, such as transplanted kidneys and artificial heart valves. New also is the administration of drugs that suppress immunity, as in the treatment of cancer with chemicals that coincidentally interfere with the growth of normal cells. The adaptability of microorganisms to change means that "the microbial world is an enduring potential source of new diseases."[18]

The most obvious way to control microorganisms that are not affected by the known antibiotics or chemical remedies is to devise new ones. Pharmaceutical companies, in particular, have accepted this challenge and have expended huge sums of money searching for new remedies and modifying existing ones. Much has been achieved by this method. Infectious diseases are not hydra-headed monsters; when one infection is cut off by a specific remedy, several do not appear to replace it. Instead, each effective remedy further increases the total number of infections over which man has control.[19] There is promise also that the role of chemotherapy will be greatly extended by the discovery of specific chemicals for the cure or prevention of viral infections. Still, the law of diminishing returns has begun to operate. The greater the number of effective remedies found, the harder it becomes to find new ones and the higher the cost in money and manpower for each new discovery. Immunization against microorganisms likewise has limits. It is impossible to vaccinate the population against every pathogenic microorganism, and even if it were, people could not be persuaded to undergo the repeated administration of vaccines that would be necessary. Clearly, investigators must look elsewhere if they are to progress much further in controlling infections.

The discovery that a particular microorganism was always associated with a certain disease led to the search for a single cause of each disease, a microorganism, and a single remedy to counteract that single cause. But infectious diseases have many causes. The tubercle bacillus alone is not responsible for tuberculosis. Poor nutrition, exhausting work, or insufficient rest can lower the body's defenses; crowded living or working conditions can increase the chances of transmission. The tubercle bacillus is thus a necessary cause but not a sufficient cause. With the realization that specific drugs and vaccines have their limitations came a shift toward research on the defenses of the host. Investigations into the functions of the cells in immunity, the production of fever in response to infections, and the ways that bacteria and viruses reach the innermost recesses of the lungs are examples of this new emphasis in research. Some of this research has already borne practical fruit. For instance, the newer knowledge of the mechanisms of immunity has aided the development of more effective vaccines, and outside the field of infectious diseases it has helped make possible the transplantation of organs. It may even supply a method for the control of cancer.

Between 1900 and 1970 the population of the United States nearly tripled, increasing from 76 million to nearly 205 million. With this growth came other changes. The nation became highly industrialized; technology and mechanization infiltrated all areas of living and working. The gross national product increased more than tenfold. A larger portion of the population were educated, over thirty times as many college degrees having been conferred in 1970 as in 1900.[20] These and other indications of growth in material and human resources can be advanced as major reasons why the nation made such rapid progress in the control of infectious diseases during this period. In addition, Americans inherited from their European forebears a will to improve the condition of life and the pragmatic attitude that good health was important for successful performance in life. From the beginning Americans had kept in close touch with the countries of northern Europe so that advances in the control of disease in those countries naturally furnished a momentum for the United States. For their own part, Americans were practical people who had struggled to wrest a living out of a new country. They wanted to see results, and soon. Thus they tended to stress research that might have a practical application. For instance, after the first sulfonamides were introduced by German and English scientists, Americans took the lead in devising new ones and in investigating techniques to facilitate their optimal use. This pattern was repeated with the antibiotics.

Other factors operating to promote success in research in America were the processes of immigration and upward mobility. The ranks of investigators could be readily filled because candidates from abroad and from the less privileged classes were continually seeking admittance. Among the three principals in the development of vaccines for poliomyelitis, for example, Enders was the son of a banker and was educated in private schools, Salk was born in a tenement in East Harlem and grew up in the Bronx, and Sabin immigrated to the United States from Poland at the age of fifteen.[21]

The United States has produced no single hero in the field of infectious diseases comparable to France's Pasteur, Germany's Koch, or England's Lister. Perhaps this is because Americans entered the competition too late, after others were already the focus of popular adulation, or because among the profusion of talented investigators no one man or woman could stand out, or because the American brand of democracy has a leveling effect on our attitudes toward scientists. But I am inclined to think it is mainly because Americans seek their heroes elsewhere. They glorify generals, martyred presidents, builders of industrial empires, entertainers, and sports figures. Even inventors such as Thomas Edison and Alexander Graham Bell are placed ahead of scientists.[22]

Certainly the list of Americans who contributed significantly to the understanding of the infectious diseases in the past century is as long as for any country. It is true that in the beginning the important new advances originated abroad: man-made vaccines in France, antitoxin and chemical specifics in Germany. As the automobile was the product of many minds, mostly European, while Americans standardized the processes and lowered the costs of manufacture so that there could be "a car for every family," so in the control of infectious diseases Americans followed the leaders in other countries.[23] They improved, amplified, extended original discoveries, and especially they applied them, often more completely and more successfully than the scientists and practitioners in the countries of origin. The popularization of diphtheria immunization, the statewide programs for the use of pneumococcal serums, the well-coordinated campaigns against tuberculosis, syphilis, and poliomyelitis, and the prompt, efficient testing of the sulfonamides and antibiotics are examples of the American genius for organization. The vast industrial and intellectual resources that grew up in the United States enabled Americans to outstrip other countries by producing more and better serums, vaccines, sulfonamides, and antibiotics after the first ones had been produced elsewhere.

As the number of investigators increased and greater emphasis was placed on basic research, Americans began to make more significant contributions in more fundamental areas. The awards of the Nobel Prize Committee in recent years illustrate this advance. From 1947, when the Nobel Prize was given to Fleming, Florey, and Chain, through 1973, eighteen investigators were honored in this way for research in infectious diseases or immunity; eleven of them were Americans.[24]

How much emphasis to place on basic and how much on applied research is always a troubling question for those who allocate resources. The story of the infectious diseases shows how Americans answered it differently at different times. Up to the twentieth century Americans allowed other countries to carry out most of the fundamental research while they applied the results. With the appointment of full-time teachers to the faculties of medical schools and the establishment of research laboratories such as the Rockefeller Institute and the National Institute of Health, interest in basic research increased. Yet acceptance of its importance was slow to come. In 1938 Abraham Flexner, in a speech entitled "The Usefulness of Useless Research," pointed out how apparently unproductive theoretical discoveries had often led to useful inventions of great benefit to mankind. Three decades later James A. Shannon, director of the National Institutes of Health, still felt it necessary to warn a subcommittee of the United States Senate that the application of knowledge should not be allowed to outrun its science base, citing the Cutter incident as a tragic example.[25]

Perhaps if more attention had been paid to basic knowledge earlier in the century, applied research in infectious diseases would have advanced more rapidly. Yet it is questionable whether the technical resources and the supply of trained personnel would have been adequate to the task. Furthermore, it is doubtful whether the American people would have contributed the large sums of money needed for fundamental research before the 1940s. They had to visualize the practical results of investigations before they would support them on a large scale. Even now one cannot be sure that the need for basic research is understood; there is still the danger that when practical results are not forthcoming in a short time, the flow of funds will be cut off.

Yet research problems cannot be sharply divided into two groups, those that seek to understand natural phenomena on one side and those that attempt to increase man's control over nature on the other. The experience of Cole's group at the Rockefeller Institute illustrates this connection. In attempting to find a cure for pneumococcal pneumonia, these investigators not only developed successful serums but also turned their interest in another direction, toward the structure of pneumococci. This led eventually to the epoch-making discovery of DNA, the substance that transmits hereditary characteristics. It had been known for some time that the capsules of pneumococci protected them from being ingested by phagocytes and thus contributed to their virulence. Pneumococci grown under conditions that produced no capsules were less virulent than fully encapsulated forms. In 1928 Fred Griffith inoculated mice with a living culture of unencapsulated pneumococci of one type and a killed culture of capsulated pneumococci of another type. To his surprise, the living bacteria changed to the same type as that of the dead bacteria. For instance, degraded, nonencapsulated, Type I pneumococci could be changed into virulent, capsulated, Type II pneumococci by some unknown substance contained in the dead Type II pneumococci. Avery's curiosity spurred him to find out what this substance was, and in 1944, with Colin M. MacLeod and Maclyn McCarty, he reported that the transforming substance was DNA, which carried the information directing the pneumococci to manufacture a capsule of a specific type and which was transmitted as a hereditary trait. Some years later an American geneticist, James D. Watson, reasoned that since "DNA was known to occur in the chromosomes of all cells, Avery's experiments strongly suggested that . . . all genes were composed of DNA." Working at Cambridge University, he and an English physicist, Francis H. C. Crick, traced out its chemical structure. In 1953 they announced that they had built a model of DNA with which they could show how it split into pairs to transmit genetic characteristics from one generation to the next. In 1962 Watson, Crick, and another English scientist, Maurice H. F. Wilkins, were awarded a Nobel Prize for this discovery, "perhaps the most famous event in biology since Darwin's book," to quote Watson again. As they received the prize, they must have silently thanked Griffith and Avery for ignoring immediately practical goals to study instead the basic biology of the pneumococcus.[26]

Research occupies a continuous spectrum from the most basic to the most applied, and investigators often move from one point on the spectrum to another as the answers begin to come in. Combining the spirit of free inquiry characteristic of basic research with the motivation that is inspired by the urge for practical application is healthy for both. In the United States, where both the freedom to explore and the desire for practical results have been a part of the national culture from the beginning, the collaboration of the two has contributed to the success of research in general and of research on infectious diseases in particular. In view of past successes, every effort should be made to keep theoretical scientists and practical physicians in contact with each other, despite the increasing size and complexity of the research enterprise.

The easiest way to lose ground in the fight against infectious diseases is to assume that they have been conquered and nothing more needs to be done. Surveys of doctors' practices show that more patients are seen because of infections than for any other group of illnesses. Among a representative sample of 1.3 billion doctor-patient contacts in homes, hospitals, and doctors' offices in the United States in 1968, 20.2 percent were for the prevention or treatment of infections or their consequences. Two studies of doctors' visits to patients' homes in the 1950s, one in Illinois and the other in Vermont, showed that 39.8 percent and 39.1 percent, respectively, of the visits were for infections. The number of deaths from infections is still by no means negligible. In the United States in 1970 infections were listed as the primary cause of death in nearly 116,000, or 6 percent, of all deaths. They were a complicating cause in many more.[27]

Most of the more common infections have now been brought under partial control, but partial control is not enough. As long as pathogenic microorganisms are around, they can counterattack whenever and wherever a group of people becomes vulnerable. A depression or a war may lower living standards, the invention of a new irritant chemical may injure the lungs of persons exposed to it, and either of these may provide the opportunity for tuberculosis to sweep through the susceptible population. The social and economic factors that favor the spread of venereal disease are not easily changed and morals cannot be legislated. Infections that have been banished from this country can return. Vaccines against syphilis and gonorrhea would make the control of venereal diseases more certain and more universal than is true today, and a more effective vaccine against tuberculosis than BCG might make it possible to stamp out the last vestige of that disease in the United States and throughout the world.

Much more research is needed on viruses and the diseases they cause. First of all, vaccines are needed to protect against viral hepatitis and encephalitis, which caused in excess of 1,000 and 300 deaths, respectively, in the United States in 1970. Some method is needed for preventing upper respiratory infections, which are mostly viral and are responsible for more patient-doctor visits than any other group of illnesses.[28] The few slow virus diseases now identified may be only the tip of the iceberg; such infections may be responsible for several diseases now considered to be noninfectious. It is known that viruses are related to some types of cancer, but how significant this relationship is and how frequently it occurs remain to be seen. Finally, as more and more persons are kept alive until old age, many having chronic diseases, measures must be found for strengthening their defenses against infections lest they become victims of the microorganisms in their day-to-day environment that have heretofore been considered harmless.

So much remains to be done that no one dares succumb to the philosophy encountered by Stokes when he began to work on syphilis in the early decades of this century, namely that "all diagnostic problems had been resolved by serologic testing, that all uncertainties of treatment were banished by the magic bullet of arsenic, and that the necessity for clinical acumen and skill had vanished."[29] Today neither the ramparts built by vaccines nor the many "magic bullets" of chemotherapy have stopped the inexorable march of the microbes. The war must go on forever, and the only way to gain and hold ground is by using past experience. The previous seventy-five years, which saw so many victories over man's minute enemies, have pointed the way to many more.

Notes

1. The Field of Battle

1. From the information available it is not possible to diagnose George Washington's last illness with certainty. It is generally believed that the inflammation was in the larnyx, although some believe it was mainly in the throat. One otolaryngologist makes a good case for epiglottitis, inflammation of the saddle-shaped plate of cartilage at the root of the tongue which covers the larynx during the act of swallowing. See H. H. E. Scheidemandel, "Did George Washington Die of Quinsy?" *Archives of Otolaryngology* 102 (Sept. 1976): 519–521. The microorganisms responsible for the inflammation may have been pneumococci (see Chapter 4), streptococci (Chapter 5), or staphylococci (Chapter 8). In any case no specific treatment would have been available before the twentieth century. Diphtheria also affects the larynx, but this diagnosis seems unlikely, because it seldom affects persons of Washington's age, because there was no mention of other cases in the vicinity, and because the symptoms of diphtheria usually come on more slowly and the local symptoms are usually less severe than in Washington's case. See also Wyndham B. Blanton, *Medicine in Virginia in the Eighteenth Century* (Richmond, Va.: Garrett and Massie, 1931), pp. 305–312; W. A. Wells, "Last Illness and Death of Washington," *Virginia Medical Monthly* 53 (Jan. 1927): 629–642; R. Reece, "George Washington: His Death and His Doctors," *Minnesota Medicine* 49 (Jul. 1966): 1185–1190.

2. James B. Herrick, *Memories of Eighty Years* (Chicago: University of Chicago Press, 1949), pp. 61, 136–154. In infectious diseases a specific remedy, or a "specific," is a substance that acts directly on the causative agent to bring about either a cure or a more rapid recovery than would otherwise occur.

3. William Bulloch, *The History of Bacteriology* (London: Oxford University Press, 1938; reprint ed., 1960), pp. 4–5; Hubert A. Lechevalier and Morris Solotorovsky, *Three Centuries of Microbiology* (New York: McGraw-Hill, 1965), pp. 1–3; Clifford Dobell, ed., *Anthony Van Leeuwenhoek and His "Little Animals"* (New York: Dover Publications, 1960), p. 243.

4. Charles Singer and E. Ashworth Underwood, *A Short History of*

Medicine, 2d ed. rev. (New York: Oxford University Press, 1962), pp. 322–342; Bulloch, *History of Bacteriology*, pp. 159–162, 59–60, 96–105.

5. Ibid., pp. 163–165, 180, 210–213, 225–229, 213–214, 217–230. Microorganisms are usually grown in the laboratory, or "cultured," by being inoculated into a liquid broth containing nutrients needed for their growth or spread across or inoculated in a semisolid medium (originally gelatin, later agar) in which the proper nutrients are dissolved. The presence of a microorganism is recognized by characteristic growth in these media. When a portion of the growth is placed on a slide, stained, and viewed under the microscope, the individual microorganisms can be seen. They can be transferred from liquid or semisolid media to other media or injected into animals for further identification and study. See A. P. Hitchens and M. C. Leikind, "The Introduction of Agar-agar into Bacteriology," *Journal of Bacteriology* 37 (May 1939): 485–493.

6. Bulloch, *History of Bacteriology*, pp. 242–247; René J. Dubos, *Louis Pasteur: Free Lance of Science* (Boston: Little, Brown, 1950), pp. 326–337.

7. Bulloch, *History of Bacteriology*, pp. 249–251. See Chapter 12 for a description of viruses.

8. J. Lister, "On a New Method of Treating Compound Fractures, Abscess, etc., with Observations on the Conditions of Suppuration: Part I. On Compound Fracture," *Lancet* 1 (Mar. 16, 1867): 326–329; Lister, "The Address in Surgery," *British Medical Journal* 2 (Aug. 26, 1871): 225–233; Singer and Underwood, *A Short History of Medicine*, pp. 354–358, 366–371.

9. Singer and Underwood, *A Short History of Medicine*, pp. 391, 411–412; Bulloch, *History of Bacteriology*, pp. 236–238, 217.

10. Bulloch, *History of Bacteriology*, pp. 255–283.

11. Frederick L. Hoffman, "American Mortality: Progress during the Last Half Century," in Mazyck P. Ravenel, ed., *A Half Century of Public Health* (New York: American Public Health Association, 1921), pp. 114–115; U.S. Bureau of the Census, *Historical Statistics of the United States, Colonial Times to 1957* (Washington, D.C.: U.S. Government Printing Office, 1960), p. 30; Robert D. Grove and Alice M. Hetzel, *Vital Statistics Rates in the United States, 1940–1960* (Washington, D.C.: U.S. Department of Health, Education, and Welfare, Public Health Service, National Center for Health Statistics, 1968), pp. 559–564.

12. Isolation is the separation of an individual ill with an infectious disease from other persons; quarantine is the separation from others for the duration of the incubation period of a healthy person who has been exposed to a person ill with the disease. See also E. H. Ackerknecht, "Anticontagionism between 1821 and 1867," *Bulletin of the History of Medicine* 22 (Sept.–Oct. 1948): 562–593; J. H. Cassedy, "The Flamboyant Colonel Waring: An Anti-contagionist Holds the American Stage in the Age of Pasteur and Koch," ibid. 36 (Mar.–Apr. 1962): 163–176.

13. Richard Harrison Shryock, *The Development of Modern Medicine: An Interpretation of the Social and Scientific Factors Involved* (New

York: Alfred A. Knopf, 1947), p. 227. The concept of the Death Registration Area was established in order to improve the completeness of reporting. The statistics when gathered became a basis for planning in public health and related areas. See J. H. Cassedy, "The Registration Area and American Vital Statistics: Development of a Health Research Resource, 1885–1915," *Bulletin of the History of Medicine* 39 (May–June 1965): 221–231; U.S. Bureau of the Census, *Historical Statistics of the United States*, p. 18.

14. George Rosen, *A History of Public Health* (New York: MD Publications, 1958), p. 248; Charles V. Chapin, *Municipal Sanitation in the United States* (Providence, R.I.: Providence Press, Snow and Farnham, 1901), p. 3; Ralph Chester Williams, *The United States Public Health Service, 1798–1950* (Washington, D.C.: Commissioned Officers Association of the United States Public Health Service, 1951), pp. 135, 162–167. The Marine Hospital Fund had been established in 1798 to care for sick sailors. In 1883 this organization, renamed the Marine Hospital Service, took over national quarantine matters, and in 1891 it was made responsible for the medical aspects of immigration.

15. Rosen, *A History of Public Health*, p. 333; Charles-Edward A. Winslow, *The Life of Hermann M. Biggs, M.D., D.Sc., LL.D.: Physician and Statesman of the Public Health* (Philadelphia: Lea & Febiger, 1929), p. 95; T. Turnbull, "Address in Medicine," *Pennsylvania Medical Journal* 4 (Sept. 1900): 181–184.

16. U.S. Bureau of the Census, *Historical Statistics of the United States*, pp. 7, 14.

17. Shryock, *The Development of Modern Medicine*, pp. 258–264, 349–350.

18. Joseph Chamberlain Furnas, *The Americans: A Social History of the United States, 1587–1914* (New York: G. P. Putnam's Sons, 1969), p. 905.

19. W. F. Norwood, "Critical Incidents in the Shaping of Medical Education in the United States," *Journal of the American Medical Association* 194 (Nov. 15, 1965): 715–718; R. P. Hudson, "Abraham Flexner in Perspective: American Medical Education, 1865–1910," *Bulletin of the History of Medicine* 46 (Nov.–Dec. 1972): 545–561; Abraham Flexner, *Medical Education in the United States and Canada: A Report to the Carnegie Foundation for the Advancement of Teaching* (New York: Carnegie Foundation for the Advancement of Teaching, 1910); Shryock, *The Development of Modern Medicine*, pp. 258–272, 349–350. Abraham Flexner was a teacher and author on education who was employed by the Carnegie Foundation to make this survey of American medical education.

20. For changes in therapy, see Harry F. Dowling, *Medicines for Man: The Development, Regulation, and Use of Prescription Drugs* (New York: Alfred A. Knopf, 1970), pp. 16–18.

21. William Osler, *Aequanimitas: With Other Addresses to Medical*

Students, Nurses, and Practitioners of Medicine, 3d ed. (Philadelphia: Blakiston, 1961), p. 430. Osler was professor of medicine at Johns Hopkins Medical School from 1889 to 1905. His textbook of medicine, which appeared in many editions, is cited frequently throughout this volume because it was the leading American textbook for nearly four decades.

2. Control by Public Health Measures

1. Richard N. Current, T. Harry Williams, and Frank Freidel, *American History: A Survey,* 2d ed. (New York: Alfred A. Knopf, 1966), pp. 578–582; Martha L. Sternberg, *George Miller Sternberg: A Biography* (Chicago: American Medical Association, 1920), p. 185; Walter Reed, Victor C. Vaughan, and Edward O. Shakespeare, *Report on the Origin and Spread of Typhoid Fever in U.S. Military Camps during the Spanish War of 1898* (Washington, D.C.: U.S. Government Printing Office, 1904), I, 656, 659–660, 663.

2. Frederick P. Gay, *Typhoid Fever Considered as a Problem of Scientific Medicine* (New York: Macmillan, 1918), pp. 1–2. For typhoid in colonial America, see John Duffy, *Epidemics in Colonial America* (1953; reprint ed., Baton Rouge: Louisiana State University Press, 1971), pp. 222–229.

3. Gay, *Typhoid Fever,* pp. 3, 14; H. C. S. Lombard, "Observations Suggested by a Comparison of the Post Mortem Appearances Produced by Typhous Fever in Dublin, Paris, and Geneva," *Dublin Journal of Medical Science* 10 (Sept. 1836): 17–24; "Second Letter from Doctor Lombard to Doctor Graves on the Subject of Typhous Fever," ibid., pp. 101–105; W. W. Gerhard, "On the Typhus Fever Which Occurred in Philadelphia in the Spring and Summer of 1836," *American Journal of the Medical Sciences* 19 (Feb. 1837): 289–322; 20 (Aug. 1837): 289–322. See also William G. MacCallum, *A Textbook of Pathology,* 1st ed. (Philadelphia: W. B. Saunders, 1916), p. 633; 2d ed. (1920), pp. 686–687.

4. Arthur L. Bloomfield, *A Bibliography of Internal Medicine: Communicable Diseases* (Chicago: University of Chicago Press, 1958), pp. 7–9; William Budd, *Typhoid Fever: Its Nature, Mode of Spreading, and Prevention* (London: Longmans, Green, 1873), pp. 190, 181. See also Charles Murchison, *A Treatise on the Continued Fevers of Great Britain,* 2d ed. (London: Longmans, Green, 1884).

5. Bloomfield, *A Bibliography of Internal Medicine,* pp. 9–10; K. W. Von Drigalski, "Ueber Ergebnisse bei der Bekämpfung des Typhus nach Robert Koch," *Zentralblatt für Bakteriologie, Parasitenkunde und Infektionskrankheiten* 35, no. 6 (1904): 776–798.

6. E. Lesky, "Viennese Serological Research about the Year 1900: Its Contribution to the Development of Clinical Medicine," *Bulletin of the New York Academy of Medicine* 49 (Feb. 1973): 100–111.

7. G. W. Fuller, "Typhoid Fever Death Rates in American Cities,"

Public Health: Papers and Reports, American Public Health Association 27 (1901): 100–102; J. Hartzell, "The Mississippi River as a Sewer," *Public Health: Papers and Reports, American Public Health Association* 21 (1896): 22–34. The mortality rate or death rate refers to the number of deaths per unit of population. The case fatality rate, or fatality rate, is the number of patients dying of a disease among a particular group of patients having the disease.

The provision of sewers and safe water by cities and towns did not eliminate typhoid immediately. In 1907 some residents of the District of Columbia preferred to drink the water from shallow wells or springs rather than the filtered city water, and in 1908 the residents of 56 of 542 houses surveyed used privies rather than water closets connected to the sewerage system. See Charles-Edward A. Winslow, *The Evolution and Significance of the Modern Public Health Campaign* (New Haven: Yale University Press, 1923), p. 38; George C. Whipple, "Fifty Years of Water Purification," in Mazyck P. Ravenel, ed., *A Half Century of Public Health* (New York: American Public Health Association, 1921), pp. 166, 172; J. B. Blake, "The Origins of Public Health in the United States," *American Journal of Public Health* 38 (Nov. 1948): 1539–1550; Allen Weir Freeman, *Five Million Patients: The Professional Life of a Health Officer* (New York: Charles Scribner's Sons, 1946), p. 63; Harvey Cushing, *The Life of Sir William Osler*, 2 vols. (Oxford: Clarendon Press, 1925), I, 380.

8. A. Taylor, Jr., G. F. Craun, G. A. Faich, L. J. McCabe, and E. J. Gangarosa, "Outbreaks of Waterborne Diseases in the United States, 1961–1970," *Journal of Infectious Diseases* 125 (Mar. 1972): 329–331.

9. Harry Wain, *A History of Preventive Medicine* (Springfield, Ill.: Charles C Thomas, 1970), pp. 252–255. In 1901 George M. Kober, professor of public health at Georgetown University, reported that in 75 percent of 195 epidemics of milkborne typhoid the disease was present in the dairy farm from which the milk came. See G. M. Kober, "Conclusions Based upon Three Hundred and Thirty Outbreaks of Infectious Diseases Spread through the Milk Supply," *American Journal of the Medical Sciences* 121 (May 1901): 552–556; M. J. Waserman, "Henry L. Coit and the Certified Milk Movement in the Development of Modern Pediatrics," *Bulletin of the History of Medicine* 46 (Jul.–Aug. 1972): 359–390; René Vallery-Radot, *The Life of Pasteur*, trans. Mrs. R. L. Devonshire (London: Constable, 1911), p. 113.

10. Waserman, "Henry L. Coit," pp. 359–390. For the quotation, see M. J. Rosenau, L. L. Lumsden, and J. H. Kastle, "Report on the Origin and Prevalence of Typhoid Fever in the District of Columbia," *Hygienic Laboratory Bulletin*, no. 35 (1907), p. 20. See also George Rosen, *A History of Public Health* (New York: MD Publications, 1958), pp. 358–360; C. E. North, "Milk and Its Relation to Public Health," in Ravenel, ed., *A Half Century of Public Health*, pp. 267–279.

11. Gay, *Typhoid Fever*, p. 20; Carl C. Dauer, Robert F. Korns, and

Leonard M. Schuman, *Infectious Diseases* (Cambridge, Mass.: Harvard University Press, 1968), p. 13.

12. Rosen, *A History of Public Health*, pp. 319–320; J. C. G. Ledingham and J. A. Arkwright, *The Carrier Problem in Infectious Diseases* (London: Edward Arnold, 1912), p. 5; E. D. W. Grieg, "Typhoid Fever Prophylaxis: Report on the Methods Employed in the Campaign against Typhoid in Germany," *Journal of the American Medical Association* 46 (Mar. 3, 1906): 673–676.

13. W. H. Park, "Typhoid Bacilli Carriers," *Journal of the American Medical Association* 51 (Sept. 19, 1908): 981–982. For the quotation, see G. A. Soper, "Typhoid Mary," *Military Surgeon* 45 (Jul. 1919): 1–15. See also G. A. Soper, "The Curious Career of Typhoid Mary," *Bulletin of the New York Academy of Medicine* 15 (Oct. 1939): 698–712; Editorial: "Healthy Disease Spreaders," *New York Times,* Jul. 1, 1909, p. 8; "Typhoid Mary," *Outlook* 109 (Apr. 7, 1915): 803–804; Letter: "Proposes a Medical Monarchy," *New York Times,* Jul. 2, 1909, p. 6; "The Germ-Carrier," *Punch* (London), 137 (Jul. 7, 1909): 2.

14. F. A. Coller and F. C. Forsbeck, "The Surgical Treatment of Chronic Biliary Typhoid Carriers," *Annals of Surgery* 105 (May 1937): 791–799; J. L. Freitag, "Treatment of Chronic Typhoid Carriers by Cholecystectomy," *Public Health Reports* 79 (Jul. 1964): 567–570.

15. Medical Research Council (Great Britain), *Diphtheria: Its Bacteriology, Pathology and Immunology* (London: His Majesty's Stationery Office, 1923), pp. 14–15.

16. For the history of diphtheria to 1923, see ibid., pp. 13–63; Bloomfield, *A Bibliography of Internal Medicine*, pp. 246–264.

17. T. M. Prudden, "On the Etiology of Diphtheria: An Experimental Study," *American Journal of the Medical Sciences* 97 (Apr. 1889): 329–350; (May 1889): 450–478; W. H. Welch and A. C. Abbott, "The Etiology of Diphtheria," *Bulletin of the Johns Hopkins Hospital* 2 (Feb.–Mar. 1891): 25–31.

18. Charles-Edward A. Winslow, *The Life of Hermann M. Biggs, M.D., D.Sc., LL.D.: Physician and Statesman of the Public Health* (Philadelphia: Lea & Febiger, 1929), esp. p. 108. See also Wade W. Oliver, *The Man Who Lived for Tomorrow: A Biography of William Hallock Park* (New York: E. P. Dutton, 1941).

19. W. H. Park, "The History of Diphtheria in New York City," *American Journal of Diseases of Children* 42 (Dec. 1931): 1439–1445; W. H. Park and A. L. Beebe, "Diphtheria and Pseudo-Diphtheria," *Medical Record* 46 (Sept. 29, 1894): 385–401; G. H. Weaver, "Diphtheria Carriers," *Journal of the American Medical Association* 76 (Mar. 26, 1921): 831–835.

20. W. H. Frost, "The Familial Aggregation of Infectious Diseases," *American Journal of Public Health* 28 (Jan. 1938): 7–13; James H. Cassedy, *Charles V. Chapin and the Public Health Movement* (Cambridge: Harvard University Press, 1962), p. 115.

21. W. H. Frost, M. Frobisher, Jr., V. A. Van Volkenburgh, and M. L. Levin, "Diphtheria in Baltimore: A Comparative Study of Morbidity, Carrier Prevalence and Antitoxic Immunity in 1921–1924 and 1933–1936," *American Journal of Hygiene* 24 (Nov. 1936): 568–586; Cassedy, *Charles V. Chapin and the Public Health Movement*, p. 74.

3. Prevention by Immunization

1. H. F. Dowling, "Human Experimentation in Infectious Diseases," *Journal of the American Medical Association* 198 (Nov. 28, 1966): 997–999; René Vallery-Radot, *The Life of Pasteur*, trans. Mrs. R. L. Devonshire (London: Constable, 1911), pp. 168–195.

2. Leonard Colebrook, *Almroth Wright: Provocative Doctor and Thinker* (London: William Heinemann Medical Books, 1954), pp. 28, 32, 36; A. E. Wright, "Association of Serous Haemorrhages with Conditions of Defective Blood Coagulability," *Lancet* 2 (Sept. 19, 1896): 807–809; R. Pfeiffer and W. Kolle, "Experimentelle Untersuchungen zur Frage des Schutzimpfung des Menschen gegen Typhus abdominalis," *Deutsche Medizinische Wochenschrift* 22 (Nov. 12, 1896): 735–737; William Derek Foster, *A History of Medical Bacteriology and Immunology* (London: William Heinemann Medical Books, 1970), p. 131.

3. Zachary Cope, *Almroth Wright: Founder of Modern Vaccine Therapy* (London: Thomas Nelson, 1966), pp. 17–32.

4. Ibid., pp. 22–32. These differences of opinion were dramatically set forth in a series of letters to the editor of the *British Medical Journal*, written by Wright and a professor of statistics at the University of London, Karl Pearson. Pearson claimed that the data obtained in the experiments on British troops, when examined by statistical methods, did not prove that vaccination prevented typhoid. He recommended, first, that the method of preparing the vaccine be improved, and, second, that more soldiers be vaccinated and the incidence of typhoid among them be compared with the incidence in a proper control group. Wright, on the other hand, believed strongly that his vaccine was effective and wanted to get on with the business of saving lives by vaccinating the entire army. He could not win the debate on statistical grounds. Instead, he contended that, in carrying out experiments involving thousands of troops under field conditions, one could not follow the rigid rules of statistics that prevailed in the laboratory. And he was convinced that, if the experiments had to be performed as Pearson recommended, hundreds of men would die unnecessarily of typhoid in the meantime. Seventy years and many clinical experiments later investigators have learned that the requirements laid down by Pearson are by no means too stringent. In fact, they could easily be employed to conduct a similar experiment today. See K. Pearson, "Report on Certain Enteric Fever Inoculation Statistics," *British Medical Journal* 2 (Nov. 5, 1904): 1243–1246; A. E. Wright, "Cor-

respondence: Antityphoid Inoculation," ibid. (Nov. 12, 1904), pp. 1343–1345; K. Pearson, "Correspondence: Antityphoid Inoculation," ibid. (Nov. 19, 1904), p. 1432; A. E. Wright, "Correspondence: Antityphoid Inoculation," ibid. (Nov. 26, 1904), pp. 1489–1491; K. Pearson, "Correspondence: Antityphoid Inoculation," ibid. (Dec. 3, 1904), pp. 1542–1543; A. E. Wright, "Correspondence: Antityphoid Inoculation," ibid. (Dec. 10, 1904), p. 1614.

5. Frederick P. Gay, *Typhoid Fever Considered as a Problem of Scientific Medicine* (New York: Macmillan, 1918), pp. 172–197; Harvey Cushing, *The Life of Sir William Osler*, 2 vols. (Oxford: Clarendon Press, 1925), II, 427–428; Cope, *Almroth Wright*, pp. 35–37; Colebrook, *Almroth Wright*, p. 44.

6. G. R. Callender and G. F. Luippold, "The Effectiveness of Typhoid Vaccine Prepared by the U.S. Army," *Journal of the American Medical Association* 123 (Oct. 9, 1943): 319–321.

7. Ibid., pp. 319–321; T. G. Duncan, J. A. Doull, E. R. Miller, and H. Bancroft, "Outbreak of Typhoid Fever with Orange Juice as the Vehicle Illustrating the Value of Immunization," *American Journal of Public Health* 36 (Jan. 1946): 34–36.

8. D. E. Marmion, G. R. E. Naylor, and I. O. Steward, "Second Attacks of Typhoid Fever," *Journal of Hygiene* 51 (June 1953): 260–267.

9. R. B. Hornick and T. E. Woodward, "Appraisal of Typhoid Vaccine in Experimentally Infected Subjects," *Transactions of the American Clinical and Climatological Association* 78 (1966): 70–78.

10. Army Medical School, Research Laboratories, *Immunization in Typhoid Fever* (Baltimore: Johns Hopkins Press, 1941); B. Cvjetanovic and K. Uemura, "The Present Status of Field and Laboratory Studies of Typhoid and Paratyphoid Vaccines with Special Reference to Studies Sponsored by the World Health Organization," *Bulletin of the World Health Organization* 32, no. 1 (1965): 29–36.

11. Francis Adams, trans., *The Genuine Works of Hippocrates* (Baltimore: Williams & Wilkins, 1939), p. 315.

12. Arthur L. Bloomfield, *A Bibliography of Internal Medicine: Communicable Diseases* (Chicago: University of Chicago Press, 1958), pp. 270–277, 254–255.

13. Smith wrote: "From the practical standpoint it [this method] offers a promising field for investigations in the active immunization of the human subject. If the latter reacts as does the guinea pig it should be an easy matter to confer a relatively high degree of active immunity lasting at least several years, without any appreciable disturbances of health." Smith's method was workable and his prophecy correct, but his own interests were in animal experimentation, and he was not closely associated with clinical investigators who might carry out his idea. T. Smith, "The Degree and Duration of Passive Immunity to Diphtheria Toxin Transmitted by Immunized Female Guinea Pigs and Their Immediate Offspring," *Journal of Medical Research* 16 (May 1907): 359–379;

Bloomfield, *A Bibliography of Internal Medicine*, p. 262; E. von Behring, "Ueber ein neues Diptherieschutzmittel," *Deutsche medizinische Wochenschrift* 39 (May 8, 1913): 873–876.

14. B. Schick, "Kutanreaktion bei Impfung mit Diphtherietoxin," *Münchener Medizinische Wochenschrift* 55 (Mar. 10, 1908): 504–506; W. H. Park, "The Schick Test and the Immunization of Children against Diphtheria: A Communication," *Medical Record* 96 (Nov. 29, 1919): 900.

15. A. T. Glenny and B. E. Hopkins, "Diphtheria Toxoid as an Immunising Agent," *British Journal of Experimental Pathology* 4 (Oct. 1923): 283–288; G. Ramon, "L'anatoxine diphthérique: Ses propriétés—ses applications," *Annales de l'Institut Pasteur* 42 (Sept. 1928): 959-1009; C. C. Slemons, "Michigan's Department of Health," *Journal of the Michigan State Medical Society* 30 (Dec. 1931): 956–958; Henry James Parish, *A History of Immunization* (Edinburgh: E. & S. Livingstone, 1965), p. 144.

16. H. N. Bundesen, W. I. Fishbein, and H. C. Niblack, "Diphtheria Control in Chicago," *Journal of the American Medical Association* 100 (Apr. 8, 1933): 1093–1097; R. C. Eelkema and E. D. Lyman, "Changing Patterns of Diphtheria in Omaha—Douglas County, 1880–1963: Comment on Family Immunization Using Leap Year as a Guide," *Nebraska Medical Journal* 49 (Nov. 1964): 612–621.

17. Reginald C. McGrane, *The Cincinnati Doctors' Forum* (Cincinnati: Academy of Medicine of Cincinnati, 1957), p. 278. The incidence of diphtheria in the United States in recent years has been summarized by K. E. Nelson, C. A. Kallick, and S. Levin, "Diphtheria in Chicago, 1960–1970," *Illinois Medical Journal* 139 (Jan. 1971): 35–39. See also U.S. Department of Health, Education, and Welfare, Public Health Service, *Reported Morbidity and Mortality in the United States, 1974: Morbidity and Mortality Weekly Report* 23, no. 53 (Atlanta: Center for Disease Control, 1975): 2.

18. Carl C. Dauer, Robert F. Korns, and Leonard M. Schuman, *Infectious Diseases* (Cambridge: Harvard University Press, 1968), p. 61. The American Medical Association was the major force in the campaign to ban the sale of fireworks. See "Third Annual Summary of Fourth of July Injuries Due to Fireworks and Explosives," *Journal of the American Medical Association* 114 (Jan. 6, 1940): 39–41; Editorial, ibid., p. 44.

19. Charles Singer and E. Ashworth Underwood, *A Short History of Medicine*, 2d ed. rev. (New York: Oxford University Press, 1962), p. 441; A. P. Long and P. E. Sartwell, "Tetanus in the United States Army in World War II," *Bulletin of the U.S. Army Medical Department* 7 (Apr. 1947): 371–385; Editorial: "Tetanus and Military Medicine," *U.S. Naval Medical Bulletin* 47 (May–June 1947): 505–506.

20. Max Sterne and W. E. van Heyningen, "The Clostridia," in René J. Dubos and James G. Hirsch, eds., *Bacterial and Mycotic Infections of Man*, 4th ed. (Philadelphia: J. B. Lippincott, 1965), p. 562; Gaston

Ramon, *Anatoxins and Anatoxic Immunizations: Results of Thirty Years' Application in the World* (Toulouse, France: Imprimerie Régionale, 1953), p. 35; S. Gottlieb, F. X. McLaughlin, L. Levine, W. C. Latham, and G. Edsall, "Long-Term Immunity to Tetanus—A Statistical Evaluation and Its Clinical Implications," *American Journal of Public Health* 54 (June 1964): 961–971.

21. Ramon, *Anatoxins and Anatoxic Immunization*, pp. 36, 38; W. W. Hall, "The U.S. Navy's War Record with Tetanus Toxoid," *Annals of Internal Medicine* 28 (Feb. 1948): 298–308; Long and Sartwell, "Tetanus in the United States Army in World War II"; Hall, "U.S. Navy's War Record with Tetanus Toxoid."

22. Ramon, *Anatoxins and Anatoxic Immunization*, p. 36; J. A. Bigler and M. Warner, "Active Immunization against Tetanus and Diphtheria in Infants and Children," *Journal of the American Medical Association* 116 (May 24, 1941): 2355–2366; U.S. Department of Health, Education, and Welfare, National Communicable Disease Center, *United States Immunization Survey—1969* (Atlanta: 1970), p. 21; U.S. Department of Health, Education, and Welfare, *Reported Morbidity and Mortality in the United States, 1974*, pp. 2, 54.

23. Roderick Heffron, *Pneumonia with Special Reference to Pneumococcus Lobar Pneumonia* (New York: Commonwealth Fund, 1939), pp. 196–198, 408–414; Benjamin White, *The Biology of Pneumococcus: The Bacteriological, Biochemical, and Immunological Characters and Activities of Diplococcus Pneumoniae* (New York: Commonwealth Fund, 1938), pp. 494–503; R. L. Cecil, "Immunization against Pneumonia," *Medicine* 4, no. 4 (1925): 395–419. Pneumonia is discussed in more detail in Chapter 4.

24. A. J. Orenstein, "Vaccine Prophylaxis in Pneumonia," *Journal of the Medical Association of South Africa* 5 (June 13, 1931): 339–346.

25. White, *The Biology of Pneumococcus*, pp. 494–499.

26. A. Sophian and J. Black, "Prophylactic Vaccination against Meningitis," *Journal of the American Medical Association* 59 (Aug. 17, 1912): 527–532.

27. H. W. Stoner, "A Resumé of Vaccine Therapy," *American Journal of the Medical Sciences* 141 (Feb. 1911): 186–213; Foster, *A History of Medical Bacteriology and Immunology*, pp. 140–147.

4. Treatment with Serum

1. E. Behring and S. Kitasato, "Ueber das Zustandekommen der Diphtherie-Immunität und der Tetanus-Immunität bei Thieren I," *Deutsche Medizinische Wochenschrift* 16 (Dec. 4, 1890): 1113–1114; E. Behring, "Ueber das Zustandekommen der Diphtherie-Immunität und Tetanus-Immunität bei Thieren II," ibid. 16 (Dec. 11, 1890): 1145–1148; A. S.

McNalty, "Emil von Behring, Born March 15, 1854," *British Medical Journal* 1 (Mar. 20, 1954): 668–670.

2. Paul Ehrlich, "The Assay of the Activity of Diphtheria-Curative Serum and Its Theoretical Basis," in F. Himmelweit, ed., *The Collected Papers of Paul Ehrlich*, 4 vols. (London: Pergamon Press, 1956–), vol. 2: *Immunology and Cancer Research*, pp. 107–125; Martha Marquardt, *Paul Ehrlich* (New York: Henry Schuman, 1951), pp. 66–69.

3. Marquardt, *Paul Ehrlich*, pp. 29–40.

4. M. E. Roux and M. L. Martin, "Contribution à l'etude de la diphthérie (serum-thérapie)," *Annales de l'Institut Pasteur* 8 (Sept. 1894): 609–639; V. C. Vaughn, "Immunization against Diphtheria by the Employment of a Toxin-Antitoxin Mixture: A Caution," *Journal of Laboratory and Clinical Medicine* 5 (Feb. 1920): 334–335.

5. Charles-Edward A. Winslow, *The Life of Hermann M. Biggs, M.D., D.Sc., LL.D.: Physician and Statesman of the Public Health* (Philadelphia: Lea & Febiger, 1929), pp. 110–111; W. H. Park, "Antitoxin Administration," *Boston Medical and Surgical Journal* 168 (Jan. 9, 1913): 73–77; Wilson G. Smillie, *Public Health, Its Promise for the Future: A Chronicle of the Development of Public Health in the United States, 1607–1914* (New York: Macmillan, 1955), pp. 394–395; W. L. Bierring, "The Modern Treatment of Diphtheria with Demonstration of Method of Preparing Antitoxin," *Journal of the Iowa State Medical Society* 15 (Apr. 1925): 171–175; Ralph Chester Williams, *The United States Public Health Service, 1798–1950* (Washington, D.C.: Commissioned Officers Association of the United States Public Health Service, 1951), pp. 179–180; Editorial: "The Antitoxin of Diphtheria at New York City—Cost of Plant," *Journal of the American Medical Association* 23 (Dec. 29, 1894): 993; Wade W. Oliver, *The Man Who Lived for Tomorrow: A Biography of William Hallock Park* (New York: E. P. Dutton, 1941), pp. 104–109.

6. Parke M. Banta, *Federal Regulation of Biologicals Applicable to the Diseases of Man*, Address before the New York State Bar Association, January 29, 1958 (Bethesda, Md.: Division of Biologics Standards, National Institutes of Health, U.S. Department of Health, Education, and Welfare, n.d.).

7. Clemens Peter von Pirquet and Bela Schick, *Serum Sickness*, trans. Bela Schick (Baltimore: Williams & Wilkins, 1951); F. J. Kojis, "Serum Sickness and Anaphylaxis Analysis of Cases of 6,211 Patients Treated with Horse Serum for Various Infections," *American Journal of Diseases of Children* 64 (Jul. 1942): 93–143; (Aug. 1942), pp. 313–350. People with allergic constitutions react with hypersensitivity to a substance that does not cause such a reaction in normal individuals.

8. R. B. Gibson, "The Concentration of Antitoxin for Therapeutic Use," *Journal of Biological Chemistry* 1 (1905–1906): 161–170; Kojis, "Serum Sickness."

9. John Fothergill, "An Account of the Sore Throat Attended with Ulcers, a Disease Which Hath of Late Years Appeared in this City, and the Parts Adjacent," *Medical Classics* 5 (Oct. 1940): 58–99; John M. T. Finney, *A Surgeon's Life: An Autobiography* (New York: G. P. Putnam's Sons, 1940), p. 329.

10. J. O'Dwyer, "Intubation of the Larynx," *New York Medical Journal* 42 (Aug. 8, 1885): 145–147.

11. A. Baginsky, "The Treatment of Diphtheria, with Special Reference to the Efficacy of Antitoxin," *Medical Record* 46 (Oct. 6, 1894): 417–418.

12. Baginsky, "The Treatment of Diphtheria with Special Reference to the Efficacy of Antitoxin," pp. 417–418; P. Ehrlich, H. Kossel, and A. Wassermann, "Ueber Gewinnung und Verwendung des Diphtherieheilserums," *Deutsche Medizinische Wochenschrift* 20 (Apr. 19, 1894): 353–355.

13. "Antitoxin, the New Cure for Diphtheria," *Ohio State Medical Journal* 37 (Dec. 1941): 1182–1183; J. H. Coughlin, "Successful Treatment of Diphtheria as Compared with Antitoxin," *Journal of the American Medical Association* 29 (Nov. 27, 1897): 1099–1104; A. Rupp, "Remarks on Antitoxin, Diphtheria, the Practitioner and History," *Medical Record* 54 (Nov. 5, 1898): 661–663.

14. Allen Weir Freeman, *Five Million Patients: The Professional Life of a Health Officer* (New York: Charles Scribner's Sons, 1946), pp. 69–70; Correspondence: "The Recent Epidemic of Diphtheria in Chicago: Results of the Serum-Antitoxin Treatment," *Medical News* 68 (Feb. 22, 1896): 216–218.

15. William Osler, *The Principles and Practice of Medicine*, 6th ed. (New York: D. Appleton, 1905), p. 261; Arthur L. Bloomfield, *A Bibliography of Internal Medicine: Communicable Diseases* (Chicago: University of Chicago Press, 1958), p. 277.

16. R. West, "Curare in Man," *Proceedings of the Royal Society of Medicine* (London) 25 (Feb. 18, 1932): 1107–1116; George B. Koelle, "Neuromuscular Blocking Agents," in Louis S. Goodman and Alfred Gilman, eds., *The Pharmacological Basis of Therapeutics*, 3d ed. (New York: Macmillan, 1965), pp. 596–597; W. A. Altemeier, W. R. Culbertson, and L. L. Gonzalez, "Clinical Experiences in the Treatment of Tetanus," *Archives of Surgery* 80 (June 1960): 977–985.

17. E. Montgomery, "A Plea for the Antiphlogistic Treatment of Disease," *Chicago Medical Journal* 29 (Aug. 1872): 474–488; R. B. Rudolph, "Bleeding in Typhoid Fever," *American Journal of the Medical Sciences* 147 (Jan. 1914): 44–56.

18. For recommendations of leading teachers, see William Osler, *The Principles and Practice of Medicine*, 4th ed. (New York: D. Appleton, 1901), pp. 41–48; James M. Anders, *A Textbook of the Practice of Medicine*, 4th ed. (Philadelphia: W. B. Saunders, 1900), p. 52. For use of drugs, see discussion of lecture by T. McCrae, "The Treatment of

Typhoid Fever," *Journal of the American Medical Association* 51 (Sept. 19, 1908): 983–987. See also Thomas McCrae, "Typhoid Fever," in William Osler and Thomas McCrae, eds., *Modern Medicine: Its Theory and Practice*, 7 vols. (Philadelphia: Lea Brothers, 1907–10), vol. 2: *Infectious Diseases*, pp. 217–229; "Hospital Notes: Treatment of Typhoid Fever in the Philadelphia Hospitals," *Medical News* 51 (Dec. 10, 1887): 676–680; "Hospital Notes: The Treatment of Typhoid Fever," ibid. (Dec. 3, 1887): 653–655; (Dec. 17, 1887): 707–710.

19. W. Coleman, "Diet in Typhoid Fever," *Journal of the American Medical Association* 53 (Oct. 9, 1909): 1145–1150; Osler, *The Principles and Practice of Medicine*, 4th ed., p. 42; W. Coleman and E. DuBois, "Calorimetric Observations on the Metabolism of Typhoid Patients with and without Food," *Archives of Internal Medicine* 15 (May 1915): 887–938.

20. Graham Lusk, "A Respiration Calorimeter for the Study of Disease," *Archives of Internal Medicine* 15 (May 1915): 793–804; W. Coleman, "The High Calory Diet in Typhoid Fever: A Study of One Hundred and Eleven Cases," *American Journal of the Medical Sciences* 143 (January 1912): 77–102; Coleman and DuBois, "Calorimetric Observations on the Metabolism of Typhoid Patients," pp. 887–938.

21. Frederick P. Gay, *Typhoid Fever Considered as a Problem of Scientific Medicine* (New York: Macmillan, 1918), pp. 217–220.

22. Osler, *Principles and Practice of Medicine*, 4th ed., p. 108; E. Ebeling, "Keilschrifttafeln medizinischen Inhalts," *Archiv für Geschichte der Medizin* 13 (May 1921): 1–42; Henry E. Sigerist, *A History of Medicine*, 2 vols. (New York: Oxford University Press, 1951–1961), vol. 1: *Primitive and Archaic Medicine*, p. 62; D. Tucher, "Incidence of Pneumonia in Two Communities in New York State," *Milbank Memorial Fund Quarterly* 30 (1952): 224–238; U.S. Bureau of the Census, *Historical Statistics of the United States, Colonial Times to 1957* (Washington, D.C.: U.S. Government Printing Office, 1960), p. 26.

23. Ralph H. Major, *Classic Descriptions of Disease* (Springfield, Ill.: Charles C Thomas, 1932), pp. 522–530; Fielding H. Garrison, *An Introduction to the History of Medicine* (Philadelphia: W. B. Saunders, 1914), pp. 366–367; F. H. Williams, "Notes on X-rays in Medicine," *Transactions of the Association of American Physicians* 11 (1896): 375–382.

24. Benjamin White, *The Biology of Pneumococcus: The Bacteriological, Biochemical, and Immunological Characters and Activities of Diplococcus Pneumoniae* (New York: Commonwealth Fund, 1938), pp. 1–21; Bloomfield, *A Bibliography of Internal Medicine*, pp. 88–94; R. Austrian, "The Gram Stain and the Etiology of Lobar Pneumonia: An Historical Note," *Bacteriological Reviews* 24 (Sept. 1960): 261–265. Doctors describe pneumococci as lancet-shaped after the instrument that was so long used to puncture the skin to let blood. Recently the technical name for the pneumococcus was officially changed to *Streptococcus pneumoniae*. Pasteur and Sternberg produced bacteremia, or the condition in

which living bacteria are present in the circulating blood, in rabbits by injecting human saliva that contained pneumococci. Pasteur reported his findings in January 1881 and Sternberg in April of the same year.

25. See citations to White, Bloomfield, and Austrian in note 24.

26. J. B. Blake, "Scientific Institutions since the Renaissance: Their Role in Medical Research," *Proceedings of the American Philosophical Society* 101 (Feb. 1957): 31–62; Simon Flexner and James Thomas Flexner, *William Henry Welch and the Heroic Age of American Medicine* (New York: Viking Press, 1941), pp. 269–278.

27. Osler, *Principles and Practice of Medicine*, 4th ed., p. 136; R. Cole, "The Treatment of Pneumonia," *Annals of Internal Medicine* 10 (Jul. 1936): 1–12; H. F. Dowling, "Frustration and Foundation: Management of Pneumonia before Antibiotics," *Journal of the American Medical Association* 220 (June 5, 1972): 1341–1345.

28. White, *The Biology of Pneumococcus*, pp. 16, 21–29, 103–109.

29. A. R. Dochez and L. J. Gillespie, "A Biological Classification of Pneumococci by Means of Immunity Reactions," *Journal of the American Medical Association* 61 (Sept. 6, 1913): 727–730; G. Cooper, M. Edwards, and C. Rosenstein, "The Separation of Types among the Pneumococci Hitherto Called Group IV and the Development of Therapeutic Antiserums for These Types," *Journal of Experimental Medicine* 49 (Mar. 1929): 461–474; G. Cooper, C. Rosenstein, A. Walter, and L. Peizer, "The Further Separation of Types among the Pneumococci Hitherto Included in Group IV and the Development of Therapeutic Antisera for These Types," ibid. 55 (Apr. 1932): 531–554; B. E. Eddy, "Cross Reactions between the Several Pneumococcus Types and their Significance in the Preparation of Polyvalent Antiserum," *Public Health Reports* 59 (Apr. 14, 1944): 485–499; E. Lund, "Laboratory Diagnosis of *Pneumococcus* Infections," *Bulletin of the World Health Organization* 23 (1960): 5–13. Many of the investigations that revealed the significance of the "higher" types of pneumococci were carried out by Maxwell Finland at the Boston City Hospital and a succession of research fellows who trained under him. See M. Finland, "The Significance of Specific Pneumococcus Types in Disease, Including Types IV to XXXII (Cooper)," *Annals of Internal Medicine* 10 (Apr. 1937): 1531–1543; M. Finland, "Recent Advances in the Epidemiology of Infections," *Medicine* 21 (Sept. 1942): 307–344.

30. R. I. Cole and A. R. Dochez, "Report on Studies on Pneumonia," *Transactions of the Association of American Physicians* 28 (1913): 606–616; E. A. Locke, "The Treatment of Type I Pneumococcus Lobar Pneumonia with Specific Serum," *Journal of the American Medical Association* 80 (May 26, 1923): 1507–1511; L. D. Felton, "A Study of the Isolation and Concentration of Specific Antibodies of Antipneumococcus Sera," *Boston Medical and Surgical Journal* 190 (May 15, 1924): 819–825. In patients with Type I pneumonia treated at the Boston City Hospital between 1924 and 1929 the fatality rate was reduced by about one-third,

as compared with patients who were not given serum, and among those admitted to the hospital within the first three days of the onset of pneumonia only 5.5 percent of patients died, as compared with 37.5 percent of the controls. In 1926 bacteriologists in Park's laboratory adopted Felton's procedure, and Park arranged to have the serums tested at Harlem Hospital by Jesse Bullowa and at Bellevue Hospital by Russell Cecil in one of the first clinical trials in which alternate patients were treated with the specific remedy while the others were given the same general care and observed as controls for statistical purposes. In the two hospitals combined the fatality rate for Type I cases was 19 percent in the treated patients compared with 33 percent in the controls. For Type II cases the difference was not as great: 35 versus 45 percent. M. Finland, "The Serum Treatment of Lobar Pneumonia," *New England Journal of Medicine* 202 (June 26, 1930): 1244–1247; W. H. Park, J. G. M. Bullowa, and M. B. Rosenbluth, "The Treatment of Lobar Pneumonia with Refined Specific Antibacterial Serum," *Journal of the American Medical Association* 91 (Nov. 17, 1928): 1503–1507.

31. F. L. Horsfall, K. Goodner, C. M. MacLeod, and A. H. Harris 2d, "Antipneumococcus Rabbit Serum as a Therapeutic Agent in Lobar Pneumonia," *Journal of the American Medical Association* 108 (May 1, 1937): 1483–1490.

32. F. Neufeld and R. Etinger-Tulczynska, "Nasale Pneumokokken-infectionen und Pneumokokkenkeimträger im Tierversuch," *Zeitschrift für Hygiene une Infektionskrankheiten* 112 (May 1931): 492–526.

33. F. H. Badger, "A Few Theories, Facts and Fallacies concerning Pneumonia Gathered from My Own Experience and Current Medical Literature," *Maine Medical Journal* 22 (Mar. 1931): 38–43; H. F. Dowling, "The Rise and Fall of Pneumonia-Control Programs," *Journal of Infectious Diseases* 127 (Feb. 1973): 201–206; M. Finland, "The Present Status of the Higher Types of Antipneumococcus Serums," *Journal of the American Medical Association* 120 (Dec. 19, 1942): 1294–1307.

34. Epidemics are discussed further in Chapter 12.

35. The clinical features of the cases in the Geneva epidemic of 1805 were described by Gaspard Vieusseux, "Mémoire sur la maladie qui a régné à Genève au printemps de 1805," *Journal de médecine, chirurgie, et pharmacie* 11 (1806): 163–182; and the pathology by A. Matthey, "Recherches sur une maladie particulière qui a régné à Genève en 1805," ibid., pp. 243–253. For a possible epidemic among the Indians in Canada in 1670, see D. J. Plazak, "Epidemic Meningitis in 1670," *Bulletin of the History of Medicine* 25 (Sept.–Oct. 1951): 457–459. For early American reports, see Elisha North, *A Treatise on a Malignant Epidemic Commonly Called Spotted Fever* (New York: T. & J. Swords, 1811). The most complete account of epidemics of this disease in the nineteenth century is in August Hirsch, *Handbook of Geographical and Historical Pathology,* trans. Charles Creighton, 3 vols. (London: New Sydenham Society, 1883–86), vol. 3: *Chronic Infective, Toxic, Parasitic, Septic and Constitutional*

Diseases, pp. 547–594. For the opinion that the disease was not contagious, see W. H. Draper, "Cerebrospinal Meningitis, or Spotted Fever,"
American Medical Times 9 (Aug. 27, 1864): 99–101; (Sept. 8, 1864):
111–114. See also A. Weichselbaum, "Ueber die Aetiologie der akuten
Meningitis cerebro-spinalis," *Fortschritte der Medizin* 5 (Sept. 15, 1887):
573–583; (Oct. 1, 1887): 620–626. After Weichselbaum's discovery, several other bacteriologists found identical cocci in cases of meningitis.
Confusion arose, however, when one reported finding cocci of similar
appearance which were gram-positive. This discrepancy may have been
the result of lack of proficiency in staining techniques in the early days
of bacteriology or the cases may have actually been caused by gram-
positive cocci, such as pneumococci or streptococci. Weichselbaum's observations were completely confirmed by several bacteriologists during a
series of epidemics that occurred around the turn of the century. See
Bloomfield, *A Bibliography of Internal Medicine,* pp. 171–172, 174–175.
See also H. Quincke, "Die Lumbalpunction des Hydrocephalus," *Berliner
Klinische Wochenschrift* 28 (Sept. 21, 1891): 929–933.

36. Bloomfield, *A Bibliography of Internal Medicine,* p. 173; E. P.
Campbell, "Meningococcemia," *American Journal of the Medical Sciences*
206 (Nov. 1943): 566–576.

37. F. W. Andrewes, "A Case of Acute Meningococcal Septicemia,"
Lancet 1 (Apr. 28, 1906): 1172–1173; R. Waterhouse, "A Case of Supra-
renal Apoplexy," ibid. 1 (Mar. 4, 1911): 577–578. Andrewes' case was
unnoticed for many years and the syndrome was named for the two
men who had popularized it, Waterhouse and Carl Friderichsen of Copenhagen. See C. Friderichsen, "Nebennierenapoplexie bei kleinen Kindern,
Jahrbuch für Kinderheilkunde 87 (Part 2, 1918): 109–125. Today it is
realized that, although hemorrhage into the adrenal glands is common,
it is not invariably present. See L. H. Dennis, R. J. Cohen, S. H. Schach-
ner, and M. E. Conrad, "Consumptive Coagulopathy in Fulminant
Meningococcemia," *Journal of the American Medical Association* 205
(Jul. 15, 1968): 179–181.

38. A. W. Hedrich, "Recent Trends in Meningococcal Disease," *Public Health Reports* 67 (May 1952): 411–420; A. W. Hedrich, "The Movements of Epidemic Meningitis, 1915–1930," *Public Health Reports* 46
(Nov. 13, 1931): 2709–2726; Hedrich, "Recent Trends in Meningococcal
Disease"; U.S. Department of Health, Education, and Welfare, Public
Health Service, *Morbidity and Mortality: Annual Supplement, Summary
1972* (Atlanta: Center for Disease Control, 1973), p. 45; Robert D. Grove
and Alice M. Hetzel, *Vital Statistics Rates in the United States, 1940–
1960* (Washington, D.C.: U.S. Department of Health, Education, and
Welfare, Public Health Service, National Center for Health Statistics,
1968), pp. 559–603; U.S. Surgeon General's Office, *Medical Department
of the United States Army in the World War,* 15 vols. (Washington,
D.C.: U.S. Government Printing Office, 1921–29), vol. 9: *Communicable
and Other Diseases,* by Joseph F. Siler and Henry C. Michie, pp. 203–

221; H. A. Feldman, "Recent Developments in the Therapy and Control of Meningococcal Infections," *Disease-a-Month*, Feb. 1966, p. 7.

39. C. Dopter, "Etude de quelques germes isolés du rhinopharynx voisins du méningocoque (paraméningococcus)," *Comptes rendus des séances de la Société de biologie et de ses filiales* 67 (Jul. 10, 1909): 74–76; Bloomfield, *A Bibliography of Internal Medicine*, pp. 177–179; M. H. Gordon and E. G. Murray, "Identification of the Meningococcus," *Journal of the Royal Army Medical Corps* 25 (Oct. 1915): 411–423; M. H. Gordon, "Identification of the Meningococcus," *Journal of Hygiene* (Cambridge, England), 17 (Jul. 1918): 290–315.

40. S. E. Branham, "Serological Relationships among Meningococci," *Bacteriological Reviews* 17 (Sept. 1953): 175–188; Feldman, "Recent Developments in the Therapy and Control of Meningococcal Infections," p. 23.

41. K. D. Blackfan, "The Treatment of Meningococcus Meningitis," *Medicine* 1 (May 1922): 139–212; G. Jochmann, "Versuche zur Serodiagnostik und Serotherapie der epidemischen Genickstarre," *Deutsche Medizinische Wochenschrift* 32 (May 17, 1906): 788–793; W. Kolle and A. von Wassermann, "Versuche zur Gewinnung und Wertbestimmung eines Meningococcenserums," ibid. (Apr. 19, 1906), pp. 609–612; Bloomfield, *A Bibliography of Internal Medicine*, p. 175; George W. Corner, *A History of the Rockefeller Institute, 1901–1953* (New York: Rockefeller Institute Press, 1964), pp. 60–62; S. Flexner, "Experimental Cerebrospinal Meningitis in Monkeys," *Journal of Experimental Medicine* 9 (Mar. 14, 1907): 142–167; S. Flexner, "Concerning a Serum-Therapy for Experimental Infection with Diplococcus Intracellularis," ibid., pp. 168–185; S. Flexner and J. W. Jobling, "An Analysis of Four Hundred Cases of Epidemic Meningitis Treated with the Anti-meningitis Serum," ibid. 10 (Sept. 5, 1908): 690–733; S. Flexner, "The Results of the Serum Treatment in Thirteen Hundred Cases of Epidemic Meningitis," ibid. 17 (May 1, 1913): 553–576.

42. William Osler and Thomas McCrae, *The Principles and Practice of Medicine*, 9th ed. (New York: D. Appleton, 1920), pp. 108–115; A. Wadsworth, "Practical Problems in the Serum Therapy of Meningococcus Meningitis," *American Journal of Public Health* 21 (Feb. 1931): 157–162; I. S. Wright, A. G. De Sanctis, and A. Sheplar, "The Determination of the Value of Serum in the Treatment for Meningococcus Meningitis," *American Journal of Diseases of Children* 38 (Oct. 1929): 730–740; K. C. Smithburn, G. F. Kempf, L. G. Zerfas, and L. H. Gilman, "Meningococcus Meningitis: A Clinical Study of One Hundred and Forty-four Epidemic Cases," *Journal of the American Medical Association* 95 (Sept. 13, 1930): 776–780.

43. Wright et al., "The Determination of the Value of Serum in the Treatment of Meningococcus Meningitis"; Wadsworth, "Practical Problems in the Serum Treatment of Meningococcus Meningitis."

44. M. Finland, "Controlling Clinical Therapeutic Experiments with

Specific Serums," *New England Journal of Medicine* 225 (Sept. 25, 1941): 495–506.

5. Streptococcal Infections

1. O. W. Holmes, "On the Contagiousness of Puerperal Fever," *New England Quarterly Journal of Medicine and Surgery* 1 (Apr. 1843): 503–530; Oliver Wendell Holmes, "The Contagiousness of Puerperal Fever," in *Medical Essays, 1842–1882* (Boston: Houghton, Mifflin, 1891), pp. 103–172 (quotations appear on pp. 103, 131, 134).

2. For an English translation of Semmelweis's report, see Ignaz Philipp Semmelweis, "The Etiology, the Concept and the Prophylaxis of Childbed Fever," ed. and trans. Frank P. Murphy, *Medical Classics* 5 (Jan.–Apr. 1941): 340–773, esp. pp. 356–357 and 392–394.

3. W. W. Ford, *Bacteriology* (New York: Hafner Publishing, 1964), pp. 83, 169; William Bulloch, *The History of Bacteriology* (London: Oxford University Press, 1938; reprint ed., 1960), pp. 149–151. "Strepto-" and "Staphylo-" were derived from Greek words meaning "twined" and "bunch of grapes." See further discussion of staphylococci in Chapter 9. Streptococci were first differentiated as hemolytic and nonhemolytic in 1903. Practically the only important infection caused by those that are not hemolytic is endocarditis. H. Schottmüller, "Die Artunterscheidung der für den Menschen pathogenen Streptokokken durch Blutagar," *Münchener Medizinische Wochenschrift* 50 (May 19, 1903): 849–853; (May 26, 1903): 909–912.

4. Arthur L. Bloomfield, *A Bibliography of Internal Medicine: Communicable Diseases* (Chicago: University of Chicago Press, 1958), pp. 108–112. This is an excellent critique of the important early discoveries in the bacteriology of scarlet fever. See also Medical Research Council (Great Britain), *A System of Bacteriology in Relation to Medicine*, 9 vols. (London: His Majesty's Stationery Office, 1929–31), II, 29–33. For the early history of scarlet fever epidemics, see John Davy Rolleston, *History of the Acute Exanthemata* (London: William Heinemann [Medical Books], 1937), pp. 47–75.

5. Bloomfield, *A Bibliography of Internal Medicine*, pp. 129–130, 115–116. Up to the mid-thirties some doctors clung to the belief that streptococci grew in the skin in scarlet fever and were unwilling to allow their patients out of isolation until peeling was complete.

6. Bloomfield, *A Bibliography of Internal Medicine*, pp. 112–113; G. F. Dick and G. H. Dick, "The Etiology of Scarlet Fever," *Journal of the American Medical Association* 82 (Jan. 26, 1924): 301–302; Dick and Dick, "A Skin Test of Susceptibility to Scarlet Fever," ibid. (Jan. 26, 1924), pp. 265–266; Dick and Dick, *Scarlet Fever* (Chicago: Year Book Publishers, 1938), pp. 62–64; A. B. Wadsworth, "The Hemolytic Streptococci and Antistreptococcus Serum in Scarlet Fever," *American Journal of Public Health* 19 (Dec. 1929): 1287–1302.

7. F. G. Blake and J. D. Trask, Jr., "Studies in Scarlet Fever: II. The Relation of the Specific Toxemia of Scarlet Fever to the Course of the Disease," *Journal of Clinical Investigation* 3 (Dec. 1926): 397–409.

8. R. C. Lancefield, "A Serological Differentiation of Human and Other Groups of Hemolytic Streptococci," *Journal of Experimental Medicine* 57 (Apr. 1933): 571–595; F. Griffith, "The Serological Classification of *Streptococcus Pyogenes*," *Journal of Hygiene* 34 (Dec. 1934): 542–584; Rebecca C. Lancefield, "Specific Relationship of Cell Composition to Biological Activity of Hemolytic Streptococci," in *The Harvey Lectures*, Series 36 (Lancaster, Pa.: Science Press Printing, 1941), pp. 251–290; Maclyn McCarty, "The Hemolytic Streptococci," in René J. Dubos and James G. Hirsch, eds., *Bacterial and Mycotic Infections of Man* (Philadelphia: J. B. Lippincott, 1958), pp. 364–365, 368–369, 376–383.

9. Anna Wessels Williams, *Streptococci in Relation to Man in Health and Disease* (Baltimore: Williams and Wilkins, 1932), pp. 152–158; Benjamin P. Thomas, *Abraham Lincoln: A Biography* (New York: Alfred A. Knopf, 1952), p. 220.

10. Dick and Dick, *Scarlet Fever*, pp. 62–64; L. W. Hunt, "The Treatment of Scarlet Fever with Antitoxin," *Journal of the American Medical Association* 101 (Nov. 4, 1933): 1444–1447.

11. A. R. Dochez and L. Sherman, "The Significance of Streptococcus Hemolyticus in Scarlet Fever," *Journal of the American Medical Association* 82 (Feb. 16, 1924): 542–544; H. F. Dowling, "Diphtheria as a Model: Introduction of Serums and Vaccines for Scarlet Fever and Pneumococcal Pneumonia," *Journal of the American Medical Association* 226 (Oct, 29, 1973): 550–553; Bloomfield, *A Bibliography of Internal Medicine*, pp. 119–120; Wadsworth, "The Hemolytic Streptococci and Antistreptococcus Serum in Scarlet Fever"; Anna M. Sexton, *A Chronicle of the Division of Laboratories and Research, New York State Department of Health: The First Fifty Years, 1914–1964* (Lunenberg, Vt.: Stinehour Press, 1967), pp. 104–109; G. H. Dick, "A Brief History of the Scarlet Fever Patent," *Journal of the American Medical Association* 113 (Jul. 22, 1939): 327–330. Many investigators believed that the restrictions imposed by the owners of the patent hampered their research. See D. T. Smith, discussion of F. L. Hutchison, "The Administration of Medical Patents for the Public Welfare," ibid., p. 334; G. W. Anderson, "Obstacles to Effective Control and Prevention of Scarlet Fever," *Journal of the Iowa State Medical Society* 30 (Dec. 1940): 580–584. Consultants to the New York State Health Department Laboratories advised against obtaining a license to produce serum by the Dicks' method because "the restrictions authorized and/or imposed by a license from the Scarlet Fever Committee would result in the production of a serum by the State Department of Health of inferior therapeutic value, would limit further development of improved methods and thereby would imperil the public health." They also recommended that the state contest any suit by the Scarlet Fever Committee for infringement of its patent. No

suit was instituted and the New York State Laboratories continued to produce antiscarlatinal serum, although it was not made available outside the state. See Sexton, *A Chronicle of the Division of Laboratories and Research*, pp. 107–108; Bloomfield, *A Bibliography of Internal Medicine*, p. 121.

12. James A. Anders, "Erysipelas," in William Osler and Thomas McCrae, eds., *Modern Medicine: Its Theory and Practice*, 7 vols. (Philadelphia: Lea Brothers, 1907–1910), vol. 2: *Infectious Diseases*, pp. 525–536; Bloomfield, *A Bibliography of Internal Medicine*, pp. 128–129; E. D. Churchill, "The Pandemic of Wound Infection in Hospitals: Studies in the History of Wound Healing," *Journal of the History of Medicine* 20 (Oct. 1965): 390–404.

13. Bloomfield, *A Bibliography of Internal Medicine*, p. 131; K. E. Birkhaug, "Studies on the Biology of the Streptococcus erysipelatis: I. Agglutination and Agglutinin Absorption with the Streptococcus erysipelatis," *Bulletin of the Johns Hopkins Hospital* 36 (Apr. 1925): 248–259; Lancefield, "A Serological Differentiation of Human and Other Groups of Hemolytic Streptococci"; Griffith, "The Serological Classification of *Streptococcus Pyogenes*."

14. D. Symmers and K. M. Lewis, "The Antitoxin Treatment of Erysipelas: Further Observations," *Journal of the American Medical Association* 99 (Sept. 24, 1932): 1082–1083; W. S. McCann, "The Serum Treatment of Erysipelas," *Journal of the American Medical Association* 91 (Jul. 14, 1928): 78–81.

15. U.S. Bureau of the Census, *Historical Statistics of the United States, Colonial Times to 1957* (Washington, D.C.: U.S. Government Printing Office, 1960), p. 257; Louis I. Dublin and Alfred J. Lotka, *Twenty-five Years of Health Progress: A Study of the Mortality Experience among the Industrial Policyholders of the Metropolitan Life Insurance Company, 1911 to 1935* (New York: Metropolitan Life Insurance, 1937), p. 353.

16. F. E. Kobrin, "The American Midwife Controversy: A Crisis of Professionalization," *Bulletin of the History of Medicine* 40 (Jul.–Aug. 1966): 350–363; P. Van Ingen, "The History of Child Welfare Work in the United States," in Mazyck P. Ravenel, ed., *A Half Century of Public Health* (New York: American Public Health Association, 1921), pp. 290–322.

17. Viola I. Paradise, "Maternity Care and the Welfare of Young Children in a Homesteading County in Montana," *U.S. Children's Bureau Publication No. 34* (Washington, D.C.: U.S. Government Printing Office, 1919), pp. 27–52.

18. J. W. Williams, "Medical Education and the Midwife Problem in the United States," *Journal of the American Medical Association* 58 (Jan. 6, 1912): 1–7; American Medical Association, *A History of the Council on Medical Education and Hospitals of the American Medical*

Association, 1904–1959 (Chicago: American Medical Association, n.d.), pp. 9, 11–14.

19. Arthur E. Hertzler, *The Horse and Buggy Doctor* (New York: Harper & Brothers, 1938), pp. 117–118.

20. George Rosen, *A History of Public Health* (New York: MD Publications, 1958), pp. 360–364; J. G. Marmol, A. L. Scriggins, and R. F. Vollman, "History of the Maternal Mortality Study Committees in the United States," *Obstetrics and Gynecology* 34 (Jul. 1969): 123–138; U.S. Bureau of the Census, *Historical Statistics of the United States*, p. 25.

21. G. F. Dick and G. H. Dick, "Scarlet Fever Toxin in Preventive Immunization," *Journal of the American Medical Association* 82 (Feb. 16, 1924): 544–545; E. S. Platou, "Present Status of Scarlet Fever Prevention and Serum Treatment," *Minnesota Medicine* 15 (Oct. 1932): 697–702. By 1926 immunization for scarlet fever was advocated by 27 percent of the health departments in 90 cities that responded to a questionnaire sent out by a committee of the American Public Health Association. See D. B. Armstrong and W. F. Walker, "Current Immunization Practice in Diphtheria, Scarlet Fever and Measles," *American Journal of Public Health* 16 (Nov. 1926): 1099–1102.

22. Bloomfield, *A Bibliography of Internal Medicine*, pp. 135–150 (for the relationship of rheumatic fever and sore throat see p. 146). See also G. E. Murphy, "The Evolution of Our Knowledge of Rheumatic Fever: An Historical Survey with Particular Emphasis on Rheumatic Heart Disease," *Bulletin of the History of Medicine* 14 (Jul. 1943): 123–147. The state of knowledge at the beginning of the twentieth century was epitomized in the 1901 edition of Osler's textbook. He remarked that tonsillitis often preceded rheumatic fever—"some would say always." He described not only the arthritis and the involvement of various parts of the heart, but also the rarer manifestations: pleurisy, pneumonia, and the appearance of nodules under the skin. He was less sure of the relationship of chorea, noting that only 88 of 554 cases of chorea analyzed were associated with rheumatic fever. Among the possible causes he gave most prominence to the infectious theory but was puzzled by the fact that each of several investigators had cultured a different microorganism from the throats of patients with rheumatic fever. See William Osler, *The Principles and Practice of Medicine*, 4th ed. (New York: D. Appleton, 1901), pp. 80, 166–175, 1079–1088. See also L. Aschoff, "Zur Myocarditisfrage," *Verhandlungen der Deutschen Gesellschaft für Pathologie* 8 (1904): 46–53.

23. For the frequency of rheumatic fever, see Alvin F. Coburn, *The Factor of Infection in the Rheumatic State* (Baltimore: Williams & Wilkins, 1931), p. 2. The deaths from acute rheumatic fever were for policyholders of the Metropolitan Life Insurance Company. See Dublin and Lotka, *Twenty-five Years of Health Progress*, p. 376. See also O. F. Hedley, "Rheumatic Heart Disease in Philadelphia Hospitals: III. Fatal

Rheumatic Heart Disease and Subacute Bacterial Endocarditis," *Public Health Reports* 55 (Sept. 20, 1940): 1707–1740.

24. Bloomfield, *A Bibliography of Internal Medicine*, pp. 154–156; H. F. Swift, "Rheumatic Fever," *Journal of the American Medical Association* 92 (June 22, 1929): 2071–2083.

25. Coburn, *The Factor of Infection in the Rheumatic State*, pp. 156–173, 211.

26. E. W. Todd, "Antihaemolysin Titres in Haemolytic Streptococcal Infections and Their Significance in Rheumatic Fever," *British Journal of Experimental Pathology* 13 (June 1932): 248–258; A. F. Coburn and R. H. Pauli, "Studies on the Relationship of Streptococcus hemolyticus to the Rheumatic Process: III. Observations on the Immunological Responses of Rheumatic Subjects to Hemolytic Streptococcus," *Journal of Experimental Medicine* 56 (Nov. 1932): 651–676; Edward E. Fischel, "The Role of Allergy in the Pathogenesis of Rheumatic Fever," *American Journal of Medicine* 7 (Dec. 1949): 772–793.

27. For the use of salicylates in rheumatic fever, see Bloomfield, *A Bibliography of Internal Medicine*, pp. 143–145.

28. C. Rammelkamp, R. S. Weaver, and J. H. Dingle, "Significance of the Epidemiological Differences between Acute Nephritis and Acute Rheumatic Fever," *Transactions of the Association of American Physicians* 65 (1952): 168–174.

6. Tuberculosis

1. Walter A. Wells, *A Doctor's Life of Keats* (New York: Vantage Press, 1959), pp. 174, 186, 211; William Hale-White, *Keats as Doctor and Patient* (London: Oxford University Press, 1938), p. 52; U.S. Bureau of the Census, *Historical Statistics of the United States, Colonial Times to 1957* (Washington, D.C.: U.S. Government Printing Office, 1960), p. 26. When community-wide surveys were made later, they showed that approximately ten persons were ill with tuberculosis for every one who died of the disease, and this ratio of 10:1 was widely used before the advent of streptomycin as an indication of the prevalence of tuberculosis.

2. Henry A. Sigerist, *A History of Medicine*, 2 vols. (New York: Oxford University Press, 1951, 1961), vol. 1: *Primitive and Archaic Medicine*, pp. 53–54; E. R. Long, "The Decline of Tuberculosis, with Special Reference to Its Generalized Form," *Bulletin of the History of Medicine* 8 (June 1940): 819–843.

3. René Jules Dubos and Jean Dubos, *The White Plague: Tuberculosis, Man, and Society* (Boston: Little, Brown, 1952), p. 8; Long, "The Decline of Tuberculosis"; R. J. Dubos, "The Tubercle Bacillus and Tuberculosis," *American Scientist* 37 (Summer 1949): 353–370, esp. p. 355.

4. Esmond R. Long, *A History of Pathology*, enlarged and corrected ed. (New York: Dover Publications, 1965), pp. 51–52, 81–82.

5. Dubos and Dubos, *The White Plague*, pp. 29–30; W. Budd, "The

Nature and the Mode of Propagation of Phthisis," *Lancet* 2 (Oct. 12, 1867): 451–452.

6. Arthur L. Bloomfield, *A Bibliography of Internal Medicine: Communicable Diseases* (Chicago: University of Chicago Press, 1958), pp. 207–208; William Bulloch, *The History of Bacteriology* (London: Oxford University Press, 1938; reprint ed., 1960), pp. 147–149; R. Koch, "Die Aetiologie der Tuberculose," *Berliner Klinische Wochenschrift* 19 (Apr. 10, 1882): 221–230, trans. W. de Rouville in Frank P. Murphy, ed. and trans., *Medical Classics* 2 (Apr. 1938): 853–880. See also Bloomfield, *A Bibliography of Internal Medicine*, pp. 212–213. The modification of Ehrlich's staining method in general use today is known as the Ziehl-Nielson method after two bacteriologists who made minor changes in the technique.

7. In addition to showing that immunity could be produced by the injection of killed bacteria (see Chapter 3), Smith had more recently solved the riddle of Texas cattle fever by demonstrating that the infectious agent was transmitted from animal to animal by ticks. Paul F. Clark, *Pioneer Microbiologists of America* (Madison, Wis.: University of Wisconsin Press, 1961), pp. 119–129; T. Smith, "A Comparative Study of Bovine Tubercle Bacilli and of Human Bacilli from Sputum," *Journal of Experimental Medicine* 3 (Jul. and Sept. 1898): 451–511; R. Koch, "The Fight against Tuberculosis in the Light of the Experience That Has Been Gained in the Successful Combat of Other Diseases," *British Medical Journal* 2 (Jul. 27, 1901): 189–193; Editorial: "The Fight against Tuberculosis," ibid., pp. 221–223; Editorial: "Professor Koch's Address," *Journal of the American Medical Association* 37 (Jul. 27, 1901): 266; M. P. Ravenel, "The Comparative Virulence of the Tubercle Bacillus from Human and Bovine Sources," *University of Pennsylvania Medical Bulletin* 14 (Sept. 1901): 238–256; Editorial: "The Royal Commission on Human and Bovine Tuberculosis," *Lancet* 2 (Jul. 15, 1911): 166–167; Lewis W. Hunt, *The People versus Tuberculosis* (Chicago: Tuberculosis Institute of Chicago and Cook County, 1966), pp. 29, 45–47.

8. Gerald B. Webb, *Tuberculosis* (New York: Paul B. Hoeber, 1936), pp. 139–143, 30–33; Fielding H. Garrison, *An Introduction to the History of Medicine* (Philadelphia: W. B. Saunders, 1914), p. 218; L. E. Henry, Jr., and R. MacKeith, "Samuel Johnson's Childhood Illnesses and the King's Evil," *Medical History* 10 (Oct. 1966): 386–389.

9. Webb, *Tuberculosis*, pp. 150–152.

10. R. Koch, "Weitere Mitteilungen über ein Heilmittel gegen Tuberculose," *Deutsche Medizinische Wochenschrift* 16 (Nov. 13, 1890): 1029–1032; J. Lister, "Lecture on Koch's Treatment of Tuberculosis," ibid. (Dec. 6, 1890), pp. 1257–1259. It was rumored that Koch at first withheld information on the method of preparing tuberculin because the German government wished to retain a monopoly on this supposed remedy. See Editorial, ibid. (Nov. 22, 1890), pp. 1107–1108.

11. An injection of tuberculin might cause a flare-up of the disease if a large quantity was injected, such as was sometimes employed in the decade or two following its introduction. There is no evidence that the small quantity of tuberculin used today in the tuberculin test has ever caused a flare-up. See Louis Hamman and Samuel Wolman, *Tuberculin in Diagnosis and Treatment* (New York: D. Appleton, 1912), esp. pp. 205–208. See also Bloomfield, *A Bibliography of Internal Medicine,* p. 219; E. L. Trudeau, "The Therapeutic Use of Tuberculin Combined with Sanatarium Treatment of Tuberculosis," *American Journal of the Medical Sciences* 132 (Aug. 1906): 175–186.

12. William Osler, *The Principles and Practice of Medicine,* 8th ed. (New York: D. Appleton, 1914), p. 227; ibid., 13th ed., rev. Henry A. Christian (New York: D. Appleton-Century, 1938), p. 261; Charles Henry May, *May's Manual of the Diseases of the Eye,* 24th ed., rev. and ed. James H. Allen (Baltimore: Williams & Wilkins, 1968), pp. 359–360.

13. Ella Marie Elizabeth Flick, *Beloved Crusader—Lawrence F. Flick* (Philadelphia: Dorrance, 1944), pp. 217–218, 321.

14. J. L. Cowan, "Climate and Consumption," *Overland Monthly* 55 (Mar. 1910): 244–253; W. G. Brown, "Some Confessions of a 'T.B.,' " *Atlantic Monthly* 113 (June 1914): 747–754; Editorial: "Indigent Consumptives in the Southwest," *American Journal of Public Health* 2 (June 1912): 449.

15. Garrison, *Introduction to the History of Medicine,* p. 651; Flick, *Beloved Crusader,* p. 55; Edward Livingston Trudeau, *An Autobiography* (Philadelphia: Lea & Febiger, 1916), pp. 77, 97, 105–108.

16. V. Y. Bowditch, "A History of the Growth of the Anti-tuberculosis Movement in Massachusetts, and the Lessons to Be Learned Therefrom," *Boston Medical and Surgical Journal* 175 (Dec. 14, 1916): 847–854; Flick, *Beloved Crusader,* pp. 156–159, 179; Webb, *Tuberculosis,* p. 167.

17. Amy Lowell, *John Keats* (Boston: Houghton Mifflin, 1925), p. 509; B. J. Kendall, "The Way to Health: II. Self-Cure with Fresh Cream," *Worlds Work* 19 (Apr. 1910): 12774–12775; Esmond R. Long, "A Pathologist's Reflections of the Control of Tuberculosis," in Dwight J. Ingle, ed., *A Dozen Doctors: Autobiographic Sketches* (Chicago: University of Chicago Press, 1963), p. 136.

18. Dubos and Dubos, *The White Plague,* p. 141; Osler, *The Principles and Practice of Medicine,* 8th ed., p. 225; ibid., 16th ed., rev. Henry A. Christian (New York: D. Appleton-Century, 1947), pp. 285–286.

19. G. Ronzoni, "Carlo Forlanini and the Contribution of the Italian School to Pulmonary Collapse Therapy," *American Review of Tuberculosis* 18 (Aug. 1928): 101–108; Lawrason Brown, *The Story of Clinical Pulmonary Tuberculosis* (Baltimore: Williams & Wilkins, 1941), pp. 236, 270–277; Osler, *Principles and Practice of Medicine,* 8th ed., p. 229; F. Tice, "A Century of Tuberculosis in Illinois," *Illinois Medical Journal* 77 (May 1940): 455–465; Henry D. Chadwick and Alton S. Pope,

The Modern Attack on Tuberculosis (New York: Commonwealth Fund, 1942), p. 45.

20. William Osler, *The Principles and Practice of Medicine*, 4th ed. (New York: D. Appleton, 1901), p. 336; E. R. Baldwin, "History of Tuberculosis Research in America," *Yale Journal of Biology and Medicine* 15 (Jan. 1943): 301–309.

21. Editorial: "The Gold Treatment of Tuberculosis," *American Journal of Public Health* 15 (Jul. 1925): 631; J. B. Amberson, B. T. McMahon, and M. Pinner, "A Clinical Trial of Sanocrysin in Pulmonary Tuberculosis," *American Review of Tuberculosis* 24 (Oct. 1931): 401–435. In 1948 the Streptomycin Committee of the British Medical Research Council stated that the controlled trial of gold treatment by Amberson and his associates was "the only report of an adequate controlled trial of tuberculosis we have been able to find in the literature." "Streptomycin Treatment of Pulmonary Tuberculosis: A Medical Research Council Investigation," *British Medical Journal* 2 (Oct. 30, 1948): 769–782. The patients did not know who was receiving the gold salt or who was receiving the distilled water. Distilled water in a case like this is called a placebo, which is a substance without pharmacological effect that is given to a subject of an experiment who supposes it to have a preventive or therapeutic effect, as the case may be.

22. U.S. Bureau of the Census, *Historical Statistics of the United States*, p. 26; Richard Harrison Shryock, *National Tuberculosis Association, 1904–1954: A Study of the Voluntary Health Movement in the United States* (New York: National Tuberculosis Association, 1957), pp. 63, 51–52, 59–62, 69–79; R. G. Paterson, "The Evolution of Official Tuberculosis Control in the United States," *Public Health Reports* 62 (Mar. 1947): 336–341.

23. Flick, *Beloved Crusader*, pp. 15–16, 205; Charles-Edward A. Winslow, *The Life of Hermann M. Biggs, M.D., D.Sc., LL.D.: Physician and Statesman of the Public Health* (Philadelphia: Lea & Febiger, 1929), pp. 135, 143–152. The quotation is cited on p. 144.

24. Shryock, *National Tuberculosis Association*, pp. 50–51; J. A. Miller, "The Beginnings of the American Antituberculosis Movement," *American Review of Tuberculosis* 48 (Dec. 1943): 361–381; Hunt, *The People versus Tuberculosis*, p. 2. The quotation is from Tice, "A Century of Tuberculosis in Illinois."

25. Brown, *Story of Clinical Pulmonary Tuberculosis*, pp. 215–217; Shryock, *National Tuberculosis Association*, pp. 167–170.

26. Charles Singer and E. Ashworth Underwood, *A Short History of Medicine*, 2d ed. rev. (New York: Oxford University Press, 1962), pp. 423–426; Louis I. Dublin, *A 40 Year Campaign against Tuberculosis* (New York: Metropolitan Life Insurance, 1952), pp. 1–36, 40, 42, 44; Hunt, *The People versus Tuberculosis*, pp. 32–34.

27. BCG stands for Bacillus of Calmette and Guérin. Singer and Un-

derwood, *A Short History of Medicine*, pp. 426–427. See also S. R. Rosenthal, E. I. Leslie, and E. Loewinsohn, "BCG Vaccination in All Age Groups," *Journal of the American Medical Association* 136 (Jan. 10, 1948): 73–79, which includes a summary of the Lübeck disaster, and Carl Muschenheim, "Tuberculosis," in Paul B. Beeson and Walsh McDermott, eds., *Textbook of Medicine*, 12th ed. (Philadelphia: W. B. Saunders, 1967), p. 294.

28. Shryock, *National Tuberculosis Association*, p. 267; H. E. Hilleboe, "Recent Developments in Tuberculosis Control," *American Review of Tuberculosis* 55 (Jan. 1947): 17–20; Godias J. Drolet and Anthony M. Lowell, *A Half Century's Progress against Tuberculosis in New York City, 1900–1950* (New York: New York Tuberculosis and Health Association, 1952), pp. iii, 23; H. R. Edwards and J. R. Drolet, "The Implications of Changing Morbidity and Mortality Rates from Tuberculosis," *American Review of Tuberculosis* 61 (Jan. 1950): 39–50; U.S. Public Health Service, Tuberculosis Control Division, *Tuberculosis in the United States*, 4 vols. (New York: National Tuberculosis Association, 1943–46), III, 5, 7; Drolet and Lowell, *A Half Century's Progress against Tuberculosis in New York City*, pp. 45–46; E. Bogen, Editorial: "Ten Beds per Death or Eradication, Not Reduction, of Tuberculosis," *American Review of Tuberculosis* 59 (Dec. 1949): 707–709.

29. Robert D. Grove and Alice M. Hetzel, *Vital Statistics Rates in the United States, 1940–1960* (Washington, D.C.: U.S. Department of Health, Education, and Welfare, Public Health Service, National Center for Health Statistics, 1968), p. 79.

7. The Venereal Diseases

1. Ralph H. Major, *Classic Descriptions of Disease* (Springfield, Ill.: Charles C Thomas, 1932), p. 35.

2. Frederick A. Pottle, ed., *Boswell's London Journal, 1762–1763* (New York: McGraw-Hill, 1950), p. 164; W. B. Ober, "Boswell's Clap," *Journal of the American Medical Association* 212 (Apr. 6, 1970): 91–95.

3. Today the latter term is used only as a vulgarism. See D'Arcy Power, "Clap and the Pox in English Literature," *British Journal of Venereal Diseases* 14 (Apr. 1938): 105–118.

4. Henry E. Sigerist, *A History of Medicine*, 2 vols. (New York: Oxford University Press, 1951, 1961), vol. 1: *Primitive and Archaic Medicine*, p. 482; Ernest Finger, *Gonorrhoea Being the Translation of Blenorrhoea of the Sexual Organs and Its Complications*, 3d ed. (New York: William Wood, 1894), p. 4. For early history of gonorrhea, see also Theodor Rosebury, *Microbes and Morals: The Strange Story of Venereal Disease* (New York: Viking Press, 1971), pp. 10–22.

5. Rosebury, *Microbes and Morals*, p. 29.

6. For Columbian theory, see William Allen Pusey, *The History and Epidemiology of Syphilis* (Springfield, Ill.: Charles C Thomas, 1933). For

modern variant of this theory, see A. W. Crosby, Jr., "The Early History of Syphilis: A Reappraisal," *American Anthropologist* 71 (Apr. 1969): 218–227. For pre-Columbian theory, see William Osler and John W. Churchman, "Syphilis," in William Osler and Thomas McCrae, eds., *Modern Medicine: Its Theory and Practice*, 7 vols. (Philadelphia: Lea Brothers, 1907–1910), vol. 3: *Infectious Diseases (continued)—Diseases of the Respiratory Tract*, pp. 436–439; R. R. Willcox, "Venereal Disease in the Bible," *British Journal of Venereal Diseases* 25 (Jul. 1949): 28–33; Thomas B. Turner, "Syphilis and the Treponematoses," in Stuart Mudd, ed., *Infectious Agents and Host Reactions* (Philadelphia: W. B. Saunders, 1970), pp. 382–385. See also Karl Sudhoff, *The Earliest Printed Literature on Syphilis*, adapted by Charles Singer (Florence: R. Lier, 1925), pp. ix–x, xii, xviii.

7. For the broader viewpoint, see E. H. Hudson, "Treponematosis and Anthropology," *Annals of Internal Medicine* 58 (June 1963): 1037–1048; Turner, "Syphilis and the Treponematoses." See also Rosebury, *Microbes and Morals*, pp. 67–93.

8. Pusey, *The History and Epidemiology of Syphilis*, pp. 37–59; Arthur L. Bloomfield, *A Bibliography of Internal Medicine: Communicable Diseases* (Chicago: University of Chicago Press, 1958), pp. 298–312.

9. John Kobler, *The Reluctant Surgeon: A Biography of John Hunter* (Garden City, N.Y.: Doubleday, Dolphin Books, 1962), pp. 179–180, 185–187, 369.

10. In 1947 investigators at the University of Chicago succeeded in producing gonococcal infection in the eyes of rabbits which closely resembled the infection in humans. M. J. Drell, C. P. Miller, and M. Bohnhoff, "Gonorrheal Iritis, Experimental Production in the Rabbit," *Archives of Ophthalmology* 38 (Aug. 1947): 221–244. Also, gonococcal urethritis has recently been produced in chimpanzees and transferred from one chimpanzee to another. C. T. Lucas, F. Chandler, Jr., J. E. Martin, Jr., and J. D. Schmale, "Transfer of Gonococcal Urethritis from Man to Chimpanzee: An Animal Model for Gonorrhea," *Journal of the American Medical Association* 216 (June 7, 1971): 1612–1614. See also J. Hill, "Experimental Infection with *Neisseria Gonorrhoeae*: I. Human Inoculations," *American Journal of Syphilis, Gonorrhea, and Venereal Diseases* 27 (November 1943): 733–771, esp. pp. 741, 745; Finger, *Gonorrhoea Being the Translation of Blenorrhoea*, pp. 4–11.

11. Bloomfield, *A Bibliography of Internal Medicine*, pp. 185–188; Hill, "Experimental Infection with *Neisseria Gonorrhoeae*." Hill accepts the experiment of Anfuso in 1891 as the first valid instance of transfer of the disease by culture (p. 745).

12. James Boswell was equally perplexed as he pondered the probable guilt of Louisa, with whom he had had a series of assignations: "What! thought I, can this beautiful, this sensible, and this agreeable woman be so sadly defiled? Can corruption lodge beneath so fair a form? Can she who professed delicacy of sentiment and sincere regard for me, use me

so very basely and so very cruelly? No, it is impossible. I have just got a gleet by irritating the parts too much with excessive venery. And yet these damned twinges, that scalding heat, and that deep-tinged loathesome matter are the strongest proofs of an infection . . . But perhaps she was ignorant of her being ill. A pretty conjecture indeed! No she could not be ignorant. Yes, yes, she intended to make the most of me." See Pottle, ed., *Boswell's London Journal*, pp. 155–156.

13. Emil Noeggerath, *Die latente Gonorrhoë im weiblichen Geschlect* (Bonn: Max Cohen & Sohn, 1872). Four years later, however, when Noeggerath spoke on the subject at a meeting of the American Gynecological Society, every doctor who discussed his paper disagreed with him. See E. Noeggerath, "Latent Gonorrhea, Especially with Regard to Its Influence on the Fertility of Women," *Transactions of the American Gynecological Society* 1 (1876): 268–293; James R. Hayden, *Venereal Diseases: A Manual for Students and Practitioners*, 3d ed. (Philadelphia: Lea Brothers, 1901), pp. 20–21; Robert W. Taylor, *The Pathology and Treatment of Venereal Diseases* (Philadelphia: Lea Brothers, 1895), p. 289.

14. H. Friedenwald, "Paths of Progress in Opthalmology," *Archives of Ophthalmology* 27 (June 1942): 1047–1096; John Vetch, *A Practical Treatise on the Diseases of the Eye* (London: G. and W. B. Whittaker, 1820), pp. 238–239, 243, 253, 256–257; Hill, "Experimental Infection with *Neisseria Gonorrhoeae*," pp. 760–762. Even though investigators usually selected patients with incurable diseases as their subjects, such studies on humans would be severely condemned by doctors and laymen alike today because there is no evidence that the experiments were explained to the patients and their consent obtained beforehand. See H. F. Dowling, "Human Experimentation in Infectious Diseases," *Journal of the American Medical Association* 198 (Nov. 28, 1966): 997–999.

15. Samuel Gross of Philadelphia, perhaps the leading American surgeon of his day, denied the relationship of the two conditions in the first edition of his textbook, published in 1859, and maintained this stand until the sixth edition of 1882. J. F. McCahey, "The History of Gonorrheal Arthritis," *British Journal of Venereal Diseases* 9 (July 1933): 194–201; Benjamin Bell, *Treatise on Gonorrhoea Virulenta and Lues Venerea*, 2 vols. (Dublin: William Jones, 1793), I, 265; Bloomfield, *A Bibliography of Internal Medicine*, pp. 189–191.

16. Bloomfield, *A Bibliography of Internal Medicine*, pp. 188–189.

17. For the reference to Arderne see Power, "Clap and the Pox." See also Jean Astruc, *A Treatise of Venereal Diseases*, 2d ed., trans. W. Barrowby (London: W. Innys and J. Richardson, J. Clarke, R. Manby, H. S. Cox, L. Davis, and C. Reymers, 1756), pp. 263–271, 287, 177–207; Bell, *Treatise on Gonorrhoea Virulenta and Lues Venerea*, p. vi; Everett M. Culver and James R. Hayden, *Manual of Venereal Diseases Being an Epitome of the Most Approved Treatment* (Philadelphia: Lea Brothers, 1891), pp. 31–34.

18. Percy Starr Pelouze, *Gonorrhea in the Male and Female*, 3d ed.

(Philadelphia: W. B. Saunders, 1939), pp. 147–150, 202–224, 384–392; George Robertson Livermore and Edward Armin Schumann, *Gonorrhea and Kindred Affections* (New York: D. Appleton, 1929), p. 5.

19. Bell, *Treatise on Gonorrhoea Virulenta and Lues Venerea*, p. 184.

20. McCahey, "History of Gonorrheal Arthritis"; Medical Research Council (Great Britain), *A System of Bacteriology in Relation to Medicine*, 9 vols. (London: His Majesty's Stationery Office, 1929–31), II, 281–282.

21. P. S. Hench, C. H. Slocumb, and W. C. Popp, "Fever Therapy: Results for Gonorrheal Arthritis, Chronic Infectious (Atrophic) Arthritis, and Other Forms of 'Rheumatism,'" *Journal of the American Medical Association* 104 (May 18, 1935): 1779–1790; S. L. Warren and K. M. Wilson, "Treatment of Gonococcal Infections by Artificial (General) Hyperthermia: A Preliminary Report," *American Journal of Obstetrics and Gynecology* 24 (Oct. 1932): 592–598; Pelouze, *Gonorrhea in the Male and Female*, pp. 266–267.

22. J. G. Clark, "A Critical Summary of the Recent Literature on Gonorrhoea in Women," *American Journal of the Medical Sciences* 119 (Jan. 1900): 73–81.

23. R. A. Benson and A. Steer, "Vaginitis in Children: A Review of the Literature," *American Journal of Diseases of Children* 53 (Mar. 1937): 806–824.

24. Fielding H. Garrison, *An Introduction to the History of Medicine* (Philadelphia: W. B. Saunders, 1914), p. 614; O. T. Schultz, "Fritz Schaudinn: A Review of His Work," *Bulletin of the Johns Hopkins Hospital* 19 (June 1908): 169–173; Editorial: "The Etiology of Syphilis," *Journal of the American Medical Association* 45 (Jul. 8, 1905): 108; Bloomfield, *A Bibliography of Internal Medicine*, pp. 322–323.

25. A. Wassermann, A. Neisser, and C. Bruck, "Eine serodiagnostische Reaktion bei Syphilis," *Deutsche Medizinische Wochenschrift* 32 (May 10, 1906): 745–746.

26. H. Noguchi, "A Method for the Pure Cultivation of *Treponema pallidum (Spirochaeta pallida),*" *Journal of Experimental Medicine* 14 (Aug. 1911): 99–108; C. C. Kast and J. A. Kolmer, "Concerning the Cultivation of *Spirocheta Pallida,*" *American Journal of Syphilis* 13 (Jul. 1929): 419–453.

27. H. Noguchi and J. W. Moore, "A Demonstration of *Treponema Pallidum* in the Brain in Cases of General Paralysis," *Journal of Experimental Medicine* 17 (Feb. 1913): 232–238; Editorial: "The Nature of General Paralysis," *Journal of the American Medical Association* 61 (Sept. 6, 1913): 776.

28. Pusey, *The History and Epidemiology of Syphilis*, p. 37; William J. Brown, James F. Donohue, Norman W. Axnick, Joseph H. Blount, Neal H. Ewen, and Oscar G. Jones, *Syphilis and Other Venereal Diseases* (Cambridge: Harvard University Press, 1970), pp. 11–14; Thomas Parran, *Shadow on the Land: Syphilis* (New York: Reynal & Hitchcock, 1937),

pp. 42–43; Pusey, *The History and Epidemiology of Syphilis,* p. 57. In his textbook of medicine, published in 1901, Osler recommended mercury either by mouth, by rubbing into the skin, by injection, or by fumigation for early syphilis. He claimed that iodides given in the late stages would cause syphilitic ulcers and tumors to "melt away." Subsequent observers found this viewpoint much too sanguine. See William Osler, *The Principles and Practice of Medicine,* 4th ed. (New York: D. Appleton, 1901), pp. 254–255. See also L. W. Harrison, "Half a Lifetime in the Management of Venereal Diseases," *Archives of Dermatology and Syphilology* 73 (May 1956): 441–454.

29. Harry F. Dowling, *Medicines for Man: The Development, Regulation, and Use of Prescription Drugs* (New York: Alfred A. Knopf, 1970), pp. 17, 38–39; P. Guttman and P. Ehrlich, "Ueber die Wirkung des Methylenblau bei Malaria," *Berliner Klinische Wochenschrift* 28 (Sept. 28, 1891): 953–956; Martha Marquardt, *Paul Ehrlich* (New York: Henry Schuman, 1951), pp. 141–157, 163–179; P. Ehrlich and S. Hata, *Die experimentelle Chemotherapie der Spirillosen* (Berlin: Julius Springer, 1910); J. J. Abraham, "Some Account of the History of the Treatment of Syphilis," *British Journal of Venereal Diseases* 24 (Dec. 1948): 153–160. Trypanosomes are oval-shaped with pointed ends and are best known as the cause of encephalitis, an inflammation of the substance of the brain. This illness is popularly known as sleeping sickness because in some forms the patient may be in coma for days or weeks. Trypanosomes also cause Chagas disease, an infection mainly involving the heart and the digestive system.

30. For a fuller account of the events following Ehrlich's discovery, see H. F. Dowling, "Comparisons and Contrasts between the Early Arsphenamine and Early Antibiotic Periods," *Bulletin of the History of Medicine* 47 (May–June 1973): 236–249.

31. In 1913 Neisser felt impelled to plead in an open letter that the price be lowered "so that the modest purse will be able to meet the expense." See A. Neisser, "Le traitement actual de la syphilis: Avantages de l'association Salvarsan et mercure; quelques principes pour l'emploi du Salvarsan," *Annales des maladies vénériennes* 8 (Oct. 1913): 721–761. See also G. W. Raiziss, "Development of the American Arsphenamine Industry," *Industrial and Engineering Chemistry* 15 (Apr. 1923): 413–415; "Trade Commission Acts on Salvarsan Patent," *American Journal of Syphilis* 2 (Jan. 1918): 202–203.

32. Dowling, "Comparisons and Contrasts"; Editorial: "The Historical Perspective in Medicine," *Journal of the American Medical Association* 63 (Oct. 10, 1914): 1300–1301; I. Rosen and N. Sobel, "Fifty Years' Progress in the Treatment of Syphilis," *New York State Medical Journal* 50 (Nov. 15, 1950): 2694–2696.

33. J. E. Kemp, "An Outline of the History of Syphilis," *American Journal of Syphilis* 24 (Nov. 1940): 759–779; John Hinchman Stokes, *Modern Clinical Syphilology: Diagnosis—Treatment—Case Studies,* 1st

ed. (Philadelphia: W. B. Saunders, 1926), pp. 947–952. The members of the Cooperative Clinical Group were: Harold N. Cole of Western Reserve University, Joseph E. Moore of Johns Hopkins University, Paul A. O'Leary of the Mayo Clinic, John H. Stokes of the University of Pennsylvania, Udo J. Wile of the University of Michigan, and Thomas Parran, then assistant surgeon general in charge of venereal diseases. See also T. Clark, T. Parran, Jr., H. N. Cole, J. E. Moore, P. A. O'Leary, J. H. Stokes, and U. J. Wile, "Clinical Studies in the Treatment of Syphilis: I. Introduction," *Venereal Disease Information* 13 (Apr. 20, 1932): 135–138; J. H. Stokes, H. N. Cole, J. E. Moore, P. A. O'Leary, U. J. Wile, and T. Parran, Jr., "Cutaneous and Mucosal Relapse in Early Syphilis and Its Differentiation from Reinfection," ibid. 12 (Feb. 20, 1931): 55–56. The sheer mass of data accumulated was almost overwhelming. The first paper on relapses after treatment covered 5,952 patients with early syphilis. A report on reactions to arsenicals involved 8,810 patients who had received 177,360 injections. For the study on syphilis and pregnancy, the records of 3,817 women were utilized. See Stokes et al., "Cutaneous and Mucosal Relapse in Early Syphilis"; H. N. Cole, J. E. Moore, P. A. O'Leary, J. H. Stokes, U. J. Wile, T. Clark, T. Parran, Jr., and L. J. Usilton, "Cooperative Clinical Studies in the Treatment of Syphilis—Arsenical Reactions," *Venereal Disease Information* 14 (Aug. 1933): 173–200; H. N. Cole, J. E. Moore, P. A. O'Leary, J. H. Stokes, U. J. Wile, T. Clark, T. Parran, Jr., R. A. Vonderlehr, and L. J. Usilton, "Cooperative Clinical Studies in the Treatment of Syphilis: Syphilis in Pregnancy," ibid. 15 (Mar. 1934): 83–107. See also H. F. Dowling, "The Emergence of the Cooperative Clinical Trial," *Transactions and Studies of the College of Physicians of Philadelphia*, 4th ser. 43 (July 1975): 22–29.

34. Joseph Earl Moore, *The Modern Treatment of Syphilis* (Springfield, Ill.: Charles C Thomas, 1933), p. 345); P. A. O'Leary, W. H. Goeckerman, and S. T. Parker, "Treatment of Neurosyphilis by Malaria: A Preliminary Report," *Archives of Dermatology* 13 (Mar. 1926): 301–318; J. R. Ewalt and F. G. Ebaugh, "Treatment of Dementia Paralytica: A Five-year Comparative Study of Artificial Fever and Therapeutic Malaria in Two Hundred and Thirty-two Cases," *Journal of the American Medical Association* 116 (May 31, 1941): 2474–2477.

35. A. L. Tatum and G. A. Cooper, "An Expermental Study of Mapharsen (meta-amino para-hydroxy phenyl arsine oxide) as an Antisyphilitic Agent," *Journal of Pharmacology and Experimental Therapeutics* 50 (Feb. 1934): 198–215; O. H. Foerster, R. L. McIntosh, L. M. Wieder, H. R. Foerster, and G. A. Cooper, "Mapharsen in the Treatment of Syphilis," *Archives of Dermatology and Syphilology* 32 (Dec. 1935): 868–889.

36. Parran, *Shadow on the Land*, pp. 48–49; S. D. Gross, "Syphilis in Its Relation to the National Health," *Transactions of the American Medical Association* 25 (1874): 249–292; Raymond A. Vonderlehr and J. R. Heller, Jr., *The Control of Venereal Disease* (New York: Reynal &

Hitchcock, 1946), p. 12; Parran, *Shadow on the Land*, p. 80; Upton Sinclair, *Sylvia: A Novel*, 2d ed. (Long Beach, Cal.: Upton Sinclair, 1927), p. 388. See also H. L. Mencken, *The American Language: An Inquiry into the Development of English in the United States* (New York: Alfred A. Knopf, 1937), pp. 300–306. Mencken gives a long list of words that were banned from the mails and from public print and of euphemisms used in their place, including an incident in which the Chattanooga police, on arresting a man for picking up a streetwalker, announced that he was charged with "walking on the street accompanied by a woman" (p. 304).

37. Pottle, ed., *Boswell's London Journal*, pp. 161, 164; Power, "Clap and the Pox in English Literature."

38. Charles Walter Clarke, *Taboo: The Story of the Progress of Social Hygiene* (Washington, D.C.: Public Affairs Press, 1961), pp. 56–63; Eugene Brieux, *Damaged Goods (Les Avaries)*, trans. John Pollock (London: Jonathan Cope, 1943), pp. 54, 57, 60, 18.

39. Brown et al., *Syphilis and Other Venereal Diseases*, p. 58; Adam J. Rapalski, "Effects of Population Concentration on the Spread of Syphilis: Military," *Proceedings of the World Forum on Syphilis and Other Treponomatoses*, Public Health Service Publication No. 997 (Washington, D.C.: U.S. Government Printing Office, 1964), pp. 134–140, the figures quoted are on p. 137. "Average strength" refers to the average number of troops in a given unit per year or other period of time. See also H. H. Young, "The Venereal Disease Probem in the A.E.F.," *Military Surgeon* 78 (Jan. 1936): 1–22; Editorial: "Model Ordinances for Venereal Disease Control," *Journal of the American Medical Association* 73 (Sept. 13, 1919): 842; Parran, *Shadow on the Land*, pp. 82–84. Seattle, Washington, was an example of a city that cooperated effectively with military authorities. W. F. Meier, "Quarantine for Social Diseases," *American City* 21 (Sept. 1919): 221–224.

40. Editorial: "Honi Soit Qui Mal y Pense," *New England Journal of Medicine* 211 (Nov. 29, 1934): 1031; Mencken, *The American Language,* p. 307; T. Parran, "The Next Great Plague to Go," *Survey Graphic* 25 (Jul. 1936): 405–411; Parran, "Why Don't We Stamp Out Syphilis," *Reader's Digest* 29 (Jul. 1936): 65–73; Parran to Mary Ross, May 20, 1936, Thomas Parran Papers, Library of the Graduate School of Public Health, University of Pittsburgh, Pittsburgh, Pa.; DeWitt Wallace to Parran, Aug. 21, 1936, ibid.; Memorandum from Survey Associates, dated Sept. 11, 1936, ibid.; "Venereal Disease Campaign," *Time* 29 (Jan. 11, 1937): 38.

41. "Shameless, Sinful," *Time* 44 (Oct. 16, 1944): 56–57; Editorial: "Catholics and Venereal Disease," *New Republic* 111 (Oct. 9, 1944): 446; Michel de Montaigne, *Selected Essays*, rev. and ed. Blanchard Bates (New York: Modern Library, 1949), p. 36; Alfred C. Kinsey, Wardell B. Pomeroy, and Clyde E. Martin, *Sexual Behavior in the Human Male* (Philadelphia: W. B. Saunders, 1948).

42. R. E. Riegel, "Changing American Attitudes toward Prostitution,"

Journal of the History of Ideas 29 (Jul.–Sept. 1968): 437–452; J. C. Burnham, "Medical Inspection of Prostitutes in America in the Nineteenth Century: The St. Louis Experiment and Its Sequel," *Bulletin of the History of Medicine* 45 (May–June 1971): 203–218; M. E. Miner, "Report of Committee on Social Hygiene," *Social Hygiene* 1 (Dec. 1914): 81–92; Abraham Flexner, *Prostitution in Europe* (New York: Century, 1914); Flexner, "The Regulation of Prostitution in Europe," *Social Hygiene* 1 (Dec. 1914): 15–28; "The Bureau of Socal Hygiene," *Outlook* 103 (Feb. 8 and 18, 1913): 287–288, 298–299; Howard B. Woolston, *Prostitution in the United States* (New York: Century, 1921), vol. 1: *Prior to the Entrance of the United States into the World War*, pp. 102–120.

43. "New Methods of Grappling with the Social Evil," *Current Opinion* 54 (Apr. 1913): 273–274; Michael Gold, *Jews without Money* (New York: Avon Book Division, The Hearst Corporation, 1965), pp. 6–8; Robert Hamlett Bremner, *From the Depths: The Discovery of Poverty in the United States* (New York: New York University Press, 1956), pp. 238–240; B. Johnson, "A Current View of Prostitution and Sex Delinquency," *Journal of Social Hygiene* 22 (Dec 1936): 389–402.

44. Parran, *Shadow on the Land*, p. 245.

45. Charles V. Chapin, "History of State and Municipal Control of Disease," in Mazyck P. Ravenel, ed., *A Half Century of Public Health* (New York: American Public Health Association, 1921), pp. 146–147; Charles-Edward A. Winslow, *The Life of Hermann M. Biggs, M.D., D.Sc., LL.D.: Physician and Statesman of the Public Health* (Philadelphia: Lea & Febiger, 1929), pp. 234–235. For the quotation on the poor treatment of patients, see H. S. Newcomer, R. Richardson, and C. Ashbrook, "One Aspect of Syphilis as a Community Problem," *American Journal of the Medical Sciences* 158 (Aug. 1919): 141–165.

46. M. A. Clark, "Venereal Disease Control: Chapter from the Forthcoming Report of the Committee on Municipal Health Department Practice of the American Public Health Association," *Journal of Social Hygiene* 9 (Jan. 1923): 27–51; U. S. Bureau of the Census, *Historical Statistics of the United States, Colonial Times to 1957* (Washington, D.C.: U.S. Government Printing Office, 1960), p. 26; T. Parran and L. J. Usilton, "The Extent of the Problem of Syphilis and Gonorrhea in the United States," *American Journal of Syphilis* 14 (Apr. 1930): 145–155; U.S. Department of Health, Education, and Welfare, Public Health Service, *V.D. Fact Sheet 1971* (Atlanta: Center for Disease Control, 1972), p. 8.

47. J. C. Funk, "The Pennsylvania Venereal Disease Program," *Journal of Social Hygiene* 9 (Mar. 1923): 150–159; J. S. Lawrence, "The New York State Venereal Disease Control Program," ibid. (May 1923), pp. 271–279; S. W. Trythall, "The Premarital Law: History and a Survey of Its Effectiveness in Michigan," *Journal of the American Medical Association* 187 (Mar. 21, 1964): 900–903.

48. A. Keidel, "Economic Aspects of the Management of Syphilis," *Archives of Dermatology and Syphilology* 25 (Mar. 1932): 470–484; Ralph Chester Williams, *The United States Public Health Service, 1798–1950* (Washington, D.C.: Commissioned Officers Association of the United States Public Health Service, 1951), pp. 153–155; "Progress in Venereal Disease Control during Fiscal Year 1939," *Venereal Disease Information* 20 (Dec. 1939): 376–378; Brown et al., *Syphilis and Other Venereal Diseases*, pp. 107–108.

49. J. C. Gebhart, "Syphilis as a Prenatal Problem," *Journal of Social Hygiene* 10 (Apr. 1924): 193–217; "Prenatal Health Laws," *Venereal Disease Information* 19 (May 1938): 128–129; Vonderlehr and Heller, *The Control of Venereal Disease*, p. 71; A. F. Brewer and F. E. Olson, "Evaluation of California's Prenatal Law Requiring a Serologic Test for Syphilis," *American Journal of Syphilis, Gonorrhea, and Venereal Diseases* 31 (Nov. 1947): 633–639; N. R. Ingraham, Jr., "The Importance of Treatment in the Control of Congenital Syphilis," *Venereal Disease Information* 19 (May 1938): 124–128.

50. C. D. Bowdoin, A. R. Remein, C. A. Henderson, J. W. Morse, and W. T. Davis, Jr., "Socioeconomic Factors in Syphilis Prevalence," *Journal of Venereal Disease Information* 30 (May 1949): 131–139; W. H. Y. Smith, "Evolution of Venereal Disease Control in Alabama," *Journal of the Medical Association of Alabama* 26 (Nov. 1956): 112–115.

51. L. J. Usilton and J. W. Morse, "Attitude of Venereal Disease Patients toward Clinics and Rapid Treatment Centers," *Journal of Venereal Disease Information* 30 (Oct. 1949): 275–280; "First," *Time* 40 (Dec. 14, 1942): 56; "Cooperating Clinics of New York and Midwestern Groups: Massive Arsenotherapy for Syphilis, United States Public Health Service Evaluation," *Journal of the American Medical Association* 126 (Oct. 28, 1944): 554–557; J. R. Heller, Jr., "Venereal Disease Control during the Postwar Period," *American Journal of Syphilis, Gonorrhea, and Venereal Diseases* 31 (Nov. 1947): 569–574.

52. U.S. Bureau of the Census, *Historical Statistics of the United States*, p. 26.

53. Friedenwald, "Paths of Progress in Ophthalmology," pp. 1047–1096; Bloomfield, *A Bibliography of Internal Medicine*, pp. 186–187. Mild silver proteinate was later found to be almost as effective for conjunctivitis as silver nitrate and less irritating. In some places public health regulations allowed doctors to substitute the milder solution; others required silver nitrate. See further discussion in Chapter 9. See also Clark, "Venereal Disease Control," p. 39; Brown et al., *Syphilis and Other Venereal Diseases*, p. 130.

54. Finger, *Gonorrhoea Being the Translation of Blenorrhoea*, p. 23; L. K. Fraenkel, "Social Hygiene and Public Health," *Journal of Social Hygiene* 11 (Apr. 1925): 210–214; L. J. Usilton, "Prevalence of Venereal Disease in the United States," *Venereal Disease Information* 11 (Dec. 20,

1930): 542–562; W. Clarke, "Some Practical Problems in the Control of Syphilis," *American Journal of Syphilis* 17 (Jan. 1933): 1–9.

8. The Sulfonamides

1. Frank Hawking, "History of Chemotherapy," in Robert J. Schnitzer and Frank Hawking, eds., *Experimental Chemotherapy*, 5 vols. (New York: Academic Press, 1963–67), I, 1–24; E. B. Chain, "Academic and Industrial Contributions to Drug Research," *Nature* (London), 200 (Nov. 2, 1963): 441–451; Iago Galdston, *Behind the Sulfa Drugs: A Short History of Chemotherapy* (New York: D. Appleton-Century, 1943).

2. Roderick Heffron, *Pneumonia with Special Reference to Pneumococcus Lobar Pneumonia* (New York: Commonwealth Fund, 1939), pp. 778–782; L. Colebrook and R. Hare, "Treatment of Puerperal Infection Due to *Streptococcus Pyogenes* by Organic Arsenical Compounds," *Lancet* 1 (Feb. 24, 1934): 388–391; Ronald Hare, *The Birth of Penicillin and the Disarming of Microbes* (London: George Allen and Unwin, 1970), pp. 52–53.

3. Frank Hawking, "Early History of Antibacterial Chemotherapy," in Schnitzer and Hawking, eds., *Experimental Chemotherapy*, II, 1–36; L. P. Garrod, "Chemicals versus Bacteria," *Proceedings of the Royal Society of Medicine* (London), 48 (Jan. 1955): 21–28; Hugh Hampton Young, *Hugh Young: A Surgeon's Autobiography* (New York: Harcourt, Brace, 1940), pp. 252–263. For the attitude of British investigators on the effectiveness of Mercurochrome, see Hare, *The Birth of Penicillin*, p. 50.

4. Thomas S. Work and Elizabeth Work, *The Basis of Chemotherapy* (Edinburgh: Oliver and Boyd, 1948), pp. 6–26. For attempts to treat bacterial infections with chemical compounds, see John A. Kolmer, *Principles and Practice of Chemotherapy with Special Reference to the Specific and General Treatment of Syphilis* (Philadelphia: W. B. Saunders, 1926), esp. pp. 98–152. Behring is quoted in Galdston, *Behind the Sulfa Drugs*, p. 139.

5. "Obituary: Gerhard Domagk," *Lancet* 1 (May 2, 1964): 992–993. In more technical language, the azo dyes were most effective when a sulphamyl group (H_2NSO_2) was introduced into an azo-dye molecule in the para position relative to the N-linkage. See Hare, *The Birth of Penicillin*, pp. 145–161. See also G. Domagk, "Ein Beitrag zur Chemotherapie der bakteriellen Infektionen," *Deutsche Medizinische Wochenschrift* 61 (Feb. 15, 1935): 250–253. The chemical name of Prontosil is para-sulphonamido-chrysoidin.

6. Perrin H. Long and Eleanor A. Bliss, *The Clinical and Experimental Use of Sulfanilamide, Sulfapyridine and Allied Compounds* (New York: Macmillan, 1939), pp. 4–5; J. Tréfouël, Mme. J. Tréfouël, F. Nitti, and D. Bovet, "Activité du p-aminophényl-sulfamide sur les infections strep-

tococciques expérimentales de la souris et du lapin," *Comptes rendus des séances de la Société de biologie et des ses filiales* 120 (1935): 756–758; Tréfouël et al., "The Contribution of the Institut Pasteur, Paris, to Recent Advances in Microbial and Functional Chemotherapy," *British Medical Bulletin* 4 (1946): 284–289. Sulfanilamide was initially spelled sulphanilamide.

7. H. Hoerlein, "The Development of Chemotherapy for Bacterial Diseases," *Practitioner* 139 (Dec. 1937): 635–649; Hare, *The Birth of Penicillin*, pp. 155–161.

8. Hare, *The Birth of Penicillin*, p. 136. D. A. E. Shephard has contended that Domagk's work remained unappreciated in England and the United States for so long because the original articles were written in German and therefore were not read widely in those countries. Shephard, "New Information in Medical Journals: Excretory or Secretory Process?" *Mayo Clinic Proceedings* 47 (June 1972): 415–423. However, Perrin Long and others in the United States read the German journals widely, as no doubt did investigators such as Leonard Colebrook in England. Besides, Hare told of a postcard sent to Colebrook in the summer of 1935 from a friend in Paris calling attention to Prontosil. Hare, *The Birth of Penicillin*, p. 136. The Anglo-American suspicion of the way in which new compounds were uncritically introduced in Germany was probably a more important factor.

9. Hare, *The Birth of Penicillin*, pp. 125–131; L. Colebrook and M. Kenny, "Treatment of Human Puerperal Infections, and of Experimental Infections in Mice, with Prontosil," *Lancet* 1 (June 6, 1936): 1279–1286; idem, "Treatment with Prontosil of Puerperal Infections Due to Haemolytic Streptococci," ibid. 2 (Dec. 5, 1936): 1319–1322.

10. Colebrook and Kenny, "Treatment with Prontosil of Puerperal Infections Due to Haemolytic Streptococci"; "The Chemotherapy of Streptococcal Infections," *Lancet* 1 (June 6, 1936): 1303–1304; Editorial: "Chemotherapy of Streptococcal Infections," ibid. 2 (Dec. 5, 1936): 1339.

11. W. R. Snodgrass and T. Anderson, "Prontosil in the Treatment of Erysipelas: A Controlled Series of 312," *British Medical Journal* 2 (Jul. 17, 1937): 101–104; G. A. H. Buttle, W. H. Gray, and D. Stephenson, "Protection of Mice against Streptococcal and Other Infections by p-Aminobenzenesulphonamide and Related Substances," *Lancet* 1 (June 6, 1936): 1286–1290; F. F. Schwentker, S. Gelman, and P. H. Long, "The Treatment of Meningococcic Meningitis with Sulfanilamide: Preliminary Report," *Journal of the American Medical Association* 108 (Apr. 24, 1937): 1407–1408.

12. Hare, *The Birth of Penicillin*, pp. 140–142; P. H. Long and E. A. Bliss, "Para-Amino-Benzene-Sulfonamide and Its Derivatives," *Journal of the American Medical Association* 108 (Jan. 2, 1937): 32–37; Long and Bliss, "The Use of Para Amino Benzene Sulphonamide (Sulphanilamide) or Its Derivatives in the Treatment of Infections Due to Beta

Hemolytic Streptococci, Pneumococci and Meningococci," *Southern Medical Journal* 30 (May 1937): 479–487; E. K. Marshall, Jr., and J. T. Litchfield, Jr., "The Determination of Sulfanilamide," *Science* 88 (Jul. 22, 1938): 85–86; E. K. Marshall, Jr., K. Emerson, and W. C. Cutting, "Para-Aminobenzenesulfonamide, Absorption and Excretion: Method of Determination in Urine and Blood," *Journal of the American Medical Association* 108 (Mar. 20, 1937): 953–957.

13. *Sulphonamides: Their Properties and Therapeutic Use* (Dagenham, England: May & Baker, 1962), pp. 4–6; L. E. H. Whitby, "Chemotherapy of Pneumococcal and Other Infections with 2 (p-Amino-benzenesulfamido-) Pyridine," *Lancet* 1 (May 28, 1938): 1210–1212; Long and Bliss, *Clinical Use of Sulfanilamide and Sulfapyridine and Allied Compounds*, pp. 231–239.

14. Within a few months of Whitby's report on the value of sulfapyridine in pneumococcal infections in animals, William Antopol and Harry Robinson of the Merck Institute for Therapeutic Research observed blocking of the urinary passages by aggregates of sulfapyridine crystals in some animals receiving this drug. Merck and Company promptly reported this to the clinicians to whom they had been supplying the drug so that they could guard against similar reactions in their patients. Draft of letter from Hans Molitor to Perrin Long, November 17, 1938, from the files of Merck Sharpe and Dohme, photocopy courtesy of Harry J. Robinson, Merck Sharpe and Dohme. See also W. Antopol and H. Robinson, "Urolithiasis and Renal Pathology after Oral Administration of 2 (sulfanilylamino) pyridine (Sulfapyridine)," *Proceedings of the Society for Experimental Biology and Medicine* 40 (Mar. 1938): 428–430; M. Finland, E. Strauss, and O. L. Peterson, "Sulfadiazine: Therapeutic Evaluation and Toxic Effects on Four Hundred and Forty-six Patients," *Journal of the American Medical Association* 116 (June 14, 1941): 2641–2647; H. F. Dowling and M. H. Lepper, "Toxic Reactions following Therapy with Sulfapyridine, Sulfathiazole, and Sulfadiazine," ibid. 121 (Apr. 10, 1943): 1190–1194; R. J. Schnitzer, R. H. K. Foster, N. Ercoli, G. Soo-Hoo, C. N. Mangieri, and M. D. Roe, "Pharmacological and Chemotherapeutic Properties of 3, 4-Dimethyl-5-Sulfanilamido-Isoxazole," *Journal of Pharmacology and Experimental Therapeutics* 88 (Sept. 1946): 47–57.

15. Harry F. Dowling, *The Acute Bacterial Diseases: Their Diagnosis and Treatment* (Philadelphia: W. B. Saunders, 1948), pp. 150, 152; A. J. Morris, R. Chamovitz, F. J. Cantazaro, and C. H. Rammelkamp, "Prevention of Rheumatic Fever by Treatment of Previous Streptococcic Infections," *Journal of the American Medical Association* 160 (Jan. 14, 1956): 114–116.

16. F. W. Denny, Jr., "The Prophylaxis of Streptococcal Infections," in Maclyn McCarty, ed., *Streptococcal Infections* (New York: Columbia University Press, 1954), pp. 176–196, esp. p. 181.

17. Ibid., p. 182; Alvin F. Coburn and Donald C. Young, *The Epide-*

miology of Hemolytic Streptococcus during World War II in the United States Navy (Baltimore: Williams and Wilkins, 1949), pp. 5–9, esp. footnote, p. 5.

18. G. M. Evans and W. F. Gaisford, "Treatment of Pneumonia with 2(p-Aminobenzenesulfonamido) Pyridine," *Lancet* 2 (Jul. 2, 1938): 14–19; J. G. M. Bullowa, N. Plummer, and M. Finland, "Sulfapyridine in the Treatment of Pneumonia," *Journal of the American Medical Association* 112 (Feb. 11, 1939): 570; E. K. Marshall, Jr., "An Unfortunate Situation in the Field of Bacterial Chemotherapy," *Journal of the American Medical Association* 112 (Jan. 28, 1939): 352–353; G. J. Langley, W. Mackay, and L. Stent, "Treatment of Lobar Pneumonia with M & B 693," *Lancet* 2 (Nov. 26, 1938): 1264–1265. The article referred to was S. C. Dyke and G. C. K. Reid, "Treatment of Lobar Pneumonia with M & B 693," ibid., Nov. 19, 1938, pp. 1157–1160. In May 1937, when the Council on Pharmacy and Chemistry of the American Medical Association accepted sulfanilamide, 13 firms were already marketing it. *Journal of the American Medical Association* 108 (May 29, 1937): 1888–1889. I have considered the subject of the clinical evaluation of drugs in more detail in Harry F. Dowling, *Medicines for Man: The Development, Regulation, and Use of Prescription Drugs* (New York: Alfred A. Knopf, 1970), esp. pp. 58–77, 268–283. Prior to the Drug Amendments Act of 1962 there was no federal law requiring that the efficacy of a drug be proved before it was marketed. Ibid., pp. 202–203.

19. M. Finland, "The Treatment of Pneumonia," *Canadian Medical Association Journal* 41 (Dec. 1939): 554–560; N. Plummer and H. K. Ensworth, "Sulfapyridine in the Treatment of Pneumonia," *Journal of the American Medical Association* 113 (Nov. 18, 1939): 1847–1853; D. S. Pepper, H. F. Flippin, L. Schwartz, and J. S. Lockwood, "The Results of Sulfapyridine Therapy in 400 Cases of Typed Pneumococcic Pneumonia," *American Journal of the Medical Sciences* 198 (Jul. 1939): 22–35; H. F. Dowling and T. J. Abernethy, "The Treatment of Pneumococcus Pnuemonia: A Comparison of the Results Obtained with Specific Serum and with Sulfapyridine," ibid. 199 (Jan. 1940): 55–62. For combined therapy, see Finland, "The Treatment of Pneumonia," p. 560; C. M. MacLeod, "Chemotherapy of Pneumonia," *Journal of the American Medical Association* 113 (Oct. 7, 1939): 1405–1410; H. F. Dowling, T. J. Abernethy, and C. R. Hartman, "Should Serum Be Used in Addition to Sulfapyridine in the Treatment of Pneumococcic Pneumonia?" ibid. 115 (Dec. 21, 1940): 2125–2128; N. Plummer, J. Liebmann, S. Solomon, W. H. Kammerer, M. Kalkstein, and H. K. Ensworth, "Chemotherapy versus Combined Chemotherapy and Serum in the Treatment of Pneumonia: A Study of 607 Alternated Cases," ibid. 116 (May 24, 1941): 2366–2371.

20. H. F. Flippin, L. Schwartz, and A. H. Domm, "Modern Treatment of Pneumococcic Pneumonia," *Journal of the American Medical Association* 121 (Jan. 23, 1943): 230–236; H. F. Dowling and Mark H. Lepper,

"The Effect of Antibiotics (Penicillin, Aureomycin, and Terramycin) on the Fatality Rate and Incidence of Complications in Pneumococcal Pneumonia: A Comparison with Other Methods of Therapy," *American Journal of the Medical Sciences* 222 (Oct. 1951): 396–403; Tom Mahoney, *The Merchants of Life: An Account of The American Pharmaceutical Industry* (New York: Harper & Brothers, 1959), pp. 169–170.

21. J. B. Neal, E. Appelbaum, and H. W. Jackson, "Sulfapyridine and Its Sodium Salt in the Treatment of Meningitis Due to the Pneumococcus and Haemophilus Influenzae," *Journal of the American Medical Association* 115 (Dec. 14, 1940): 2055–2058.

22. Buttle, Gray, and Stephenson, "Protection of Mice against Streptococcal and Other Infections by p-Aminobenzenesulphonamide and Related Substances," pp. 1286–1290; Schwentker, Gelman, and Long, "The Treatment of Meningococcic Meningitis with Sulfanilamide," pp. 1407–1408; J. M. Waghelstein, "Sulfanilamide in the Treatment of 106 Patients with Meningococcic Infections," *Journal of the American Medical Association* 111 (Dec. 10, 1938): 2172–2174.

23. P. B. Beeson and E. Westerman, "Cerebrospinal Fever: Analysis of 3,575 Case Reports with Special Reference to Sulphonamide Therapy," *British Medical Journal* 1 (Apr. 24, 1943): 497–500; U.S. Department of Commerce, Bureau of the Census, *Vital Statistics of the United States, 1944* (Washington, D.C.: U.S. Government Printing Office, 1946), part 1, p. 9; Dowling, *The Acute Bacterial Diseases*, p. 216; W. B. Daniels, "Cause of Death in Meningococcic Infection: Analysis of 300 Fatal Cases," *American Journal of Medicine* 8 (Apr. 1950): 468–473.

24. S. E. Branham and S. M. Rosenthal, "Studies in Chemotherapy: V. Sulphanilamide, Serum and Combined Drug Therapy in Experimental Meningitis and Pneumococcus Infections in Mice," *Public Health Reports* 52 (May 28, 1937): 685–695; H. S. Banks, "Serum and Sulphanilamide in Acute Meningococcal Meningitis: A Preliminary Survey Based on 113 Cases," *Lancet* 2 (Jul. 2, 1938): 7–13.

25. The quotation is from Annotations: "Treatment of Meningococcal Meningitis," *Lancet* 2 (July 2, 1938): 29. For the early prophylactic trials, see J. F. Meehan and C. R. Merrillees, "An Outbreak of Cerebro-spinal Meningitis in a Foundling Hospital: The Treatment of Carriers with 'M & B 693,' " *Medical Journal of Australia* 2 (Jul. 27, 1940): 84–90; S. E. Seid, "Meningitis Epidemic among Navaho Indians," *Journal of the American Medical Association* 115 (Sept. 14, 1940): 923–924; R. W. Fairbrother, "Cerebrospinal Meningitis: The Use of Sulphonamide Derivatives in Prophylaxis," *British Medical Journal* 2 (Dec. 21, 1940): 859–862; F. C. Gray and J. Gear, "Sulphapyridine M & B 693 as a Prophylactic against Cerebrospinal Meningitis," *South African Medical Journal* 15 (Apr. 12, 1941): 139–140.

26. For the quotation, see Fairbrother, "Cerebrospinal Meningitis," pp. 859–862. See also D. M. Kuhns and H. A. Feldman, "Laboratory Methods Used in Determining the Value of Sulfadiazine as a Mass

Prophylactic against Meningococcic Infections," *American Journal of Public Health* 33 (Dec. 1943): 1461–1465; D. M. Kuhns, C. T. Nelson, H. A. Feldman, and L. R. Kuhn, "The Prophylactic Value of Sulfadiazine in the Control of Meningococcic Meningitis," *Journal of the American Medical Association* 123 (Oct. 9, 1943): 335–339; F. X. Cheever, B. B. Breese, and H. E. Upham, "The Treatment of Meningococcus Carriers with Sulfadiazine," *Annals of Internal Medicine* 19 (Oct. 1943): 602–608.

27. Harry A. Feldman, "Sulfadiazine Resistant Meningococci," dittoed memorandum, May 22, 1963; J. W. Millar, E. E. Siess, H. A. Feldman, C. Silverman, and P. Frank, "In Vivo and In Vitro Resistance to Sulfadiazine in Strains of *Neisseria Meningitidis*," *Journal of the American Medical Association* 186 (Oct. 12, 1963): 139–141; H. A. Feldman, "Some Recollections of the Meningococcal Diseases," ibid. 220 (May 22, 1972): 1107–1112; Feldman, "Recent Developments in the Therapy and Control of Meningococcal Infections," *Disease-a-Month*, Feb. 1966, p. 21; J. A. Jacobson, R. E. Weaver, and C. Thornsberry, "Trends in Meningococcal Disease, 1974," *Journal of Infectious Diseases* 132 (Oct. 1975): 480–484.

28. Feldman, "Some Recollections of the Meningococcal Diseases," p. 1109.

29. J. E. Dees and J. A. C. Colston, "The Use of Sulfanilamide in Gonococcic Infections," *Journal of the American Medical Association* 108 (May 29, 1937): 1855–1858; Editorial: "Gonorrhea and Sulfanilamide," ibid. 110 (Jan. 1, 1938): 51; J. F. Mahoney, C. J. Van Slyke and R. R. Wolcott, "Sulfathiazole Treatment of Gonococcal Infections in Men and Women, Results in 360 Patients," *Venereal Disease Information* 22 (Dec. 1941): 425–431; J. A. Robinson, H. L. Hirsh, W. W. Zeller, and H. F. Dowling, "Gonococcal Arthritis: A Study of 202 Patients Treated with Penicillin, Sulfonamides or Fever Therapy," *Annals of Internal Medicine* 30 (June 1949): 1212–1223; P. H. Futcher and V. C. Scott, "Four Cases of Gonococcal Endocarditis Treated with Sulfanilamide, with Recovery of One," *Bulletin of the Johns Hopkins Hospital* 65 (Nov. 1939): 377–391.

30. For the quotation, see Editorial: "Gonorrhea Gets a Place in the Venereal Disease Program," *American Journal of Public Health* 32 (April 1942): 413–414. See also "State and Territorial Health Officers Confer on Health Defenses," *Public Health Reports* 56 (June 6, 1941): 1194–1217, remarks of R. A. Vonderlehr, pp. 1209–1211.

31. J. A. Loveless and W. Denton, "The Oral Use of Sulfathiazole as a Prophylaxis for Gonorrhea: Preliminary Report," *Journal of the American Medical Association* 121 (Mar. 13, 1943): 827–828; J. L. Rice, "Stamp Out Gonorrhea Now!" *American Journal of Public Health* 32 (Feb. 1942): 129–130.

32. H. Felke, "Die Wirkung der Sulfonamidverbindungen auf die Erreger der Gonorrhoë," *Klinische Wochenschrift* 17 (Jan. 1, 1938): 13–16; M. Meads and M. Finland, "Penicillin in the Treatment of Gono-

coccal Infections: Analysis of the Results Reported in the Literature through 1945," *American Journal of Syphilis, Gonorrhea, and Venereal Diseases* 30 (Nov. 1946): 586–609.

33. R. S. Mueller and J. E. Thompson, "The Local Use of Sulfanilamide in the Treatment of Peritoneal Infections," *Journal of the American Medical Association* 118 (Jan. 17, 1942): 189–193; H. C. Jackson and F. A. Coller, "The Use of Sulfanilamide in the Peritoneum: Experimental and Clinical Observations," ibid., pp. 194–199.

34. Council on Pharmacy and Chemistry, "Sulfonamides for Local Application Deleted from N.N.R.," *Journal of the American Medical Association* 135 (Sept. 20, 1947): 157–158.

35. J. S. Lockwood, "Sulfanilamide in Surgical Infections: Its Possibilities and Limitations," *Journal of the American Medical Association* 115 (Oct. 5, 1940): 1190–1195; D. D. Woods, "The Biochemical Mode of Action of the Sulphonamide Drugs," *Journal of General Microbiology* 29 (Dec. 1962): 687–702; P. Fildes, "A Rational Approach to Research in Chemotherapy," *Lancet* 1 (May 25, 1940): 955–957.

36. E. K. Marshall, Jr., A. C. Bratton, H. J. White, and J. T. Litchfield, Jr., "Sulfanilylguanadine: A Chemotherapeutic Agent for Intestinal Infections," *Bulletin of the Johns Hopkins Hospital* 67 (Sept. 1940): 163–188; E. K. Marshall, Jr., A. C. Bratton, L. B. Edwards, and E. Walker, "Sulfanilylguanadine in the Treatment of Acute Bacillary Dysentery in Children," ibid. 68 (Jan. 1941): 94–111; H. J. Smyly, "Bacillary Dysentery," in Russell L. Cecil, ed., *A Textbook of Medicine*, 7th ed. (Philadelphia: W. B. Saunders, 1947), pp. 240–247; A. V. Hardy, W. Burns, and T. De Capito, "Studies of Acute Diarrheal Diseases: XA. Cultural Observations on the Relative Efficacy of Sulfonamides in *Shigella dysenteriae* Infections," *Public Health Reports* 58 (Apr. 30, 1943): 689–693.

37. Carl C. Dauer, Robert F. Korns, and Leonard M. Schuman, *Infectious Diseases* (Cambridge: Harvard University Press, 1968), p. 26; B. F. Garfinkel, G. M. Martin, J. Watt, F. J. Payne, R. P. Mason, and A. V. Hardy, "Antibiotics in Acute Bacillary Dysentery: Observations in 1,408 Cases with Positive Cultures," *Journal of the American Medical Association* 151 (Apr. 4, 1953): 1157–1159.

38. E. J. Poth, "Succinylsulfathiazole: An Adjuvant in Surgery of the Large Bowel," *Journal of the American Medical Association* 120 (Sept. 26, 1942): 265–268; "Appendicitis Mortality near Vanishing Point," *Statistical Bulletin of the Metropolitan Life Insurance Co.* 28 (Jan. 1947): 7–8. The figures refer to industrial policyholders of the Metropolitan Life Insurance Company.

39. J. A. Kolmer, "Progress in Chemotherapy of Bacterial and Other Diseases with Special Reference to the Prontosils, Sulfanilamide, and Sulfapyridine," *Archives of Internal Medicine* 65 (Apr. 1940): 671–743.

40. "Army Surgeon Reports Success with Sulfanilamide," *New York Times*, Nov. 20, 1938, p. 5; "Medical Triumphs of 1939," ibid., Dec. 31,

1939, sec. 2, p. 7; H. A. Reimann, "Progress in Internal Medicine: Ninth Annual Review of Significant Publications," *Archives of Internal Medicine* 72 (Sept. 1943): 388–426.

41. W. D. Sutliff, M. Helpern, G. Griffin, and H. Brown, "Sulfonamide Toxicity as a Cause of Death in New York City in 1941," *Journal of the American Medical Association* 121 (Jan. 30, 1943): 307–312; Reimann, "Progress in Internal Medicine"; R. V. Lee, "Reactions following Mass Administration of Sulfadiazine," *Journal of the American Medical Association* 126 (Nov. 4, 1944): 630–631.

42. J. H. Young, "The 'Elixir Sulfanilamide' Disaster," *Emory University Quarterly* 14 (Dec. 1958): 230–237.

43. A. Loubatières, "The Discovery of Hypoglycemic Sulfonamides and Particularly of Their Action Mechanism," *Acta Diabetologica Latina*, supplement 1, 6 (Sept. 1969): 20–56.

44. R. F. Pitts, "Some Reflections on Mechanisms of Action of Diuretics," *American Journal of Medicine* 24 (May 1958): 745–763.

9. Penicillin

1. S. A. Waksman, "History of the Word 'Antibiotic,'" *Journal of the History of Medicine and Allied Sciences* 28 (Jul. 1973): 284–286; H. W. Florey, "Steps Leading to the Therapeutic Application of Microbial Antagonisms," *British Medical Bulletin* 4 (1946): 248–258; Lloyd G. Stevenson, "Antibacterial and Antibiotic Concepts in Early Bacteriological Studies and in Ehrlich's Chemotherapy," in Iago Galdston, ed., *The Impact of the Antibiotics on Medicine and Society* (New York: International Universities Press, 1958; first I.U.P. paperback ed., 1969), pp. 38–57; Florey, "Steps Leading to the Therapeutic Application of Microbial Antagonisms," pp. 243, 254; Stevenson, "Antibacterial and Antibiotic Concepts," pp. 48–49. Once penicillin had proved its success, many practices were dredged up from medicine's shadowy past which purported to show that some people had previously been aware of penicillin's action on infections. Folk remedies used at various times as a dressing for wounds and infections included a substance extracted from wallflowers, moldy bread, and an ointment made from a moss that grew on dead men's skulls. There is no way of telling whether any of these practices actually involved the application of an antibiotic substance to an infection for which it was a specific cure. Even if some did, such occasions must have been relatively few, given the inaccuracy of diagnosis and the crudeness and inconsistency of the method by which these substances were prepared. Most of the "cures" could be attributed to magic or other forms of suggestion. More important, the use of these preparations led nowhere. They did not result in the growth of knowledge or the formation of scientific concepts. This subject is covered more fully in H. W. Florey, E. Chain, N. G. Heatley, M. A. Jennings, A. G. Sanders, E. P. Abraham, and M. E. Florey, *Antibiotics: A Survey of Penicillin, Streptomycin, and*

Other Antimicrobial Substances from Fungi, Actinomycetes, Bacteria, and Plants, 2 vols. (London: Oxford University Press, 1949), I, 1–73; Stevenson, "Antibacterial and Antibiotic Concepts."

2. V. D. Allison, "Sir Alexander Fleming, 1881–1955," *Journal of General Microbiology* 14 (Feb. 10, 1956): 1–13; André Maurois, *The Life of Sir Alexander Fleming, Discoverer of Penicillin,* trans. Gerard Hopkins (New York: E. P. Dutton, 1959), pp. 1–108, 109–122; A. Fleming, "On the Antibacterial Action of Cultures of a Penicillium, with Special Reference to Their Use in the Isolation of *B. Influenzae,*" *British Journal of Experimental Pathology* 10 (June 1929): 226–236.

3. Fleming, "On the Antibacterial Action of Cultures of a Penicillium," p. 226. Inhibitory means that an agent slows but does not completely stop the growth of microorganisms; bactericidal is the property of killing bacteria; and bacteriolytic is the property of dissolving bacteria. Penicillin is sometimes inhibitory and sometimes bactericidal, depending upon the concentration of the antibiotic, the number of bacteria present, and other factors. From the experience of many investigators it now seems more likely that penicillin, rather than having lysed the colonies of staphylococci, inhibited their growth so that a clear zone, free of colonies, appeared around the mold.

4. Ibid., pp. 230, 232, 236.

5. Maurois, *Life of Alexander Fleming,* pp. 131–136; A. Fleming, "Penicillin: The Robert Campbell Oration," *Ulster Medical Journal* 13 (Nov. 1944): 95–108.

6. P. W. Clutterbuck, R. Lovell, and H. Raistrick, "Studies in the Biochemistry of Micro-organisms: XXVI. The Formation from Glucose by Members of the *Penicillium Chrysogenum* Series of a Pigment, an Alkali Soluble Protein, and Penicillin—the Antibacterial Substance of Fleming," *Biochemical Journal* 26 (1932): 1907–1918; Lennard Bickel, *Rise Up to Life: A Biography of Howard Walter Florey, Who Gave Penicillin to the World* (New York: Charles Scribner's Sons, 1972), p. 85.

7. E. P. Abraham, "Howard Walter Florey: Baron Florey of Adelaide and Marston, 1898–1968," *Biographical Memoirs of Fellows of the Royal Society* 17 (Nov. 1971): 255–302. This brief biography is an authentic account of Florey's scientific achievements written by one of his associates.

8. R. J. Dubos, "The Effect of Specific Agents Extracted from Soil Microörganisms upon Experimental Bacterial Infections," *Annals of Internal Medicine* 13 (May 1940): 2025–2037.

9. L. A. Falk, "The History of Penicillin," *Journal of the American Medical Association* 124 (Apr. 22, 1944): 1219. Leslie Epstein later changed his name to Falk. See Bickel, *Rise Up to Life,* pp. 177–178.

10. Abraham, "Howard Walter Florey," pp. 264–265; N. G. Heatley, "In Memoriam, H. W. Florey: An Episode," *Journal of General Microbiology* 61 (Jul. 1970): 289–299; Bickel, *Rise Up to Life,* pp. 95–97. For the quotation, see Florey, "Steps Leading to the Therapeutic Application of Microbial Antagonisms," p. 255. See also E. Chain, H. W. Florey,

A. D. Gardner, N. G. Heatley, M. A. Jennings, J. Orr-Ewing, and A. G. Sanders, "Penicillin as a Chemotherapeutic Agent," *Lancet* 2 (Aug. 24, 1940): 226–228.

11. Bickel, *Rise Up to Life*, pp. 112–115; E. P. Abraham, E. Chain, C. M. Fletcher, H. W. Florey, A. D. Gardner, N. G. Heatley, and M. A. Jennings, "Further Observations on Penicillin," *Lancet* 2 (Aug. 16, 1941): 177–188, esp. p. 178; Bickel, *Rise Up to Life*, p. 121. The unit that they arbitrarily defined remained the standard for assaying penicillin until the antibiotic was obtained in pure form. Then it was found that a unit was equivalent to 0.6 micrograms, or less than one-millionth of a gram.

12. Heatley, "In Memoriam, H. W. Florey," p. 294; Bickel, *Rise Up to Life*, pp. 120–123; Heatley, "In Memoriam, H. W. Florey," pp. 294–295.

13. Heatley, "In Memoriam, H. W. Florey," p. 295; Abraham et al., "Further Observations on Penicillin," pp. 177–188; Abraham, "Howard Walter Florey," p. 266.

14. Bickel, *Rise Up to Life*, pp. 124–129.

15. Ibid., pp. 140–149; H. W. Florey, "Penicillin in Perspective," *Antibiotics Annual, 1958–1959* (New York: Medical Encyclopedia, 1959), pp. 12–20.

16. A. N. Richards, "Production of Penicillin in the United States (1941–1946)," *Nature* (London), 201 (Feb. 1, 1964): 441–445; Bickel, *Rise Up to Life*, pp. 153–155.

17. Richards, "Production of Penicillin in the United States," pp. 442–444; C. S. Keefer, "Dr. Richards as Chairman of the Committee on Medical Research," *Annals of Internal Medicine* 71, supp. 8 (Nov. 1969): 61–70; Richards, "Production of Penicillin in the United States," pp. 443, 442.

18. C. F. Schmidt, "Alfred Newton Richards, 1876–1966," *Annals of Internal Medicine* 71, supp. 8 (Nov. 1969): 15–27; G. Urdang, "The Antibiotics and Pharmacy," *Journal of the History of Medicine* 6 (Summer 1951): 392.

19. C. S. Keefer, F. G. Blake, E. K. Marshall, Jr., J. S. Lockwood, and W. B. Wood, Jr., "Penicillin in the Treatment of Infections: A Report of 500 Cases," *Journal of the American Medical Association* 122 (Aug. 28, 1943): 1217–1224.

20. C. Lyons, "Penicillin Therapy of Surgical Infections in the U.S. Army," *Journal of the American Medical Association* 123 (Dec. 18, 1943): 1007–1018.

21. J. F. Mahoney, R. C. Arnold, and A. Harris, "Penicillin Treatment of Early Syphilis: A Preliminary Report," *American Journal of Public Health* 33 (Dec. 1943): 1387–1391; J. E. Moore, J. F. Mahoney, W. Schwartz, T. Sternberg, and W. B. Wood, Jr., "The Treatment of Early Syphilis with Penicillin: A Preliminary Report of 1,418 Cases," *Journal of the American Medical Association* 126 (Sept. 9, 1944): 67–73.

22. W. S. Middleton, "Continuing Education in Medical Rambling," *Pharos* 28 (Jan. 1965): 22; Florey et al., *Antibiotics*, II, 651–667; M. E.

Florey and H. W. Florey, "General and Local Administration of Penicillin," *Lancet* 1 (Mar. 27, 1943): 387–397; Heatley, "In Memoriam, H. W. Florey," p. 297; "Distribution of Penicillin," *British Medical Journal* 1 (Apr. 7, 1945): 493.

23. Bickel, *Rise Up to Life*, pp. 146–148, 238–240; C. Thom, "Mycology Presents Penicillin," *Mycologia* 37 (Jul.–Aug. 1945): 460–475; David Wilson, *In Search of Penicillin* (New York: Alfred A. Knopf, 1976), p. 194; Heatley, "In Memoriam, H. W. Florey," p. 296.

24. Bickel, *Rise Up to Life*, pp. 166–168, 171–172; Hare, *The Birth of Penicillin*, p. 186; Florey and Florey, "General and Local Administration of Penicillin"; A. Wright, Letter: "Penicillin," *Times* (London), Aug. 31, 1942, p. 5. For the quotation, see B. Seeman, "Penicillin's Unfolding Drama," *New York Times*, May 24, 1959, magazine section, p. 40. See also Hare, *The Birth of Penicillin*, pp. 79–81.

25. Abraham, "Howard Walter Florey," pp. 268–269; Hare, *The Birth of Penicillin*, pp. 79, 105–109; René Dubos to Harry F. Dowling, June 27, 1975; Bickel, *Rise Up to Life*, p. 107n.

26. A. Fleming, "Penicillin—Its Discovery, Development, and Uses in the Field of Medicine and Surgery: Lecture I—Discovery and Development of Penicillin," *Journal of the Royal Institute of Public Health and Hygiene* 8 (Feb. 1945): 36–49; Bickel, *Rise Up to Life*, pp. 81–99; Abraham, "Howard Walter Florey," p. 269. None of the three Nobel Laureates claimed the major credit for the discovery of penicillin and its value; as is usually the case, such claims were made by others.

27. Bickel, *Rise Up to Life*, pp. 81–89; Abraham, "Howard Walter Florey," pp. 269–276.

28. W. S. Tillett, M. J. Cambier, and J. E. McCormack, "The Treatment of Lobar Pneumonia and Pneumococcal Empyema with Penicillin," *Bulletin of the New York Academy of Medicine* 20 (Mar. 1944): 142–178. Doctors at Fort Bragg, North Carolina, later found that patients given less than 40,000 units a day often failed to respond or else the disease relapsed. J. M. Kinsman, W. B. Daniels, S. Cohen, J. P. McCracken, C. A. D'Alonzo, S. P. Martin, and W. M. M. Kirby, "The Treatment of Pneumonia with Sulfonamides and Penicillin," *Journal of the American Medical Association* 128 (Aug. 25, 1945): 1219–1224.

29. Abraham et al., "Further Observations on Penicillin," pp. 183–184; W. McDermott, P. A. Bunn, M. Benoit, R. DuBois, and M. E. Reynolds, "The Absorption, Excretion, and Destruction of Orally Administered Penicillin," *Journal of Clinical Investigation* 25 (Mar. 1946): 190–210; J. A. Robinson, H. L. Hirsh, and H. F. Dowling, "Oral Penicillin in the Treatment of Various Bacterial Infections," *American Journal of Medicine* 4 (May 1948): 716–723.

30. Karl H. Beyer, *Pharmacological Basis of Penicillin Therapy* (Springfield, Ill.: Charles C Thomas, 1950), pp. 115–131.

31. Harry F. Dowling, *The Acute Bacterial Infections: Their Diagnosis and Treatment* (Philadelphia: W. B. Saunders, 1948), p. 121; H. F. Dow-

ling, H. H. Hussey, H. L. Hirsh, and F. Wilhelm, "Penicillin and Sulfadiazine, Compared with Sulfadiazine Alone, in the Treatment of Pneumococcic Pneumonia," *Annals of Internal Medicine* 25 (Dec. 1946): 950–956.

32. Bickel, *Rise Up to Life*, p. 167; E. Dumoff-Stanley, H. F. Dowling, and L. K. Sweet, "The Absorption into and Distribution of Penicillin in the Cerebrospinal Fluid," *Journal of Clinical Investigation* 25 (Jan. 1946): 87–93; G. X. Schwemlein, R. L. Barton, T. J. Bauer, L. Loewe, H. N. Bundesen, and R. M. Craig, "Penicillin in Spinal Fluid after Intravenous Administration," *Journal of the American Medical Association* 130 (Feb. 9, 1946): 340–341; H. F. Dowling, L. K. Sweet, J. A. Robinson, W. W. Zeller, and H. L. Hirsh, "The Treatment of Pneumococcic Meningitis with Massive Doses of Systemic Penicillin," *American Journal of the Medical Sciences* 217 (Feb. 1949): 149–156.

33. M. H. Lepper, H. F. Dowling, P. F. Wehrle, N. H. Blatt, H. W. Spies, and M. Brown, "Meningococcic Meningitis: Treatment with Large Doses of Penicillin Compared to Treatment with Gantrisin," *Journal of Laboratory and Clinical Medicine* 40 (Dec. 1952): 891–900.

34. M. H. Dawson and T. H. Hunter, "The Treatment of Subacute Bacterial Endocarditis with Penicillin: Results in Twenty Cases," *Journal of the American Medical Association* 127 (Jan. 20, 1945): 129–137; M. E. Florey, "Clinical Uses of Penicillin," *British Medical Bulletin* 2 (1944): 9–13.

35. L. Loewe, P. Rosenblatt, H. J. Greene, and M. Russell, "Combined Penicillin and Heparin Therapy of Subacute Bacterial Endocarditis," *Journal of the American Medical Association* 124 (Jan. 15, 1944): 144–149; Paul de Kruif, *Life among the Doctors* (New York: Harcourt, Brace, 1949), pp. 212–231; Dawson and Hunter, "The Treatment of Subacute Bacterial Endocarditis with Penicillin"; J. R. Goerner, A. J. Geiger, and F. G. Blake, "Treatment of Subacute Bacterial Endocarditis with Penicillin: Report of Cases Treated without Anticoagulant Agents," *Annals of Internal Medicine* 23 (Oct. 1945): 491–519. The skepticism regarding the ability of penicillin to cure endocarditis was widespread and did not subside immediately. Dr. Thomas Hunter recalls that when he showed the first patient to recover from endocarditis to his department head, he was told: "Bring me back in six months and if the patient is still alive, I'll begin to pay attention." Hunter to Harry F. Dowling, Jan. 9, 1974. In mid-1945 Edward B. Krumbhaar, then editor of the *American Journal of the Medical Sciences*, rejected my manuscript reporting the recovery of seven of ten patients after penicillin therapy with the comment that no therapeutic agent would produce recovery in this disease.

36. Stephen D. Elek, *Staphylococcus Pyogenes and Its Relation to Disease* (Edinburgh: E. & S. Livingstone, 1959), pp. 4–6; William Bulloch, *The History of Bacteriology* (London: Oxford University Press, 1938), pp. 149–151.

37. Keefer et al., "Penicillin in the Treatment of Infections"; D. Skin-

ner and C. S. Keefer, "Significance of Bacteremia Caused by Staphylococcus Aureus," *Archives of Internal Medicine* 68 (Nov. 1941): 851–875; W. W. Spink and W. H. Hall, "Penicillin Therapy at the University of Minnesota Hospitals, 1942–1944," *Annals of Internal Medicine* 22 (Apr. 1945): 510–525.

38. Elek, *Staphylococcus Pyogenes*, pp. 451–455; W. M. M. Kirby, "Extraction of a Highly Potent Penicillin Inactivator from Penicillin Resistant Staphylococci," *Science* 99 (June 2, 1944): 452–453; E. P. Abraham and E. Chain, "An Enzyme from Bacteria Able to Destroy Penicillin," *Nature* (London), 146 (Dec. 28, 1940): 837; H. F. Dowling, M. H. Lepper, and G. G. Jackson, "Clinical Significance of Antibiotic-Resistant Bacteria," *Journal of the American Medical Association* 157 (Jan. 22, 1953): 327–331.

39. M. Barber, "Staphylococcal Infection Due to Penicillin-Resistant Strains," *British Medical Journal* 2 (Nov. 29, 1947): 863–865; M. Barber and M. Rozwadowska-Dowzenko, "Infection by Penicillin-Resistant Staphylococci," *Lancet* 2 (Oct. 23, 1948): 641–644; M. Finland and T. H. Haight, "Antibiotic Resistance of Pathogenic Staphylococci," *Archives of Internal Medicine* 91 (Feb. 1953): 143–158; W. W. Spink, "The Clinical Problem of Antimicrobial Resistant Staphylococci," *Annals of the New York Academy of Sciences* 65, art. 3 (Aug. 31, 1956): 175–190.

40. James H. Cassedy, *Charles V. Chapin and the Public Health Movement* (Cambridge: Harvard University Press, 1962), pp. 74–77; Elek, *Staphylococcus Pyogenes*, pp. 508–529; R. Dubos, Editorial: "Staphylococci and Infection Immunity," *American Journal of Diseases of Children* 105 (June 1963): 643–645.

41. Keefer et al., "Penicillin in the Treatment of Infections"; Dowling, *The Acute Bacterial Diseases*, pp. 230–232.

42. T. M. Gocke, C. Wilcox, and M. Finland, "Antibiotic Spectrum of the Gonococcus," *American Journal of Syphilis, Gonorrhea, and Venereal Diseases* 34 (May 1950): 265–272; A. G. Franks, "Successful Combined Treatment of Penicillin-Resistant Gonorrhea," *American Journal of the Medical Sciences* 211 (May 1946): 553–555; Editorial: "Alleged Penicillin-Resistant Gonorrhea," *New England Journal of Medicine* 239 (Sept. 9, 1948): 411–413; John F. Mahoney, "The Effects of the Antibiotics on the Concepts and Practices of Public Health," in Galdston, ed., *The Impact of the Antibiotics on Medicine and Society*, p. 213. A microgram is one-thousandth of a milligram or one-millionth of a gram.

43. "Resistance of Gonococci to Penicillin: Interim Report by a Medical Research Council Working Party Appointed to Examine the Resistance of Gonococci to Penicillin," *Lancet* 2 (Jul. 29, 1961): 226–230; L. C. Sabath and J. J. Kivlahan, "Dosage of Penicillin for Acute Gonorrhea of Males," *American Journal of the Medical Sciences* 240 (Dec. 1961): 663–672; J. E. Martin, Jr., A. Lester, E. V. Price, and J. D. Schmale, "Comparative Study of Gonococcal Susceptibility to Penicillin in the United States," *Journal of Infectious Diseases* 122 (Nov. 1970): 459–461.

44. B. Sokoloff and H. Goldstein, "Clinicolaboratory Study of Acute Gonorrhea in Men in El Paso, Tex., Area," *Journal of the American Medical Association* 184 (Apr. 20, 1963): 197–200; U.S. Department of Health, Education, and Welfare, Public Health Service, *Morbidity and Mortality Weekly Report* 25, no. 33 (Atlanta: Center for Disease Control, 1976): 261; Ian Phillips, "B-Lactamase-Producing, Penicillin-Resistant Gonococcus," *Lancet* 2 (Sept. 25, 1976): 656–657; Winston A. Ashford, R. G. Golash, and V. G. Hemming, "Penicillinase-Producing Neisseria Gonorrhoeae," ibid., pp. 657–658.

45. Editorial: "The Prophylaxis of Gonorrheal Ophthalmia Neonatorum," *American Journal of Obstetrics and Gynecology* 65 (May 1953): 1155–1159; Editorial: "Opthalmia Neonatorum Prophylaxis," *Journal of Pediatrics* 50 (June 1957): 789–790; M. Greenberg and J. E. Vandow, "Ophthalmia Neonatorum: Evaluation of Different Methods of Prophylaxis in New York City," *American Journal of Public Health* 51 (June 1961): 836–845; American Social Health Association, *Today's VD Control Program—1971* (New York: American Social Health Association, 1971), p. 9; P. C. Barsam, "Specific Prophylaxis of Gonorrheal Ophthalmia Neonatorum: A Review," *New England Journal of Medicine* 274 (Mar. 31, 1966): 731–734; U.S. Department of Health, Education, and Welfare, *Venereal Disease Control Laws—Summary* (Atlanta: Center for Disease Control, 1972).

46. Joseph Earle Moore, *Penicillin in Syphilis* (Springfield, Ill.: Charles C Thomas, 1947), p. 4; J. E. Moore, J. F. Mahoney, W. Schwartz, T. Sternberg, and W. B. Wood, Jr., "The Treatment of Early Syphilis with Penicillin: A Preliminary Report of 1,418 Cases," *Journal of the American Medical Association* 126 (Sept. 9, 1944): 67–73; J. H. Stokes, J. H. Sternberg, W. H. Schwartz, J. F. Mahoney, J. E. Moore, and W. B. Wood, Jr., "The Action of Penicillin in Late Syphilis Including Neurosyphilis, Benign Late Syphilis and Late Congenital Syphilis: Preliminary Report," ibid., pp. 73–79; William J. Brown, James F. Donohue, Norman W. Axnick, Joseph H. Blount, Neal H. Ewen, and Oscar G. Jones, *Syphilis and Other Venereal Diseases* (Cambridge: Harvard University Press, 1970), p. 33.

47. U.S. Department of Health, Education, and Welfare, Public Health Service, *VD Fact Sheet 1971* (Atlanta: Center for Disease Control, 1972), pp. 5, 9.

48. Brown et al., *Syphilis and Other Venereal Diseases*; R. H. Kampmeier, "Comments on the Present-day Management of Syphilis," *Southern Medical Journal* 46 (Mar. 1953): 226–237; J. E. Moore, Editorial: "The Impending Loss of a Partly Won War against Venereal Disease," *American Journal of Syphilis, Gonorrhea, and Venereal Diseases* 38 (May 1954): 237–238.

49. U.S. Department of Health, Education, and Welfare, *VD Fact Sheet 1971*, p. 9.

50. American Social Health Association, *Today's VD Control Program—1971*, p. 21.

51. Ibid., pp. 57, 19–20.

52. U.S. Department of Health, Education, and Welfare, *VD Fact Sheet 1971*, p. 9.

53. L. Baumgartner, "Syphilis Eradication—A Plea for Action Now," in U.S. Department of Health, Education, and Welfare, *Proceedings of World Forum on Syphilis and Other Treponematoses, Washington, D.C., September 4–8, 1962* (Washington, D.C.: U.S. Government Printing Office, 1964), pp. 26–32.

54. Keefer et al., "Penicillin in the Treatment of Infections."

55. M. Meads, M. E. Flipse, Jr., M. W. Barnes, and M. Finland, "Penicillin Treatment of Scarlet Fever: Bacteriologic Study of the Nose and Throat of Patients Treated Intramuscularly or by Spray with Penicillin and a Comparison with Sulfadiazine," *Journal of the American Medical Association* 129 (Nov. 17, 1945): 785–789; H. L. Hirsh, G. Rotman-Kavka, H. F. Dowling, and L. K. Sweet, "Penicillin Therapy of Scarlet Fever: Comparison with Antitoxin and Symptomatic Therapy," ibid. 133 (Mar. 1947): 657–661.

56. G. H. Stollerman, "The Use of Antibiotics for the Prevention of Rheumatic Fever," *American Journal of Medicine* 17 (Dec. 1954): 757–767; G. H. Stollerman, J. H. Rusoff, and I. Hirschfeld, "Prophylaxis against Group A Streptococci in Rheumatic Fever: The Use of Single Monthly Injections of Benzathine Penicillin G," *New England Journal of Medicine* 252 (May 12, 1955): 787–792.

57. L. W. Wannamaker, C. H. Rammelkamp, Jr., F. W. Denny, W. R. Brink, H. B. Houser, and E. O. Hahn, "Prophylaxis of Acute Rheumatic Fever by Treatment of the Preceding Streptococcal Infection with Various Amounts of Depot Penicillin," *American Journal of Medicine* 10 (June 1951): 673–695.

58. Since viruses contain either deoxyribonucleic acid (DNA) or ribonucleic acid (RNA) but not both, as do bacteria and other true microorganisms, they cannot reproduce themselves except by entering a living cell and using some of the components of the cell in the process. Viruses produce a wide spectrum of diseases in humans, ranging from smallpox and yellow fever to the common cold.

59. P. A. L. Chapple, L. M. Franklin, J. D. Paulett, E. Tuckman, J. T. Woodall, A J. H. Tomlinson, and J. C. McDonald, "Treatment of Acute Sore Throat in General Practice: Therapeutic Trial, with Observations on Symptoms and Bacteriology," *British Medical Journal* 1 (Mar. 31, 1956): 705–708.

60. B. B. Breese, "Treatment of Beta Hemolytic Streptococcic Infections in the Home," *Journal of the American Medical Association* 152 (May 2, 1953): 10–14.

61. B. Phibbs, D. Becker, C. R. Lowe, R. Holmes, R. Fowler, O. K. Scott, K. Roberts, W. Watson, and R. Malott, "The Casper Project—An Enforced Mass-Culture Streptococcic Control Program: 1. Clinical Aspects," *Journal of the American Medical Association* 166 (Mar. 8, 1958): 1113–1119.

62. R. Whittemore, D. E. Brinsfield, and K. D. Brownell, "The Community Service Aspects of Rheumatic Fever and Rheumatic Heart Disease," in E. Cowles Andrus, ed., *The Heart and Circulation: Second National Conference on Cardiovascular Diseases, Washington, D.C., 1964,* 2 vols. (Bethesda, Md.: Federation of American Societies for Experimental Biology, 1965), vol. 2: *Community Services and Education,* pp. 719–730. The American Heart Association also included in its recommendations the administration of penicillin prior to the extraction of teeth in patients with deformities of the heart valves, based on the observations of several investigators that penicillin would prevent or greatly diminish the bacteremia that often follows removal of a tooth. See Committee on Prevention of Rheumatic Fever and Endocarditis, American Heart Association, "Prevention of Rheumatic Fever and Bacterial Endocarditis through Control of Streptococcal Infections," *Circulation; Journal of the American Heart Association* 11 (Feb. 1955): 317–320.

63. E. F. Bland, "Declining Severity of Rheumatic Fever: A Comparative Study of the Past Four Decades," *New England Journal of Medicine* 262 (Mar. 24, 1960): 597–599; Rheumatic Fever and Rheumatic Heart Disease Study Group, "Prevention of Rheumatic Fever and Rheumatic Heart Disease," *Circulation; Journal of the American Heart Association* 41 (May 1970): A1–A15. The Good Samaritan Hospital was a special hospital for patients with rheumatic fever or its sequels.

64. Phibbs et al., "The Casper Project"; C. A. Stetson, C. H. Rammelkamp, Jr., R. M. Krause, R. J. Kohen, and W. D. Perry, "Epidemic Acute Nephritis: Studies on Etiology, Natural History and Prevention," *Medicine* 34 (Dec. 1955): 431–450.

65. U.S. Bureau of the Census, *Historical Statistics of the United States, Colonial Times to 1957* (Washington, D.C.: U.S. Government Printing Office, 1960), p. 25; National Center for Health Statistics, *Facts of Life and Death,* DHEW Publication no. (HRA) 74–1222 (Rockville, Md.: U.S. Department of Health, Education, and Welfare, Public Health Service, Health Resources Administration, 1974), p. 9; W. R. McCabe and A. A. Abrams, "An Outbreak of Streptococcal Puerperal Sepsis," *New England Journal of Medicine* 272 (Mar. 25, 1965): 615–618; J. F. Jewett, D. E. Reid, L. E. Safon, and C. L. Easterday, "Childbed Fever— A Continuing Entity," *Journal of the American Medical Association* 206 (Oct. 7, 1968): 344–350.

66. L. Weinstein, "The Treatment of Acute Diphtheria and the Chronic Carrier State with Penicillin," *American Journal of the Medical Sciences* 213 (Mar. 1947): 308–314; A. Akkoyunlu, "Treatment of Diphtheria with Penicillin Alone," *Archives françaises de pediatrie* 13, no. 10 (1956): 1065–1067.

67. H. W. Florey and H. Cairns, *Investigation of War Wounds: Penicillin—A Preliminary Report to the War Office and the Medical Research Council in Investigations Concerning The Use of Penicillin in War Wounds* (London: War Office [A.M.D. 7], 1943), esp. p. 7; Lyons, "Peni-

cillin Therapy of Surgical Infections in the U.S. Army"; Bickel, *Rise Up to Life*, p. 187.

68. Bickel, *Rise Up to Life*, pp. 159, 193–207; Florey and Cairns, *Investigations of War Wounds: Penicillin*.

69. Albert Baird Hastings, Oral History Transcript, 1969, National Library of Medicine, Bethesda, Md., pp. 201–202.

70. *Time* magazine told its readers about the "marvelous mold that saves lives when the sulfa drugs fail." See "Mold for Infections," *Time* 38 (Sept. 15, 1941): 55. A medical journal headlined an editorial, "Let's Be Getting Ready for Penicillin." See *Journal of Iowa State Medical Society* 33 (Oct. 1943): 475–476. See also "Rush on Penicillin," *Time* 42 (Aug. 30, 1943): 44, 46; F. J. Stock, "Penicillin Production and Distribution," *Journal of the American Pharmaceutical Association*, Practical Pharmacy Ed. 6 (Apr. 1945): 110–114; R. D. Coghill and R. S. Koch, "Penicillin: A Wartime Accomplishment," *Chemical and Engineering News* 23 (Dec. 1945): 2310–2316.

71. Coghill and Koch, "Penicillin: A Wartime Accomplishment"; Federal Trade Commission, *Economic Report on Antibiotics Manufacture, June 1958* (Washington, D.C.: U.S. Government Printing Office, 1958), pp. 67, 148–149.

72. Abraham et al., "Further Observations on Penicillin," p. 183; M. Ethel Florey, *The Clinical Application of Antibiotics: Penicillin* (London: Oxford University Press, 1952), p. 7.

73. Keefer et al., "Penicillin in the Treatment of Infections," pp. 22–23; P. H. Long, "Fatal Anaphylactic Reactions to Penicillin," *Antibiotics Annual, 1953–1954* (New York: Medical Encyclopedia, 1953), pp. 35–37; H. Welch, C. N. Lewis, H. I. Weinstein, and B. B. Boeckman, "Severe Reactions to Penicillin: A Nationwide Survey," *Antibiotics Annual, 1957–1958* (New York: Medical Encyclopedia, 1958), pp. 296–309.

74. L. Galton, "Penicillin Turns Killer," *Coronet* 37 (Nov. 1954): 17–21; Queries and Minor Notes: "Routine Use of Penicillin," Letter to the Editor, *Journal of the American Medical Association* 156 (Sept. 4, 1954): 92; *Registrar General's Statistical Review of England and Wales for the Year 1958, Part III: Commentary* (London: Her Majesty's Stationery Office, 1960), cited in Editorial: "Fatal Reaction to Drugs," *British Medical Journal* 2 (Nov. 5, 1960): 1373–1374; M. H. Lepper, H. F. Dowling, J. A. Robinson, T. E. Stone, R. L. Brickhouse, E. R. Caldwell, Jr., and R. L. Whelton, "Studies on Hypersensitivity to Penicillin: I. Incidence of Reactions in 1303 Patients," *Journal of Clinical Investigation* 28 (Sept. 1949): 826–831.

10. Streptomycin and the Chemotherapy of Tuberculosis

1. Hubert A. Lechevalier and Morris Solotorovsky, *Three Centuries of Microbiology* (New York: McGraw-Hill, 1965), pp. 475–478; S. A.

Waksman, "Autobiographic Sketch," *Perspectives in Biology and Medicine* 7 (Summer 1964): 377–398.

2. Lechevalier and Solotorovsky, *Three Centuries of Microbiology*, pp. 477–478; Waksman, "Autobiographic Sketch"; R. J. Dubos, "The Effect of Specific Agents Extracted from Soil Microörganisms upon Experimental Bacterial Infections," *Annals of Internal Medicine* 13 (May 1940): 2030–2034; R. Dubos, "Medicine's Living History," *Medical World News* 9 (May 5, 1975): 77–79, 81–82, 85, 87. See also H. W. Florey, E. Chain, N. G. Heatley, M. A. Jennings, A. G. Sanders, E. P. Abraham, and M. E. Florey, *Antibiotics: A Survey of Penicillin, Streptomycin, and Other Antimicrobial Substances from Fungi, Actinomycetes, Bacteria, and Plants*, 2 vols. (London: Oxford University Press, 1949), I, 38.

3. Selman A. Waksman, *The Conquest of Tuberculosis* (Berkeley: University of California Press, 1964), pp. 113–115; A. Schatz, E. Bugie, and S. A. Waksman, "Streptomycin, a Substance Exhibiting Activity against Gram-Positive and Gram-Negative Bacteria," *Proceedings of the Society for Experimental Biology and Medicine* 55 (Jan. 1944): 66–69; W. H. Feldman and H. C. Hinshaw, "Successful Treatment of Tuberculosis in Guinea Pigs," *Proceedings of the Staff Meetings of the Mayo Clinic* 19 (Dec. 1944): 593–599.

4. W. H. Feldman, "Tuberculosis Chemotherapy: Reminiscence of In Vivo Research," *Scandinavian Journal of Respiratory Diseases* 50, no. 3 (1969): 186–196; A. R. Rich and R. H. Follis, Jr., "The Inhibitory Effect of Sulfanilamide on the Development of Experimental Tuberculosis in the Guinea Pig," *Bulletin of the Johns Hopkins Hospital* 62 (Jan. 1938): 77–84; W. H. Feldman, "Streptomycin: Some Historical Aspects of Its Development as a Chemotherapeutic Agent in Tuberculosis," *American Review of Tuberculosis* 69 (June 1954): 859–868. Sulfones are a class of organic compounds containing the bivalent group, —SO_2—, attached to two hydrocarbon groups, that is, groups containing only hydrogen and carbon. A report by Feldman and his associates in 1940 on the use of a sulfone, Promin, in animals attracted the attention of leprologists at the leprosarium of the Public Health Service in Carville, Louisiana. They obtained a supply from the manufacturer, Parke, Davis and Company, and tried it in patients with leprosy. This was the start of the sulfone therapy of leprosy, which revolutionized the treatment of that disease. See Feldman, "Tuberculosis Chemotherapy," pp. 188–189.

5. Schatz, Bugie, and Waksman, "Streptomycin"; Feldman, "Tuberculosis Chemotherapy"; S. A. Waksman, "Tenth Anniversary of the Discovery of Streptomycin, the First Chemotherapeutic Agent Found to Be Effective against Tuberculosis in Humans," *American Review of Tuberculosis* 70 (Jul. 1954): 1–8; Feldman, "Tuberculosis Chemotherapy," the quotation is on p. 195.

6. G. Birath, "Introduction of Para-amino-salicylic Acid and Streptomycin in the Treatment of Tuberculosis," *Scandinavian Journal of*

Respiratory Diseases 50 (1969): 204–209; letter and accompanying documents, H. Corwin Hinshaw to Harry F. Dowling, Aug. 26, 1975.

7. Waksman, "Autobiographic Sketch"; Waksman, *The Conquest of Tuberculosis,* pp. 115–117; R. Silman and J. Lear, "Boswell of the Microbes: Dr. Selman A. Waksman," *Saturday Review of Literature* 41 (Sept. 6, 1958): 56–57; René Dubos to Harry F. Dowling, June 27 and Jul. 18, 1975.

8. Silman and Lear, "Boswell of the Microbes"; "Streptomycin Profit Asked by Ex-Student," *New York Times,* Mar. 11, 1950, p. 16; Lechevalier and Solotorovsky, *Three Centuries of Microbiology,* pp. 486–487; "Streptomycin Suit Is Labeled 'Baseless,'" *New York Times,* Mar. 13, 1950, p. 25; "Dr. Schatz Wins 3% of Royalty; Named Co-finder of Streptomycin," ibid., Dec. 30, 1950, p. 1 (the quotation is from this article). See also Edward Robert Isaacs to Harry F. Dowling, Mar. 3, 1975.

9. Birath, "Introduction of Para-amino-salicylic Acid and Streptomycin in the Treatment of Tuberculosis." Waksman later stated that he and Schatz had reported that streptomycin was highly effective against *Mycobacterium tuberculosis* in April 1944 and that they had also discovered its effectiveness in animals. These were apparently erroneous recollections since the report in question did not appear until November 1944 and the first announcement of the value of streptomycin in animal tuberculosis was published by Feldman and Hinshaw the following month. See Waksman, *The Conquest of Tuberculosis,* p. 121; Discussion by S. A. Waksman, in National Association for the Prevention of Tuberculosis, *Tuberculosis in the Commonwealth, 1947: Complete Transactions of the Commonwealth and Empire Health and Tuberculosis Conference, London, July, 1947* (London: National Association for the Prevention of Tuberculosis, 1947), pp. 175–176; A. Schatz and S. A. Waksman, "Effect of Streptomycin and Other Antibiotic Substances upon *Mycobacterium tuberculosis* and Related Organisms," *Proceedings of the Society of Experimental Biology and Medicine* 57 (Nov. 1944): 244–248; W. H. Feldman and H. C. Hinshaw, "Successful Treatment of Tuberculosis in Guinea Pigs"; letter and accompanying documents, H. Corwin Hinshaw to Harry F. Dowling, Aug. 26, 1975.

10. Feldman, "Tuberculosis Chemotherapy"; Feldman, "Streptomycin: Some Historical Aspects"; K. H. Pfeutze, M. M. Pyle, H. C. Hinshaw, and W. H. Feldman, "The First Clinical Trial of Streptomycin in Human Tuberculosis," *American Review of Tuberculosis and Pulmonary Diseases* 71 (May 1955): 752.

11. H. C. Hinshaw, "Tuberculosis Chemotherapy: Reminiscences of Early Clinical Trials," *Scandinavian Journal of Respiratory Diseases* 50, no. 3 (1969) 186–196; H. C. Hinshaw, W. H. Feldman, and K. H. Pfeutze, "Treatment of Tuberculosis with Streptomycin: A Summary of Observations on One Hundred Cases," *Journal of the American Medical Association* 132 (Nov. 30, 1946): 778–782.

12. Chester S. Keefer and William L. Hewitt, *The Therapeutic Value of Streptomycin: A Study of 3,000 Cases* (Ann Arbor, Mich.: J. W. Edwards, 1948), pp. v–viii.

13. U.S. Federal Trade Commission, *Economic Report on Antibiotics Manufacture* (Washington, D.C.: U.S. Government Printing Office, 1958), pp. 67, 93; H. C. Hinshaw, "Historical Notes on Earliest Use of Streptomycin in Clinical Tuberculosis," *American Review of Tuberculosis* 70 (Jul. 1954): 9–14.

14. W. McDermott, C. Muschenheim, S. J. Hadley, P. A. Bunn, and R. V. Gorman, "Streptomycin in the Treatment of Tuberculosis in Humans: I. Meningitis and Generalized Hematogenous Tuberculosis," *Annals of Internal Medicine* 27 (Nov. 1947): 769–822.

15. H. McLeod Riggins and H. Corwin Hinshaw, eds., *Streptomycin and Dihydrostreptomycin in Tuberculosis* (New York: National Tuberculosis Association, 1949), pp. vii–viii, 29–35, 372–402; J. B. Barnwell, Obituary: "Arthur Meeker Walker, 1896–1955," *American Review of Tuberculosis and Pulmonary Diseases* 73 (May 1956): 790–794.

16. P. R. Hawley, "The Tuberculosis Program of the Veterans' Administration," *American Review of Tuberculosis* 55 (Jan. 1947): 1–7; William B. Tucker, "The Evolution of the Cooperative Studies in the Chemotherapy of Tuberculosis of the Veterans' Administration and Armed Forces of the U.S.A.," *Advances in Tuberculosis Research* (Basel), 10 (1960): 1–68, esp. pp. 6–7.

17. "Careless Care for Veterans," *Time* 45 (May 7, 1945): 88–92; "Inspired Choice," ibid. (June 18, 1945), p. 13; P. R. Hawley, "Progress in Reorganization of Medical Services in the Veterans' Administration," *Journal of the Medical Society of New Jersey* 43 (Oct. 1946): 405–409; Barnwell, Obituary: "Arthur Meeker Walker"; James Phinney Baxter 3d, *Scientists against Time* (Boston: Little, Brown, 1946), pp. 337–359, for penicillin in syphilis see pp. 358–359; H. F. Dowling, "The Emergence of the Cooperative Clinical Trial," *Transactions and Studies of the College of Physicians of Philadelphia*, 4th ser. 43 (Jul. 1975): 20–29.

18. Council on Pharmacy and Chemistry, American Medical Association, "Streptomycin in the Treatment of Tuberculosis: Current Status," *Journal of the American Medical Association* 138 (Oct. 23, 1948): 584–593. These conferences were published as the Minutes of the VA Streptomycin Conferences, 1–8, 1946–1949; as the Transactions of the Conferences on the Chemotherapy of Tuberculosis, at first VA, then VA-Army, then VA-Armed Forces, 9–21, 1950-1962; and as VA-Armed Forces, Transactions of Research Conferences on Pulmonary Diseases, 22–, 1963–.

19. Council on Pharmacy and Chemistry, "Streptomycin in the Treatment of Tuberculosis."

20. George Rosen, *A History of Public Health* (New York: MD Publications, 1958), pp. 259–264; A. Bradford Hill to Harold Schoolman, Nov. 8, 1970; "Streptomycin Treatment of Pulmonary Tuberculosis: A Medical

Research Council Investigation," *British Medical Journal* 2 (Oct. 30, 1948): 769–782.

21. E. R. Long and S. H. Ferebee, "A Controlled Study of Streptomycin Treatment in Pulmonary Tuberculosis," *Public Health Reports* 65 (Nov. 3, 1950): 1421–1451.

22. G. P. Youmans, E. H. Williston, W. H. Feldman, and H. C. Hinshaw, "Increase in Resistance of Tubercle Bacilli to Streptomycin: A Preliminary Report," *Proceedings of the Staff Meetings of the Mayo Clinic* 21 (Mar. 20, 1946): 126–127.

23. J. Lehmann, "The Treatment of Tuberculosis in Sweden with Para-Aminosalicylic Acid (PAS): A Review," *Diseases of the Chest* 16 (Dec. 1949): 684–703; The Therapeutic Trials Committee of the Swedish National Association against Tuberculosis, "Para-Aminosalicylic Acid in Pulmonary Tuberculosis: Comparison between 94 Treated and 82 Untreated Cases," *American Review of Tuberculosis* 61 (May 1950): 597–612.

24. Tucker, "Evolution of the Cooperative Studies in the Chemotherapy of Tuberculosis," pp. 46, 51, 65; Veterans' Administration, Minutes of the Eighth Streptomycin Conference, Nov. 10–13, 1949 (Washington, D.C.: Veterans' Administration Area Office, 1949), p. 343.

25. Domagk's compound was given the trade name of Conteben by the original manufacturer and officially named thiacetazone in the United States. See W. McDermott, "The Story of INH," *Journal of Infectious Diseases* 119 (June 1969): 678–683; H. C. Hinshaw and W. McDermott, "Thiosemicarbazone Therapy of Tuberculosis in Humans," *American Review of Tuberculosis* 61 (Jan. 1950): 145–157.

26. *American Review of Tuberculosis* 65 (Apr. 1952): 357–442, 484–485; "TB Milestone," *Life* 32 (Mar. 3, 1952): 20–21; E. H. Robitzek and I. J. Selikoff, "Hydrazine Derivatives of Isonicotinic Acid (Rimifon, Marsalid) in the Treatment of Active Progressive Caseous-Pulmonic Tuberculosis: A Preliminary Report," *American Review of Tuberculosis* 65 (Apr. 1952): 402–428; McDermott, "Story of INH," p. 682. For the quotation, see A. Shamaskin, in Veterans' Administration, Minutes of the Sixth Streptomycin Conference, Oct. 21–24, 1948 (St. Paul, Minn.: Veterans' Administration Branch Office, 1948), p. 129.

27. "Treatment of Pulmonary Tuberculosis with Isoniazid: Interim Report to Medical Research Council by Their Tuberculosis Chemotherapy Trials Committee," *British Medical Journal* 2 (Oct. 4, 1952): 735–746; F. W. Mount and S. H. Ferebee, "Control Study of Comparative Efficacy of Isoniazid, Streptomycin-Isoniazid, and Streptomycin-Para-Aminosalicylic Acid in Pulmonary Tuberculosis Therapy: I. Report on Twelve-week Observations on 526 Patients," *American Review of Tuberculosis* 66 (Nov. 1952): 632–635.

28. American Medical Association, Department of Drugs, *AMA Drug Evaluations* (Acton, Mass.: Publishing Sciences Group, 1973), pp. 581–589; R. Newman, B. E. Doster, F. J. Murray, and S. F. Woolpert,

"Rifampin in Initial Treatment of Pulmonary Tuberculosis," *American Review of Respiratory Diseases* 109 (Feb. 1974): 216–232.

29. Shirley H. Ferebee, "Controlled Chemoprophlaxis Trials in Tuberculosis: A General Review," *Advances in Tuberculosis Research* (Basel), 17 (1969): 28–106, esp. pp. 30–32, 49–50. This experiment was "double blind," that is, neither the doctors evaluating the children for the presence of tuberculosis nor the children themselves and their parents knew which ones were receiving isoniazid and which placebos. In this way, the planners of the study excluded as much as possible all personal bias in interpreting the results.

30. Ibid., pp. 49–50, 52–54, 56–61. Administration of isoniazid for prophylaxis was later found to cause injury to the liver in some persons over the age of twenty-seven. Serious injury can be largely avoided by not administering it to alcoholics and persons with chronic liver disease and by testing periodically for disturbances in liver function. See R. A. Garibaldi, R. E. Drusin, S. H. Ferebee, and M. B. Gregg, "Isoniazid-Associated Hepatitis: Report of an Outbreak," *American Review of Respiratory Diseases* 106 (Sept. 1972): 357–365; American Thoracic Society, "Preventive Therapy of Tuberculous Infection," ibid. 110 (Sept. 1974): 371–374.

31. U.S. Department of Health, Education, and Welfare, Public Health Service, *Tuberculosis Beds in Hospitals, June 30, 1971* (Atlanta: Center for Disease Control, 1971), p. 2; Brightman and Hilleboe, "The Present Status of Tuberculosis Control"; "A Victim of Progress," *Life* 27 (Dec. 27, 1954): 76–79; "Oldest TB Haven Will Close Dec. 1," *New York Times,* Oct. 13, 1954, p. 23.

32. T. C. Doege, "Tuberculosis Mortality in the United States, 1900 to 1960," *Journal of the American Medical Association* 192 (June 21, 1965): 1045–1048.

33. Brightman and Hilleboe, "The Present Status of Tuberculosis Control."

34. F. L. Soper, "Problems to Be Solved If the Eradication of Tuberculosis Is to Be Realized," *American Journal of Public Health* 52 (May 1962): 734–745.

35. D. Jones, H. J. Metzger, A. Schatz, and S. A. Waksman, "Control of Gram-Negative Bacteria in Experimental Animals by Streptomycin," *Science* 100 (Aug. 4, 1944): 103–105; Keefer and Hewitt, *The Therapeutic Value of Streptomycin,* esp. p. 75; T. F. Paine, R. Murray, and M. Finland, "Streptomycin: II. Clinical Uses," *New England Journal of Medicine* 236 (May 15, 1947): 748–760.

36. G. W. McCoy and C. W. Chapin, "Further Observations on a Plague-like Disease of Rodents with a Preliminary Note on the Causative Agent, Bacterium Tularense," *Journal of Infectious Diseases* 10 (Jan. 1912): 61–72. Later the causative microorganism was named *Pasteurella tularensis* and finally *Francisella tularensis* after Edward Francis, another Public Health Service bacteriologist who contributed to the knowledge of

the disease and popularized it among the medical and public health professions. See also Lee Foshay, "Tularemia," *Annual Review of Microbiology* 4 (1950): 313–330; Karl F. Meyer, "Pasteurella and Francisella," in René J. Dubos and James G. Hirsch, eds., *Bacterial and Mycotic Infections of Man*, 4th ed. (Philadelphia: J. B. Lippincott, 1965), pp. 681–684; Louis Weinstein and N. Joel Ehrenkranz, *Streptomycin and Dihydrostreptomycin* (New York: Medical Encyclopedia, 1958), pp. 67–69.

37. Keefer and Hewitt, *Therapeutic Value of Streptomycin*; Paine, Murray, and Finland, "Streptomycin: II. Clinical Uses."

11. Other Diseases and Other Antibiotics

1. John C. Snyder, "Typhus Fever Rickettsiae," in Frank L. Horsfall, Jr., and Igor Tamm, eds., *Viral and Rickettsial Infections of Man*, 4th ed. (Philadelphia: J. B. Lippincott, 1965), pp. 1059–1061. See also Hans Zinsser, *Rats, Lice and History* (Boston: Little, Brown, 1935); T. E. Woodward, "A Historical Account of the Rickettsial Diseases with a Discussion of Unsolved Problems," *Journal of Infectious Diseases* 127 (May 1973): 583–594.

2. Zinsser, *Rats, Lice and History*, p. 285.

3. N. E. Brill, "An Acute Infectious Disease of Unknown Origin: A Clinical Study Based on 221 Cases," *American Journal of the Medical Sciences* 139 (Jul. 1910): 484–502; Zinsser, *Rats, Lice and History*, pp. 233–234; Snyder, "Typhus Fever Rickettsiae," pp. 1078–1087.

4. Theodore E. Woodward and Elizabeth B. Jackson, "Spotted Fever Rickettsiae," in Horsfall and Tamm, eds., *Viral and Rickettsial Infections of Man*, pp. 1095–1122.

5. Joseph E. Smadel and Bennett L. Elisberg, "Scrub Typhus Rickettsia," in Horsfall and Tamm, eds., *Viral and Rickettsial Infections of Man*, pp. 1130–1143.

6. Richard A. Ormsbee, "Q Fever Rickettsia," in Horsfall and Tamm, eds., *Viral and Rickettsial Infections of Man*, pp. 1144–1160.

7. Hubert A. Lechevalier and Morris Solotorovsky, *Three Centuries of Microbiology* (New York: McGraw-Hill, 1965), pp. 327–332; Ormsbee, "Q Fever Rickettsia," p. 1144.

8. Herald R. Cox, "The Preparation and Standardization of Rickettsial Vaccines," in Forest Ray Moulton, ed., *Rickettsial Diseases of Man* (Washington, D.C.: American Association for the Advancement of Science, 1948), pp. 203–214; John C. Snyder, "The Treatment of the Rickettsial Diseases of Man," in ibid., pp. 169–177.

9. William S. Jordan, Jr., and John H. Dingle, "Mycoplasma Pneumoniae Infections," in René J. Dubos and James G. Hirsch, eds., *Bacterial and Mycotic Infections of Man*, 4th ed. (Philadelphia: J. B. Lippincott, 1965), pp. 810–824.

10. Karl F. Meyer, "Psittacosis-Lymphogranuloma Venereum Agents,"

in Horsfall and Tamm, eds., *Viral and Rickettsial Infections of Man*, pp. 1006–1041; Ernest Jawetz and Phillips Thygeson, "Trachoma and Inclusion Conjunctivitis Agents," in ibid., pp. 1042–1058.

11. Sanford S. Elberg, "The Brucellae," in Dubos and Hirsch, eds., *Bacterial and Mycotic Infections of Man*, pp. 698–723; Wesley W. Spink, *The Nature of Brucellosis* (Minneapolis: University of Minnesota Press, 1956), pp. 3–25.

12. See note 11. See also I. Forest Huddleson, *Brucellosis in Man and Animals* (New York: Commonwealth Fund, 1939), pp. 1–3.

13. Spink, *The Nature of Brucellosis*; U.S. Department of Health, Education, and Welfare, Public Health Service, *Reported Morbidity and Mortality in the United States, 1973: Morbidity and Mortality Weekly Report* 22, no. 53 (Atlanta: Center for Disease Control, 1974): 2, 38.

14. Harry F. Dowling, *Medicines for Man: The Development, Regulation, and Use of Prescription Drugs* (New York: Alfred A. Knopf, 1970), pp. 39–42.

15. J. Ehrlich, Q. R. Bartz, R. M. Smith, D. A. Joslyn, and P. R. Burkholder, "Chloromycetin, a New Antibiotic from a Soil Actinomycete," *Science* 106 (Oct. 31, 1947): 417; letter and accompanying documents, H. E. Carnes to Harry F. Dowling, Aug. 8, 1958. Chloramphenicol is an example of simultaneous discovery by different groups. Herbert E. Carter and his associates at the University of Illinois obtained the same antibiotic from a fungus grown from soil on the farm of the Illinois Agricultural Experiment Station in Urbana, Illinois, and announced their findings only three months after the first publication from the Parke, Davis laboratories. H. E. Carter, D. Gottlieb, and H. W. Anderson, "Chloromycetin and Streptothricin," *Science* 107 (Jan. 30, 1948): 113.

16. H. E. Carnes to Harry F. Dowling, Aug. 8, 1958; Theodore E. Woodward to Harry F. Dowling, Jul. 18, 1958; W. H. Mohrhoff and W. D. Mogerman, "Chloromycetin: Another Weapon for the Doctor's Arsenal," *Process Industries Quarterly* 12, no. 1 (1949): 2–14; H. L. Ley, Jr., and J. E. Smadel, "Antibiotic Therapy of Rickettsial Diseases," *Antibiotics and Chemotherapy* 4 (Jul. 1954): 792–802.

17. Theodore E. Woodward to Harry F. Dowling, Jul. 18, 1958, and Nov. 6, 1967.

18. James H. Williams to Harry F. Dowling, Aug. 18, 1958.

19. Harry F. Dowling, "History of the Broad-Spectrum Antibiotics," *Antibiotics Annual, 1958–1959* (New York: Medical Encyclopedia, 1959), pp. 39–44; "Aureomycin—A New Antibiotic," *Annals of the New York Academy of Sciences* 51, art. 2 (Nov. 30, 1948): 175–342.

20. J. H. Kane, A. C. Finlay, and B. A. Sobin, "Antimicrobial Agents from Natural Sources," *Annals of the New York Academy of Sciences* 53, art. 2 (Sept. 15, 1950): 226–228; Harry F. Dowling, *Tetracycline* (New York: Medical Encyclopedia, 1955), pp. 14, 49–51.

21. U.S. Federal Trade Commission, *Economic Report on Antibiotics Manufacture* (Washington, D.C.: U.S. Government Printing Office, 1958),

pp. 248–257; Dowling, *Medicines for Man*, pp. 91–101; U.S. Senate, Committee on the Judiciary, Subcommittee on Antitrust and Monopoly, *Study of Administered Prices in the Drug Industry*, 87th Cong. 1st Sess., pp. 66–69; "Charge Drug Monopoly," *Science News Letter*, Aug. 16, 1958, p. 102.

22. Theodore E. Woodward and Charles L. Wisseman, Jr., *Chloromycetin (Chloramphenicol)* (New York: Medical Encyclopedia, 1958), p. 54.

23. V. Knight, F. Ruiz-Sanchez, A. Ruiz-Sanchez, and W. McDermott, "Aureomycin in Typhus and Brucellosis," *American Journal of Medicine* 6 (Apr. 1949): 407–416.

24. Y. Kneeland, H. M. Rose, and C. D. Gibson, "Aureomycin in the Treatment of Primary Atypical Pneumonia," *American Journal of Medicine* 6 (Jan. 1949): 41–50; E. B. Schoenbach and M. S. Bryer, "Treatment of Primary Atypical Non-bacterial Pneumonia with Aureomycin," *Journal of the American Medical Association* 139 (Jan. 29, 1949): 275–280; G. Meiklejohn and R. I. Shragg, "Aureomycin in Primary Atypical Pneumonia: A Controlled Evaluation," ibid. 140 (May 28, 1949): 391–396; S. H. Walker, "Ineffectiveness of Aureomycin in Primary Atypical Pneumonia: A Controlled Study of 212 Cases," *American Journal of Medicine* 15 (Nov. 1953): 593–602; J. R. Kingston, R. M. Chanock, M. A. Mufson, L. P. Hellman, W. D. James, H. H. Fox, M. A. Manko, and J. Boyers, "Eaton Agent Pneumonia," *Journal of the American Medical Association* 176 (Apr. 15, 1961): 118–123. Later, erythromycin was shown to be as effective as the tetracyclines in mycoplasma pneumonia.

25. Meyer, "Psittacosis-Lymphogranuloma Venereum Agents," p. 1021; P. T. Durfee and R. M. Moore, Jr., "Human Psittacosis in the United States, 1971–1973," *Journal of Infectious Diseases* 131 (Feb. 1975): 193–194.

26. Tom Mahoney, *The Merchants of Life: An Account of the American Pharmaceutical Industry* (New York: Harper & Brothers, 1959), pp. 67–68, 178.

27. E. Chain, in "Round Table: Are New Antibiotics Needed?" *Antimicrobial Agents and Chemotherapy—1965: Proceedings of the Fifth Interscience Conference on Antimicrobial Agents and Chemotherapy and IVth International Congress of Chemotherapy, Washington, D.C., 17–21 October 1965* (Ann Arbor, Mich.: American Society for Microbiology, 1966), p. 1112; Gordon T. Stewart, *The Penicillin Group of Drugs* (Amsterdam: Elsevier, 1965), pp. 17–28.

28. A. B. Christie, "Treatment of Typhoid Carriers with Ampicillin," *British Medical Journal* 1 (June 20, 1964): 1609–1611.

29. H. W. Florey, "Antibiotic Products of a Versatile Fungus," *Annals of Internal Medicine* 43 (Sept. 1955): 480–490; Lennard Bickel, *Rise Up to Life: A Biography of Howard Walter Florey, Who Gave Penicillin to the World* (New York: Charles Scribner's Sons, 1972), pp. 255–261;

Eli Lilly and Company, *Profile of an Antibiotic: Keflin (Sodium Cephalothin)* (Indianapolis: Eli Lilly, 1966), pp. 7–10.

30. H. F. Dowling, "Present Status of Therapy with Combinations of Antibiotics," *American Journal of Medicine* 39 (Nov. 1965): 796–803. Recently recovery without relapse has been reported in all of 100 patients with endocarditis caused by penicillin-sensitive streptococci treated with a combination of penicillin and streptomycin. J. C. Wolfe and W. D. Johnson, Jr., "Penicillin-Sensitive Streptococcal Endocarditis: In-vitro and Clinical Observations of Penicillin-Streptomycin Therapy," *Annals of Internal Medicine* 81 (Aug. 1974): 178–181.

31. Dowling, "Present Status of Therapy with Combinations of Antibiotics"; J. F. Wallace, R. H. Smith, M. Garcia, and R. G. Petersdorf, "Studies on the Pathogenesis of Meningitis: VI. Antagonism between Penicillin and Chloramphenicol in Experimental Pneumococcal Meningitis," *Journal of Laboratory and Clinical Medicine* 70 (Sept. 1967): 408–418.

32. "Shock Treatment for Parke, Davis," *Fortune* 48 (Sept. 1953): 108–113; P. de Kruif, "God's Gift to the Doctors," *Reader's Digest* 55 (Jul. 1949): 49–52; H. E. Simmons and P. D. Stolley, "This Is Medical Progress? Trends and Consequences of Antibiotic Use in the United States," *Journal of the American Medical Association* 227 (Mar. 4, 1974): 1023–1028. The figures are for medicinal antibiotics for both man and animals, but there is no reason to believe that the amounts used in animals increased disproportionately.

33. W. E. Scheckler and J. V. Bennett, "Antibiotic Usage in Seven Community Hospitals," *Journal of the American Medical Association* 213 (Jul. 13, 1970): 264–267; L. G. Seidl, G. F. Thornton, and L. E. Cluff, "Epidemiological Studies of Adverse Drug Reactions," *American Journal of Public Health* 55 (Aug. 1965): 1170–1175.

34. Simmons and Stolley, "This Is Medical Progress?"; J. E. McGowan and M. Finland, "Usage of Antibiotics in a General Hospital: Effect of Requiring Justification," *Journal of Infectious Diseases* 130 (Aug. 1974): 165–168.

35. C. N. Lewis, L. E. Putnam, F. D. Hendricks, I. Kerlan, and H. Welch, "Chloramphenicol (Chloromycetin) in Relation to Blood Dyscrasias with Observations on Other Drugs," *Antibiotics and Chemotherapy* 2 (Dec. 1952): 601–609; Editorial: "Blood Dyscrasias Following the Use of Chloramphenicol," *Journal of the American Medical Association* 149 (June 28, 1952): 840; U.S. Senate, Select Committee on Small Business, Subcommittee on Monopoly, *Hearings on Competitive Problems in the Drug Industry, 1968,* 90th Cong., 1st and 2d Sess., part 6, pp. 2473–2474, 2477, 2541–2543, 2618, 2630, 2649.

36. U.S. Senate, Select Committee on Small Business, Subcommittee on Monopoly, *Competitive Problems in the Drug Industry: Summary and Analysis,* 92d Cong., 2d Sess., Nov. 2, 1972, p. 62; Dowling, *Medicines for Man,* pp. 196–202.

37. H. F. Dowling, M. H. Lepper, and G. G. Jackson, "Clinical Significance of Antibiotic-Resistant Bacteria," *Journal of the American Medical Association* 157 (Jan. 22, 1955): 327–331.

38. T. Watanabe, "Infective Heredity of Multiple Drug Resistance in Bacteria," *Bacteriological Reviews* 27 (Mar. 1963): 87–115.

39. D. E. Rogers, "The Changing Pattern of Life-threatening Microbial Disease," *New England Journal of Medicine* 261 (Oct. 1, 1959): 677–683; M. Finland, "Changing Ecology of Bacterial Infections as Related to Antibacterial Therapy," *Journal of Infectious Diseases* 122 (Nov. 1970): 419–431.

40. "Rising Threat of Hospital Infection Spurs Tightening of Control Methods," *Medical Tribune,* Jan. 25, 1971, pp. 1, 17; U.S. Senate, *Hearings on Competitive Problems,* part 6, pp. 2464–2468.

41. H. F. Dowling and M. H. Lepper, "The Effect of Antibiotics (Penicillin, Aureomycin, and Terramycin) on the Fatality Rate and Incidence of Complications in Pneumococcic Pneumonia: A Comparison with Other Methods of Therapy," *American Journal of the Medical Sciences* 222 (Oct. 1951): 396–403; Wolfe and Johnson, "Penicillin-Sensitive Streptococcal Endocarditis"; Carl C. Dauer, "A Demographic Analysis of Recent Changes in Mortality, Morbidity, and Age Group Distribution in Our Population," in Iago Galdston, ed., *The Impact of Antibiotics on Medicine and Society* (New York: International Universities Press, 1958; first I.U.P. paperback ed., 1969), p. 116.

12. The Newer Vaccines

1. Herald R. Cox, "The Preparation and Standardization of Rickettsial Vaccines," in Forest Ray Moulton, ed., *Rickettsial Diseases of Man* (Washington, D.C.: American Association for the Advancement of Science, 1948), pp. 203–214; Hans Zinsser and Stanhope Bayne-Jones, *A Textbook of Bacteriology,* 8th ed. (New York: D. Appleton-Century, 1939), pp. 668–670; Frank L. Horsfall, Jr., and Igor Tamm, eds., *Viral and Rickettsial Infections of Man* (Philadelphia: J. B. Lippincott, 1965), pp. 1077, 1087, 1116. Another factor accounting partly for the infrequency of typhus during World War II was the use of DDT (dichloro-diphenyl-trichlorethane) to eliminate lice from soldiers' bodies and clothing. James Phinney Baxter 3d, *Scientists against Time* (Boston: Little, Brown, 1946), pp. 368–372.

2. See Hubert A. Lechevalier and Morris Solotorovsky, *Three Centuries of Microbiology* (New York: McGraw-Hill, 1965), pp. 280–301.

3. Theophilus Thompson, *Annals of Influenza or Epidemic Catarrhal Fever in Great Britain from 1510 to 1837* (London: Sydenham Society, 1852), pp. 11–12.

4. J. F. Townsend, "History of Influenza Epidemics," *Annals of Medical History,* n.s. 5 (Nov. 1933): 533–547.

5. August Hirsch, *Handbook of Geographical and Historical Pathol-*

ogy, trans. Charles Creighton, 3 vols. (London: New Sydenham Society, 1883–1886), vol. 1: *Acute Infective Diseases*, pp. 7, 18; Thomas Francis, Jr., and Hunein F. Maasab, "Influenza Viruses," in Horsfall and Tamm, eds., *Viral and Rickettsial Infections of Man*, p. 689. For early epidemics in the United States, see John Duffy, *Epidemics in Colonial America* (Baton Rouge, La.: Louisiana State University Press, 1953); Ernest Caulfield, *The Pursuit of Pestilence* (Worcester, Mass.: American Antiquarian Society, 1950). For the epidemic of 1918, see R. E. Shope, "Influenza: History, Epidemiology, and Speculation," *Public Health Reports* 73 (February 1958): 165–178. For the quotation, see Simon Flexner and James Thomas Flexner, *William Henry Welch and the Heroic Age of American Medicine* (New York: Viking Press, 1941), pp. 376–377. Influenza has frequently been called by the name of the country or region in which it was supposed to have originated. The pandemic of 1889 was called Russian influenza and Siberian fever, the pandemic of 1918 the Spanish influenza, and those of 1957 and 1968 the Asian and Hong Kong influenzas, respectively. Other diseases have been similarly named. Syphilis was called the French disease when it first broke out in virulent form in the fifteenth century, and German measles is still the popular name for rubella.

6. Edwin F. Hirsch, "The Influenza Pandemic of Fifty Years Ago," *Chicago Medicine* 71 (Dec. 7, 1968): 957–960; U.S. Bureau of the Census, *Historical Statistics of the United States, Colonial Times to 1957* (Washington, D.C.: U.S. Government Printing Office, 1960), p. 26; W. H. Frost, "The Epidemiology of Influenza," *Journal of the American Medical Association* 73 (Aug. 2, 1919): 313–318; Shope, "Influenza: History, Epidemiology, and Speculation," p. 166.

7. D. D. Eisenhower, "Ike Tells of Camp's Flu Epidemic," *Washington Post*, Apr. 15, 1967, p. A–15; *Washington Wife: Journal of Ellen Maury Slayden from 1897–1919* (New York: Harper & Row, 1963), p. 344.

8. S. Galishoff, "Newark and the Great Influenza Pandemic of 1918," *Bulletin of the History of Medicine* 43 (May–June 1969): 246–258.

9. James H. Cassedy, *Charles V. Chapin and the Public Health Movement* (Cambridge: Harvard University Press, 1962), p. 188.

10. F. G. Blake and R. L. Cecil, "Studies in Experimental Pneumonia: IX. Production in Monkeys of an Acute Respiratory Disease Resembling Influenza by Inoculation with Bacillus Influenzae," *Journal of Experimental Medicine* 32 (Dec. 1920): 691–717. For the quotation, see Francis Blake to Dorothy Blake, Aug. 29, 1919, in the possession of the family. See also J. W. Nuzum, I. Pilot, F. H. Stangl, and B. E. Bonar, "Pandemic Influenza and Pneumonia in a Large Civil Hospital," *Journal of the American Medical Association* 71 (Nov. 9, 1918): 1562–1565; C. P. Brown and F. W. Palfrey, "Influenza Pneumonia at Camp Greene, N.C.," *New York Medical Journal* 110 (Aug. 23, 1919): 316–321; Arthur L. Bloomfield, *A Bibliography of Internal Medicine: Communicable Diseases* (Chicago: University of Chicago Press, 1958), pp. 392–393.

11. Shope, "Influenza: History, Epidemiology, and Speculation"; W. Smith, C. H. Andrewes, and P. P. Laidlaw, "A Virus Obtained from Influenza Patients," *Lancet* 2 (Jul. 8, 1933): 66–68; Christopher Andrewes, *The Common Cold* (New York: W. W. Norton, 1965), pp. 40–41; C. Andrewes, "Recollections," in *International Virology I: Proceedings of the First International Congress for Virology, Helsinki, 1968* (Basel: S. Karger, 1969), pp. 220–228. The illness in the ferrets turned out not to be influenza after all but a form of distemper which ferrets sometimes catch from dogs. Andrewes, *The Common Cold*, p. 41.

12. T. Francis, Jr., "New Type of Virus from Epidemic Influenza," *Science* 92 (Nov. 1, 1940): 405–408; T. P. Magill, "Virus from Cases of Influenza-like Upper Respiratory Infection," *Proceedings of the Society for Experimental Biology and Medicine* 45 (Oct. 1940): 162–164. The antigens of the influenza viruses are proteins on the surface of the viruses, where they are in a position when they enter a human or other host to stimulate the host's cells to form antibodies. These facts were unknown when Francis and Magill discovered the new type. At that time little more was known about viruses than that they passed through small-pore filters and induced certain infections. Knowledge about them was built up gradually, at first by immunological methods, then by chemical and physical experiments, and finally by the use of the electron microscope. See Francis and Maasab, "Influenza Viruses," pp. 690–702.

13. See G. Loosli, "Influenza and the Interaction of Viruses and Bacteria in Respiratory Infections," *Medicine* 52 (Sept. 1973): 369–384.

14. Francis and Maasab, "Influenza Viruses," pp. 699, 716–720; V. Knight in Tinsley Randolph Harrison, Maxwell M. Wintrobe, George W. Thorn, Raymond M. Adams, Eugene Braunwald, Kurt J. Isselbacher, and Robert G. Petersdorf, *Harrison's Principles of Internal Medicine*, 7th ed., 2 vols. (New York: McGraw-Hill, 1974), I, 938. The official designations of the Type A influenza viruses are more complicated, being based on the characteristics of the two known surface antigens, hemagglutinin (H) and neuraminidase (N), both of which change from time to time. Thus the A0 viruses are officially H0N1, the A1 virus H1N1; the A2 viruses causing epidemics from 1957 through 1967 are designated as H2N2, and those causing epidemics from 1968 to 1972 as H3N2. G. G. Jackson and R. L. Muldoon, "Viruses Causing Common Respiratory Infections in Man: V. Influenza A (Asian)," *Journal of Infectious Diseases* 131 (Mar. 1975): 311.

15. C. C. Dauer and R. E. Serfling, "Mortality from Influenza, 1957–1958 and 1959–1960," International Conference on Asian Influenza, Bethesda, Md., Feb. 17–19, 1960, *American Review of Respiratory Diseases* 83 (Feb. 1961, part 2): 15–28.

16. T. Francis and T. P. Magill, "Vaccination of Human Subjects with Virus of Human Influenza," *Proceedings of the Society for Experimental Biology and Medicine* 33 (Jan. 1936): 604–606; T. Francis, "Vaccination against Influenza," *Bulletin of the World Health Organization* 8, nos. 5, 6 (1953): 725–741; Andrewes, *The Common Cold*, pp. 46–47; F. M.

Davenport, "Inactivated Influenza Virus Vaccines: Past, Present, and Future," International Conference on Asian Influenza, *American Review of Respiratory Diseases* 83 (Feb. 1961, part 2): 146–150.

17. Francis and Maasab, "Influenza Viruses," pp. 720–729; Davenport, "Inactivated Influenza Virus Vaccines," pp. 146–150; C. H. Andrewes, "Diagnostic Laboratory Organization: Future," International Conference on Asian Influenza, *American Review of Respiratory Diseases* 83 (Feb. 1961, part 2): 142–143.

18. "The Asiatic Flu," *Life* 43 (Sept. 2, 1957): 113–121; "U.S. Finds Flu Put 50% in Nation to Bed," *New York Times*, Dec. 11, 1957, p. 9.

19. "The Life of A2/Hong Kong/68," Medical News Section, *Journal of the American Medical Association* 207 (Mar. 17, 1969): 2017.

20. Vernon Knight, "Influenza," *Disease-a-Month*, Aug. 1976, pp. 46–47; National Institute of Allergy and Infectious Diseases of the National Institutes of Health, the Center for Disease Control, and the Bureau of Biologics of the Food and Drug Administration, "Summary of Clinical Trials of Influenza Vaccines," *Journal of Infectious Diseases* 134 (Jul. 1976): 100; U.S. Department of Health, Education, and Welfare, Public Health Service, *Morbidity and Mortality Weekly Report* 25, no. 6 (Atlanta: Center for Disease Control, 1976): 47–48.

21. The new strain was designated A/New Jersey/76 and officially named Hsw1 N1. See National Institute of Allergy and Infectious Diseases, "Summary of Clinical Trials of Influenza Vaccines"; U.S. Department of Health, Education, and Welfare, *Morbidity and Mortality Weekly Report* 25, no. 21: 165–171; no. 28: 221–227; no. 51: 416; no. 52: 430.

22. Jean Gould, *A Good Fight: The Story of F.D.R.'s Conquest of Polio* (New York: Dodd, Mead, 1960), p. 10.

23. Elisabeth F. Hutchin, "Historical Summary," in International Committee for the Study of Infantile Paralysis, *Poliomyelitis* (Baltimore: William & Wilkins, 1932), pp. 1–22; John R. Paul, *A History of Poliomyelitis* (New Haven: Yale University Press, 1971), pp. 12–25, 30–33. Paul's book is a comprehensive and highly readable history of poliomyelitis from early times to the present. I have drawn upon it extensively because the author played a major part in determining the manner in which the virus spreads and produces disease, and because he has written an impartial account of the controversial issues surrounding the introduction of vaccines for poliomyelitis.

24. Paul, *A History of Poliomyelitis*, pp. 38–47, 74–75.

25. Ibid., pp. 75–78, 88–92.

26. Ibid., pp. 79–83; C.-E. A. Winslow, Wilson G. Smillie, James A. Doull, and John E. Gordon, *The History of American Epidemiology*, ed. Franklin H. Top (St. Louis: C. V. Mosby, 1952), p. 87.

27. Paul, *A History of Poliomyelitis*, pp. 54–58, 98–100; K. Landsteiner and E. Popper, "Mikroscopische Präparate von einem menschlichen und zwei Affenrückenmarken," *Wiener Klinische Wochenschrift*

21 (Dec. 18, 1908): 1830. The name poliovirus was not adopted until later, but it is used throughout this chapter for convenience.

29. Quoted in Paul, *A History of Poliomyelitis*, p. 18.

30. Ibid., pp. 64–65, 69, 190–198; S. D. Kramer, W. L. Aycock, C. I. Solomon, and C. L. Thenebe, "Convalescent Serum Therapy in Preparalytic Poliomyelitis," *New England Journal of Medicine* 206 (Mar. 3, 1932): 432–435.

31. Robert W. Lovett, *The Treatment of Infantile Paralysis*, 2d ed. (Philadelphia: P. Blakiston's Sons, 1917), pp. 31–32; Paul, *A History of Poliomyelitis*, pp. 337–339.

32. "Paralysis Treatment Boosted," *Newsweek* 18 (Sept. 8, 1941): 62. "Sister" is a term applied to a head nurse in British countries.

33. Paul, *A History of Poliomyelitis*, pp. 338–344; R. M. Yoder, "Healer from the Outback," *Saturday Evening Post* 214 (Jan. 17, 1942): 18–19; "Sister Kenny Fights On," *Time* 45 (Apr. 2, 1945): 54.

34. Paul, *A History of Poliomyelitis*, pp. 324–334.

35. Ibid., pp. 274–278.

36. Gould, *A Good Fight*, p. 281, italics mine; Saul Benison, *Tom Rivers: Reflections on a Life in Medicine and Science* (Cambridge: M.I.T. Press, 1967), pp. 229–232. This volume is a fascinating oral history of the life of Thomas M. Rivers, a leading virologist of the Rockefeller Institute, who became the chief scientific counselor of the National Foundation for Infantile Paralysis and played a large part in the development of poliomyelitis vaccines.

37. Paul, *A History of Poliomyelitis*, pp. 142–143, 357–368.

38. Ibid., pp. 360–363.

39. T. M. Rivers, "Immunological and Serological Phenomena in Poliomyelitis," in National Foundation for Infantile Paralysis, *Poliomyelitis* (New York: National Foundation for Infantile Paralysis, 1941), p. 71; Paul, *A History of Poliomyelitis*, pp. 89–92, 126–136, 279–280, 131.

40. Helen Harrington, "Etiology," in International Committee for the Study of Infantile Paralysis, *Poliomyelitis*, pp. 256–259; Paul, *A History of Poliomyelitis*, pp. 248–251; A. B. Sabin and R. Ward, "The Natural History of Poliomyelitis: I. Distribution of Virus in Nervous and Nonnervous Tissues," *Journal of Experimental Medicine* 73 (June 1941): 771–793; H. A. Howe and D. Bodian, "Poliomyelitis in the Chimpanzee: A Clinical-Pathological Study," *Journal of Experimental Medicine* 69 (Aug. 1941): 149–181.

41. David Bodian and Dorothy M. Horstmann, "Polio-viruses," in Horsfall and Tamm, eds., *Viral and Rickettsial Infections of Man*, p. 439.

42. D. Bodian, I. M. Morgan, and H. A. Howe, "Differentiation of Types of Poliomyelitis Viruses: III. The Grouping of Fourteen Strains into Three Basic Immunological Types," *American Journal of Hygiene* 49 (Jan. 1949): 234–245; Paul, *A History of Poliomyelitis*, pp. 232–235.

43. Benison, *Tom Rivers*, pp. 184–190; Paul, *A History of Poliomyelitis*, pp. 254–262.

44. J. F. Enders, T. H. Weller, and F. C. Robbins, "Cultivation of the Lansing Strain of Poliomyelitis Virus in Cultures of Various Human Embryonic Tissues," *Science* 109 (Jan. 7, 1949): 85.

45. Benison, *Tom Rivers*, p. 493; Paul, *A History of Poliomyelitis*, pp. 412–416.

46. Paul, *A History of Poliomyelitis*, p. 442; I. M. Morgan, "Immunization of Monkeys with Formalin-Inactivated Poliomyelitis Viruses," *American Journal of Hygiene* 48 (Nov. 1948): 394–405; H. A. Howe, "Antibody Response of Chimpanzees and Human Beings to Formalin-Inactivated Trivalent Poliomyelitis Vaccine," *American Journal of Hygiene* 56 (Nov. 1952): 265–286.

47. J. E. Salk, "Studies in Human Subjects on Active Immunization against Poliomyelitis: 1. A Preliminary Report of Experiments in Progress," *Journal of the American Medical Association* 151 (Mar. 28, 1953): 1081–1098. An emulsion is a preparation of two liquids in which one is dispersed as minute globules throughout the other.

48. Paul, *A History of Poliomyelitis*, pp. 419–420; P. J. Fisher, *The Polio Story* (London: Heinemann, 1967), p. 66, cited in ibid., p. 312.

49. Paul, *History of Poliomyelitis*, pp. 419–425; Benison, *Tom Rivers*, pp. 509–512; J. P. L[eake]., "Joseph Asbury Bell, 1904–1968: A Biographical Appreciation," *American Journal of Epidemiology* 90 (Dec. 1969): 466.

50. Paul, *A History of Poliomyelitis*, pp. 428–429; T. Francis, Jr., and R. F. Korns, "Evaluation of 1954 Field Trial of Poliomyelitis Vaccine: Synopsis of Summary Report," *American Journal of the Medical Sciences* 229 (June 1955): 608–612.

51. Paul, *A History of Poliomyelitis*, pp. 432–435; N. Nathanson and A. D. Langmuir, "The Cutter Incident: Poliomyelitis following Formaldehyde-Inactivated Poliovirus Vaccination in the United States during the Spring of 1955: I. Background," *American Journal of Hygiene* 78 (Jul. 1963): 16–28; A. D. Langmuir, "The Surveillance of Communicable Diseases of National Importance," *New England Journal of Medicine* 268 (Jan. 24, 1963): 182–192.

52. Paul, *A History of Poliomyelitis*, pp. 437–438; Richard Carter, *Breakthrough: The Saga of Jonas Salk* (New York: Trident Press, 1966), p. 326.

53. Carter, *Breakthrough*, pp. 293–300, 398; "Salk Award," *Time* 67 (Feb. 6, 1956): 74.

54. Langmuir, "The Surveillance of Communicable Diseases of National Importance"; J. R. Paul, "Status of Vaccination against Poliomyelitis, with Particular Reference to Oral Vaccination," *New England Journal of Medicine* 264 (Mar. 30, 1961): 651–658.

55. Paul, *A History of Poliomyelitis*, p. 442; A. B. Sabin, "Oral Poliovirus Vaccine: History of Its Development and Prospects for Eradication of Poliomyelitis," *Journal of the American Medical Association* 194 (Nov. 22, 1965): 872–876. This contest is described in detail from a

popular point of view in John Rowan Wilson, *Margin of Safety* (Garden City, N.Y.: Doubleday, 1963). See also A. B. Sabin, "Status of Field Trials with an Orally Administered, Live Attenuated Poliovirus Vaccine," *Journal of the American Medical Association* 171 (Oct. 17, 1959): 863–868.

56. Paul, "Status of Vaccination against Poliomyelitis"; V. J. Cabasso, E. Jungherr, A. W. Moyer, M. Roca-Garcia, and H. R. Cox, "Oral Polio-myelitis Vaccine, Lederle: Thirteen Years of Laboratory and Field Investigation: An Interim Review," *New England Journal of Medicine* 263 (Dec. 29, 1960): 1321–1330; Luther L. Terry, *The Association of Cases of Poliomyelitis with the Use of Type III Oral Poliomyelitis Vaccines: A Technical Report* (Washington, D.C.: U.S. Government Printing Office, 1962); Wilson, *Margin of Safety*, pp. 216–223, 247. See also Paul, *A History of Poliomyelitis*, pp. 441–464.

57. "Oral Poliomyelitis Vaccines: Report of Special Advisory Committee on Oral Poliomyelitis Vaccines to the Surgeon General of the Public Health Service," *Journal of the American Medical Association* 190 (Oct. 5, 1964): 49–51; C. C. Hopkins, W. E. Dismukes, T. H. Glick, and R. J. Warren, "Surveillance of Paralytic Poliomyelitis in the United States: 1966 and 1967 Cases, and 1965–1967 Cases Associated with Oral Poliovirus Vaccine," *Journal of the American Medical Association* 210 (Oct. 27, 1969): 694–700.

58. U.S. Department of Health, Education, and Welfare, Public Health Service, *Reported Morbidity and Mortality in the United States, 1974: Morbidity and Mortality Weekly Report* 23, no. 53 (Atlanta: Center for Disease Control, 1975): 2; Paul, *A History of Poliomyelitis*, pp. 463–464, 467–468.

59. Bloomfield, *A Bibliography of Internal Medicine*, pp. 432, 437–440; F. G. Blake and J. D. Trask, Jr., "Studies in Measles: I. Susceptibility of Monkeys to the Virus of Measles," *Journal of Experimental Medicine* 33 (Mar. 1921): 385–412.

60. J. F. Enders, "Vaccination against Measles: Francis Home Redivivus," *Yale Journal of Biology and Medicine* 34 (Dec.–Feb. 1961–1962): 239–260; J. F. Enders and T. C. Peebles, "Propagation in Tissue Cultures of Cytopathogenic Agents from Patients with Measles," *Proceedings of the Society for Experimental Biology and Medicine* 86 (June 1954): 277–286; P. J. Landrigan and J. L. Conrad, "Current Status of Measles in the United States," *Journal of Infectious Diseases* 124 (Dec. 1971): 620–623; U.S. Department of Health, Education, and Welfare, Health Services and Mental Health Administration, *Morbidity and Mortality: Annual Supplement, Summary 1971* (Atlanta: Center for Disease Control, 1972), p. 4; P. J. Imperato, L. Pincus, C. L. Hwa, and A. D. Chaves, "The Control of Measles in New York City," *Bulletin of the New York Academy of Medicine* 50 (May 1974): 602–619.

61. John Ruhräh, "Measles, Rubella, The Fourth Disease, Erythema Infectiosum," in William Osler and Thomas McCrae, eds., *Modern*

Medicine: Its Theory and Practice, 7 vols. (Philadelphia: Lea Brothers, 1907–10), vol. 2: *Infectious Diseases*, pp. 393–394; D. M. Horstmann, "Rubella: The Challenge of Its Control," *Journal of Infectious Diseases* 123 (June 1971): 640–654.

62. L. Z. Cooper and S. Krugman, "The Rubella Problem," *Disease-a-Month*, Feb. 1969, pp. 4–5, 7–10, 18–30.

63. Horstmann, "Rubella"; U.S. Department of Health, Education, and Welfare, *Reported Morbidity and Mortality in the United States, 1974*, p. 2.

64. J. Stokes, "Recent Advances in Immunization against Viral Diseases," *Annals of Internal Medicine* 73 (Nov. 1970): 829–840; Werner Henle and John F. Enders, "Mumps Virus," in Horsfall and Tamm, eds., *Viral and Rickettsial Infections of Man*, pp. 755–765; Dorothy Horstmann to Harry F. Dowling, Jul. 28, 1975.

65. George Gee Jackson, "The Common Cold," in Paul B. Beeson and Walsh McDermott, eds., *Cecil-Loeb Textbook of Medicine*, 12th ed. (Philadelphia: W. B. Saunders, 1967), pp. 10–11.

66. Andrewes, *The Common Cold*, pp. 51–64; W. P. Rowe, R. J. Huebner, J. W. Hartley, T. G. Ward, and R. H. Parrott, "Studies of the Adenoidal-Pharyngeal-Conjunctival (APC) Group of Viruses," *American Journal of Hygiene* 61 (Mar. 1955): 197–218.

67. R. M. Chanock, H. W. Kim, A. J. Vargosko, A. Deleva, K. M. Johnson, C. Cumming, and R. H. Parrott, "Respiratory Syncytial Virus: I. Virus Recovery and Other Observations during 1960 Outbreak of Bronchitis, Pneumonia, and Minor Respiratory Diseases in Children," *Journal of the American Medical Association* 176 (May 27, 1961): 647–653; R. H. Parrott, A. Vargosko, A. Luckey, H. W. Kim, C. Cumming, and R. Chanock, "Clinical Features of Infection with Hemadsorption Viruses," *New England Journal of Medicine* 260 (Apr. 9, 1959): 731–738; H. F. Dowling and L. B. Lefkowitz, Jr., "Problems in Definition of Respiratory Diseases and Respiratory Disease Agents: Clinical Syndromes in Adults Caused by Respiratory Diseases," Conference on Newer Respiratory Diseases, Bethesda, Md., Oct. 3–5, 1962, *American Review of Respiratory Diseases* 88 (Sept. 1963, part 2): 61–72; R. H. Parrott, A. J. Vargosko, H. W. Kim, and R. M. Chanock, "Clinical Syndromes among Children," ibid., pp. 73–76.

68. W. H. Price, H. Emerson, I. Ibler, R. Lachaine, and A. Terrell, "Studies of the JH and 2060 Viruses and Their Relationship to Mild Upper Respiratory Disease in Humans," *American Journal of Hygiene* 69 (May 1959): 224–249; D. A. J. Tyrell and M. L. Bynoe, "Some Further Virus Isolations from Common Colds," *British Medical Journal* 1 (Feb. 11, 1961): 393–397; Dowling and Lefkowitz, "Clinical Syndromes in Adults Caused by Respiratory Diseases"; Parrott et al., "Clinical Syndromes among Children."

69. Andrewes, *The Common Cold*, pp. 139–143; G. G. Jackson and H. F. Dowling, "Transmission of the Common Cold to Volunteers under

Controlled Conditions: IV. Specific Immunity to the Common Cold," *Journal of Clinical Investigation* 38 (May 1959): 762–769; L. B. Lefkowitz, G. G. Jackson, and H. F. Dowling, "The Role of Immunity in the Common Cold and Related Viral Respiratory Infections," *Medical Clinics of North America* 47 (Sept. 1963): 1171–1184.

70. W. Paul Havens, Jr., and John R. Paul, "Infectious Hepatitis and Serum Hepatitis," in Horsfall and Tamm, eds., *Viral and Rickettsial Infections of Man*, pp. 965–966; D. J. Gocke, "New Faces of Viral Hepatitis," *Disease-a-Month*, Jan. 1973, pp. 5–6.

71. T. C. Chalmers, "Viral Hepatitis on the Threshold of Control," *American Journal of the Medical Sciences* 270 (Jul.–Aug. 1975): 3–6; U.S. Department of Health, Education, and Welfare, Public Health Service, *Reported Morbidity and Mortality in the United States, 1973: Morbidity and Mortality Weekly Report* 22, no. 53 (Atlanta: Center for Disease Control, 1974): 2; U.S. Department of Health, Education, and Welfare, Public Health Service, *Reported Morbidity and Mortality in the United States, 1974*, pp. 2, 4.

72. F. O. MacCallum, "Early Studies of Viral Hepatitis," *British Medical Bulletin* 28 (May 1972): 105–108; S. Krugman, J. P. Giles, and J. Hammond, "Infectious Hepatitis: Evidence for Two Distinctive Clinical, Epidemiological, and Immunological Types of Infection," *Journal of the American Medical Association* 200 (May 1, 1967): 365–373.

73. B. S. Blumberg, B. J. S. Gerstley, D. A. Hungerford, W. T. London, and A. I. Sutnick, "A Serum Antigen (Australia Antigen) in Down's Syndrome, Leukemia, and Hepatitis," *Annals of Internal Medicine* 66 (May 1967): 924–931. Blumberg was given the Nobel Prize for this discovery in 1976. See *Time*, Oct. 25, 1976, p. 72. See also F. Deinhardt, A. W. Holmes, R. B. Capps, and H. Popper, "Studies on the Transmission of Human Viral Hepatitis to Marmoset Monkeys: I. Transmission of Disease, Serial Passages and Description of Liver Lesions," *Journal of Experimental Medicine* 125 (Apr. 1967): 673–688; F. Deinhardt, D. Peterson, G. Cross, L. Wolfe, and A. W. Holmes, "Hepatitis in Marmosets," *American Journal of the Medical Sciences* 270 (Jul.–Aug. 1975): 73–80. Recent evidence indicates that a third virus, labeled C, is responsible for some cases of hepatitis. A. M. Prince, B. Brotman, G. F. Grady, W. J. Kuhns, C. Hazzi, R. W. Levine, and S. J. Millian, "Long-incubation Post-transfusion Hepatitis without Serological Evidence of Exposure to Hepatitis-B Virus," *Lancet* 2 (Aug. 3, 1974): 241–246.

74. Gocke, "New Faces of Viral Hepatitis," p. 30; L. B. Seeff, H. J. Zimmerman, E. C. Wright, B. F. Felsher, J. D. Finkelstein, P. Garcia-Pont, H. B. Greenlee, A. A. Dietz, J. Hamilton, R. S. Koff, C. M. Leevy, T. Kiernan, C. H. Tamburro, E. R. Schiff, Z. Vlahcevic, R. Zemel, D. S. Zimmon, and N. Nath, "Efficacy of Hepatitis B Immune Serum Globulin after Accidental Exposure: Preliminary Report of the Veterans Administration Cooperative Study," *Lancet* 2 (Nov. 15, 1975): 939–941; Chalmers, "Viral Hepatitis on the Threshold of Control," pp. 3–6.

75. M. Heidelberger and O. T. Avery, "The Soluble Specific Substance of the Pneumococcus," *Journal of Experimental Medicine* 38 (July 1923): 73–85; S. Mudd, "Sequences in Medical Microbiology: Some Observations over Fifty Years," *American Review of Microbiology* 23 (1969): 1–28; Benjamin White, *The Biology of Pneumococcus: The Bacteriological, Biochemical, and Immunological Characters and Activities of Diplococcus Pneumoniae* (New York: Commonwealth Fund, 1938), pp. 243–244.

76. White, *The Biology of Pneumococcus,* pp. 251–254, 344–346; H. F. Dowling, "The Pneumococcus and the Human Host," *Journal of Infectious Diseases* 125 (Mar. 1972 supp.): S8–S9.

77. P. Kaufman, "Pneumonia in Old Age: Active Immunization against Pneumonia with Pneumococcus Polysaccharide: Results of a Six Year Study," *Archives of Internal Medicine* 79 (May 1947): 518–531; C. M. MacLeod, R. G. Hodges, M. Heidelberger, and W. G. Bernhard, "Prevention of Pneumococcal Pneumonia by Immunization with Specific Capsular Polysaccharides," *Journal of Experimental Medicine* 82 (Dec. 1945): 445–465.

78. R. Austrian and J. Gold, "Pneumococcal Bacteremia with Especial Reference to Bacteremic Pneumococcal Pneumonia," *Annals of Internal Medicine* 60 (May 1964): 759–766; Robert Austrian, "Of Man and Pneumococcus: An Historical Paradox," in Jacques Monod and Ernest Borek, eds., *Of Microbes and Life* (New York: Columbia University Press, 1971), pp. 165–173.

79. E. C. Gotschlich, T. Y. Liu, and M. S. Artenstein, "Human Immunity to the Meningococcus: IV. Immunogenicity of Group A and Group C Meningococcal Polysaccharides in Human Volunteers," *Journal of Experimental Medicine* 129 (June 1969): 1367–1384; M. Finland, "Revival of Antibacterial Immunization: Meningococcal Vaccines Prove Promising," *Journal of Infectious Diseases* 121 (Apr. 1970): 445–448; M. S. Artenstein, "The Current Status of Bacterial Vaccines," *Hospital Practice* 8 (June 1973): 49–56; U.S. Department of Health, Education, and Welfare, *Morbidity and Mortality Weekly Report* 24, no. 45: 381–382.

13. The Continuing War

1. National Center for Health Statistics, *Facts of Life and Death,* DHEW Publication no. (HRA) 74–1222 (Rockville, Md.: U.S. Department of Health, Education, and Welfare, Public Health Service, Health Resources Administration, 1974), pp. 15–16. For syphilis, see Robert D. Grove and Alice M. Hetzel, *Vital Statistics Rates in the United States, 1940–1960* (Washington, D.C.: U.S. Department of Health, Education, and Welfare, Public Health Service, National Center for Health Statistics, 1968), p. 560; National Center for Health Statistics, *Facts of Life and Death,* p. 39. For appendicitis, see Grove and Hetzel, *Vital Statistics,* p.

562; National Center for Health Statistics, *Facts of Life and Death*, p. 39. For tetanus, see Grove and Hetzel, *Vital Statistics*, p. 561; National Center for Health Statistics, *Facts of Life and Death*, p. 16.

2. Grove and Hetzel, *Vital Statistics*, p. 79; National Center for Health Statistics, *Facts of Life and Death*, p. 39. Deaths from pneumonia include those listed as caused by influenza, since pneumonia is almost invariably the cause of death when it is attributed by the attending physician to influenza. No comparison can be made for infections of the kidney between 1900 and 1970 because they were not reported as such in 1900.

3. P. Diosi, "Long Term Changes in the Natural History of Infectious Diseases," *Centaurus* 9 (1963): 288–292; E. H. Kass, "Infectious Diseases and Social Change," *Journal of Infectious Diseases* 123 (Jan. 1971): 110–114; Thomas McKeown, "Medical Issues in Historical Demography," in Edwin Clarke, ed., *Modern Methods in the History of Medicine* (London: Athlone Press of the University of London, 1971), pp. 57–74.

4. U.S. Bureau of the Census, *Historical Statistics of the United States, Colonial Times to 1957* (Washington, D.C.: U.S. Government Printing Office, 1960), p. 30.

5. S. D. Collins, "Diphtheria Incidence and Trends in Relation to Artificial Immunization with Some Comparative Data for Scarlet Fever," *Public Health Reports* 61 (Feb. 15, 1946): 203–240; Graham S. Wilson and Ashley Miles, *Principles of Bacteriology, Virology and Immunity*, 6th ed., 2 vols. (Baltimore: Williams & Wilkins, 1975), II, 1825; H. J. Parish, *A History of Immunization* (Edinburgh: E. & S. Livingstone, 1965), p. 157.

6. U.S. Bureau of the Census, *Historical Statistics of the United States*, pp. 26, 30; C. C. Dauer, "Trends in Age Distribution of Diphtheria in the United States," *Public Health Reports* 65 (Sept. 22, 1950): 1209–1218; Percival Hartley, Manuel Anderson, James Grant, Charles Neubauer, Richard Norton, William John Tulloch, William Anderson Davidson, William Maxwell Jamieson, and George Henderson Robertson, *A Study of Diphtheria in Two Areas of Great Britain*, Medical Research Council, Special Report Series #272 (London: His Majesty's Stationery Office, 1950).

7. U.S. Bureau of the Census, *Historical Statistics of the United States*, pp. 26, 30. For England and Wales, see McKeown, "Medical Issues in Historical Demography," p. 57–74.

8. Anthony M. Lowell, *Tuberculosis in New York City, 1955: Prevalence of the Disease in an Urban Environment* (New York: Tuberculosis and Health Association, 1956), chart following p. 5. The increases in the prevalence of new cases shown in the figure can be attributed to increased reporting at certain times as well as to stepped-up public health campaigns. For the United States as a whole, see U.S. Bureau of the Census, *Historical Statistics of the United States*, p. 26; National Center for Health Statistics, *Facts of Life and Death*, p. 39.

9. J. E. Moore, "An Evaluation of Public Health Measures for the Control of Syphilis: An Epidemiologic Study," *American Journal of Syphilis, Gonorrhea, and Venereal Diseases* 35 (Mar. 1951): 101–137.

10. Figures for 1900–1902 are for the Death Registration Area; in 1971 for the entire United States. (See Chapter 1 for further explanation of the Death Registration Area.) Louis I. Dublin, Alfred J. Lotka, and Mortimer Spiegelman, *Length of Life*, rev. ed. (New York: Ronald Press, 1949), p. 35; National Center for Health Statistics, *Facts of Life and Death*, p. 7.

11. Grove and Hetzel, *Vital Statistics*, p. 79. There is evidence, at least in some cancers of animals, that viruses can induce changes in cells that can lead to cancer.

12. Grove and Hetzel, *Vital Statistics*, p. 776; National Center for Health Statistics, *Facts of Life and Death*, p. 3. An increasing life span has not been the only factor affecting the age distribution of the American people. Changes in the birth rate and immigration have also produced shifts in the age distribution at various times.

13. The idea and the quotation are from G. Rosen, "People, Disease, and Emotion: Some Newer Problems for Research in Medical History," *Bulletin of the History of Medicine* 41 (Jan.–Feb. 1967): 5–23, esp. p. 19.

14. Job 5:26.

15. Richard Harrison Shryock, *The Development of Modern Medicine: An Interpretation of the Social and Scientific Factors Involved* (New York: Alfred A. Knopf, 1947), pp. 248–272, 336–355.

16. See Charles A. Beard, Introduction to *The Idea of Progress: An Inquiry into Its Origin and Growth*, by John Bagnell Bury (New York: Macmillan, 1932), esp. pp. xxxii–xxxv.

17. D. C. Gajdusek, "Slow-Virus Infections of the Nervous System," *New England Journal of Medicine* 276 (Feb. 16, 1967): 392–400; S. L. Katz and John F. Griffith, "Slow Virus Infections," *Hospital Practice* 6 (Mar. 1971): 64–74.

18. M. E. Roux, M. L. Martin, and M. A. Chaillou, "Trois cents cas de diphthérie traités par le sérum antidiphthérique," *Annales de L'Institut Pasteur* 8 (Sept. 1894): 640–661, esp. 661n2; Paul B. Beeson and Walsh McDermott, eds., *Cecil-Loeb Textbook of Medicine*, 14th ed. (Philadelphia: W. B. Saunders, 1967), p. 182.

19. Case fatality rates cited in this volume for patients treated with antibiotics include those patients who failed to recover either because of adverse reactions, a complicating infection, or the appearance of resistant strains of the originally infecting microorganism. Thus, the figure of 5 percent cited for deaths among patients who received antibiotic therapy for pneumococcal pneumonia includes those who died from a complicating pneumonia caused by other microorganisms or from adverse reactions to any drugs received or procedures performed. In other words, among 100 patients with pneumococcal pneumonia treated with the

proper antibiotics, approximately 70 would have recovered without specific therapy, 25 will be saved by the medication, and five will die.

20. U.S. Bureau of the Census, *Statistical Abstract of the United States, 1974*, 95th ed. (Washington, D.C.: U.S. Government Printing Office, 1974), pp. 5, 374, 138; U.S. Bureau of the Census, *Historical Statistics of the United States*, pp. 139, 212.

21. Nobel Foundation, *Nobel Lectures, Including Presentation Speeches and Laureates' Biographies: Physiology or Medicine, 1942–1962* (Amsterdam: Elsevier, 1964), p. 468; Richard Carter, *Breakthrough: The Saga of Jonas Salk* (New York: Trident Press, 1966), p. 29; John R. Paul, *A History of Poliomyelitis* (New Haven: Yale University Press, 1971), p. 445.

22. See Dixon Wecter, *The Hero in America: A Chronicle of Hero-Worship* (New York: Charles Scribner's Sons, 1941).

23. Harold Underwood Faulkner, *The Quest for Social Justice, 1898–1914* (New York: Macmillan, 1931), p. 131.

24. *Les Prix Nobel [1947–1973]* (Stockholm: Imiprimerie Royale, P. A. Norstedt & Soner, 1949–1974); Theodore L. Sourkes, *Nobel Prize Winners in Medicine and Physiology, 1901–1965* (London: Abelard-Schuman, 1966).

25. A. Flexner, "The Usefulness of Useless Knowledge," in *The Field and the Work of the Squibb Institute for Medical Research* (New Brunswick, N.J.: Squibb Institute for Medical Research, 1938), pp. 19–23; James A. Shannon, "NIH: Present and Potential Contribution to Application of Biomedical Knowledge," in U.S. Senate, Committee on Government Operations, Subcommittee on Government Research, *Research in the Service of Man: Biomedical Knowledge, Development, and Use* (Washington, D.C.: U.S. Government Printing Office, 1967), pp. 72–85.

26. F. Griffith, "The Significance of Pneumococcal Types," *Journal of Hygiene* 27 (Jan. 1928): 113–159; O. T. Avery, C. M. MacLeod, and M. McCarty, "Studies on the Chemical Nature of the Substance Inducing Transformation of Pneumococcal Types: Induction of Transformation by a Deoxyribonucleic Acid Fraction Isolated from Pneumococcus Type III," *Journal of Experimental Medicine* 79 (Feb. 1944): 137–158; René J. Dubos, *The Professor, the Institute, and DNA* (New York: Rockefeller University Press, 1976), pp. 139–149; J. D. Watson and F. H. C. Crick, "Molecular Structure of Nucleic Acids: A Structure for Deoxyribose Nucleic Acid," *Nature* 171 (Apr. 25, 1953): 737–738; James D. Watson, *The Double Helix: A Personal Account of the Discovery of the Structure of DNA* (New York: Atheneum, 1968), pp. 14, 222.

27. "Leading Diagnoses and Reasons for Patient Visits," *NDTI* [National Disease and Therapeutic Index] *Review* 1 (Mar. 1970): 18–23; W. T. Couter, A. T. Held, and C. L. York, "Analysis of One Thousand Consecutive Residence Visits to Acutely Ill Medical Patients," *Journal of the American Medical Association* 152 (Aug. 29, 1953): 1704–1706;

R. B. Davis and D. J. Kresge, "Domiciliary Medical Practice in an Indigent Population," ibid. 161 (Aug. 25, 1956): 1617–1624; National Office of Vital Statistics, *Table 1–23. Deaths and Death Rates for Each Cause, by Color and Sex: United States, 1970* (Rockville, Md.: U.S. Department of Health, Education, and Welfare, Public Health Service, Health Resources Administration, n.d.). Comparable data are not available for the incidence of various illnesses until recent years, but it is interesting to compare the studies of Couter et al. and Davis and Kresge with the patients John Burke saw in their homes in New York City on one day in December 1866. Twenty-one of the 24 patients were suffering from infections. C. Rosenberg, "The Practice of Medicine in New York a Century Ago," *Bulletin of the History of Medicine* 41 (May–June 1967): 223–253.

28. National Center for Health Statistics, *Facts of Life and Death*, p. 15; "Leading Diagnoses and Reasons for Patient Visits."

29. "In Honor of Dr. John H. Stokes," *American Journal of Syphilis, Gonorrhea, and Venereal Diseases* 31 (Mar. 19, 1947): 102–108.

Index

325

DNA, 247; for discovery of hepatitis virus, 319
Noeggerath, Emil, 87
Noguchi, Hideyo, 91–92
Northern Regional Research Laboratories (Peoria, Ill.), 131, 134
Nosocomial infections, 192

Obsolescence, 114
Obstetrics, 64
O'Connor, Basil, 208, 214
O'Dwyer, Joseph, 40
Office of Scientific Research and Development, 131, 155
Ogston, Alexander, 57, 141
Ornithosis, 177, 183
Osler, William, 10, 42, 64, 75, 96, 253–254
Osteopathy, 9
Otitis media, 153
Oxford group. *See* Florey, Howard Walter
Oxytetracycline, 181

Pamaquine, 105
Para-aminobenzenesulphonamide. *See* Sulfanilamide
Para-aminobenzoic acid, 119–120, 176, 180
Para-aminosalicylic acid (PAS, PASA), 167
Parainfluenza virus, 222
Parasyphilis, 92
Paresis, 92, 95, 147
Park, William Hallock, 8, 17, 20, 30. *See also* New York City Health Department Bacteriological Laboratories
Parke, Davis and Company, 179–180, 184, 189
Parran, Thomas, 94, 97–98, 99
Parrott, Robert H., 222
PAS. *See* Para-aminosalicylic acid
PASA. *See* Para-aminosalicylic acid
Pasteur, Louis, 158; studies on fermentation by, 3; attenuation of microorganisms by, 4, 24; and rabies vaccination, 4–5; and preservation of wine, 15; and identification of pneumococcus, 46, 263; and antibiotics, 125; and cultivation of staphylococci, 141

Pasteur Institute, 47
Pasteurization. *See* Milk
Patents: on scarlet fever antitoxin, 61, 269; on arsphenamine, 93; on penicillin, 134; on streptomycin, 161; on antibiotics, 182
Pathogenic microorganisms, 2
Paul, John R., 209, 216
Payne, Eugene H., 179–180
Pearson, Karl, 166, 257
Peebles, Thomas C., 219–220
Penicillin: adverse reactions to, 41, 130, 156–157; discovery of, 125–128, 134–136, 293; testing for, 128–129; extraction of, 129; production of, 129–132, 133, 156; unit of, 130, 294; in treatment of streptococcal infections, 130–131, 132–133, 134, 149–153; in treatment of staphylococcal infections, 130–131, 132–133, 141–144; in treatment of endocarditis, 130–131, 140; in treatment of war wounds, 131, 133, 154–155; cost of, 131, 156; synthesis of, 131–132; acceptance of, 132, 154–157, 296; in treatment of gonococcal infections, 132–133, 144–145; in treatment of clostridial infections, 133; in treatment of syphilis, 133, 146–148, 238–239; patents on, 134, 185; dosage of, 136, 138, 145–146; in treatment of pneumococcal infections, 136–137, 190, 295; absorption, distribution, and excretion of, 137–139; in treatment of meningococcal infections, 139; staphylococci resistant to, 141–144; gonococci resistant to, 144–146; in prevention of gonococcal ophthalmitis, 146; in prevention of rheumatic fever, 150–153; in prevention of nephritis, 153; in treatment of diphtheria carriers and actinomycosis, 154; allergic reactions to, 156–157; molecular modifications of, 184–186; in prevention of endocarditis, 300
Penicillinase, 142, 145–146, 185–186
Penicillium notatum, 127
Perkin, William Henry, 92